Real-Time
Digital Signal Processing

Real-Time Digital Signal Processing

Implementations, Applications, and Experiments with the TMS320C55X

Sen M Kuo
Northern Illinois University, DeKalb, Illinois, USA

Bob H Lee
Texas Instruments, Inc., Schaumburg, Illinois, USA

JOHN WILEY & SONS, LTD.
Chichester · New York · Weinheim · Brisbane · Singapore · Toronto

Other Wiley Editorial Offices

John Wiley & Sons, Inc., 605 Third Avenue,
New York, NY 10158–0012, USA

Wiley-VCH Verlag GmbH
Pappelallee 3, D-69469 Weinheim, Germany

Jacaranda Wiley Ltd, 33 Park Road, Milton,
Queensland 4064, Australia

John Wiley & Sons (Canada) Ltd, 22 Worcester Road
Rexdale, Ontario, M9W 1L1, Canada

John Wiley & Sons (Asia) Pte Ltd, 2 Clementi Loop #02–01,
Jin Xing Distripark, Singapore 129809

Library of Congress Cataloging-in-Publication Data

Kuo, Sen M. (Sen-Maw)
 Real-time digital signal processing: implementations, applications, and experiments
 with the TMS320C55x / Sen M. Kuo, Bob H. Lee
  p. cm.
 Includes bibliographical references and index.
 ISBN 0–470–84137–0
 1. Signal processing—Digital techniques. 2. Texas Instruments TMS320 series
microprocessors. I. Lee, Bob H. II. Title.

 TK5102.9 .K86 2001
 621.382′2—dc21   2001026651

British Library Cataloguing in Publication Data

A catalogue record for this book is available from the British Library

ISBN 0 470 84137 0

Typeset by Kolam Information Services Pvt. Ltd, Pondicherry, India
Printed and bound in Great Britain by Antony Rowe Ltd
This book is printed on acid-free paper responsibly manufactured from sustainable forestry, in which at least two trees are planted for each one used for paper production

To my wife Paolien, and children Jennifer, Kevin, and Kathleen.

– Sen M. Kuo

To my dear wife Vikki and daughter Jenni.

– Bob H. Lee

Contents

Preface   **xv**

1 Introduction to Real-Time Digital Signal Processing   **1**
   1.1 Basic Elements of Real-Time DSP Systems   2
   1.2 Input and Output Channels   3
       1.2.1 Input Signal Conditioning   3
       1.2.2 A/D Conversion   4
       1.2.3 Sampling   5
       1.2.4 Quantizing and Encoding   7
       1.2.5 D/A Conversion   9
       1.2.6 Input/Output Devices   9
   1.3 DSP Hardware   11
       1.3.1 DSP Hardware Options   11
       1.3.2 Fixed- and Floating-Point Devices   13
       1.3.3 Real-Time Constraints   14
   1.4 DSP System Design   14
       1.4.1 Algorithm Development   14
       1.4.2 Selection of DSP Chips   16
       1.4.3 Software Development   17
       1.4.4 High-Level Software Development Tools   18
   1.5 Experiments Using Code Composer Studio   19
       1.5.1 Experiment 1A – Using the CCS and the TMS320C55x Simulator   20
       1.5.2 Experiment 1B – Debugging Program on the CCS   25
       1.5.3 Experiment 1C – File Input and Output   28
       1.5.4 Experiment 1D – Code Efficiency Analysis   29
       1.5.5 Experiment 1E – General Extension Language   32
   References   33
   Exercises   33

2 Introduction to TMS320C55x Digital Signal Processor   **35**
   2.1 Introduction   35
   2.2 TMS320C55x Architecture   36
       2.2.1 TMS320C55x Architecture Overview   36
       2.2.2 TMS320C55x Buses   39
       2.2.3 TMS320C55x Memory Map   40

2.3   Software Development Tools   40
   2.3.1 C Compiler   42
   2.3.2 Assembler   44
   2.3.3 Linker   46
   2.3.4 Code Composer Studio   48
   2.3.5 Assembly Statement Syntax   49
2.4   TMS320C55x Addressing Modes   50
   2.4.1 Direct Addressing Mode   52
   2.4.2 Indirect Addressing Mode   53
   2.4.3 Absolute Addressing Mode   56
   2.4.4 Memory-Mapped Register Addressing Mode   56
   2.4.5 Register Bits Addressing Mode   57
   2.4.6 Circular Addressing Mode   58
2.5   Pipeline and Parallelism   59
   2.5.1 TMS320C55x Pipeline   59
   2.5.2 Parallel Execution   60
2.6   TMS320C55x Instruction Set   63
   2.6.1 Arithmetic Instructions   63
   2.6.2 Logic and Bits Manipulation Instructions   64
   2.6.3 Move Instruction   65
   2.6.4 Program Flow Control Instructions   66
2.7   Mixed C and Assembly Language Programming   68
2.8   Experiments – Assembly Programming Basics   70
   2.8.1 Experiment 2A – Interfacing C with Assembly Code   71
   2.8.2 Experiment 2B – Addressing Mode Experiments   72
References   75
Exercises   75

3 DSP Fundamentals and Implementation Considerations   **77**

3.1   Digital Signals and Systems   77
   3.1.1 Elementary Digital Signals   77
   3.1.2 Block Diagram Representation of Digital Systems   79
   3.1.3 Impulse Response of Digital Systems   83
3.2   Introduction to Digital Filters   83
   3.2.1 FIR Filters and Power Estimators   84
   3.2.2 Response of Linear Systems   87
   3.2.3 IIR Filters   88
3.3   Introduction to Random Variables   90
   3.3.1 Review of Probability and Random Variables   90
   3.3.2 Operations on Random Variables   92
3.4   Fixed-Point Representation and Arithmetic   95
3.5   Quantization Errors   98
   3.5.1 Input Quantization Noise   98
   3.5.2 Coefficient Quantization Noise   101
   3.5.3 Roundoff Noise   102
3.6   Overflow and Solutions   103
   3.6.1 Saturation Arithmetic   103
   3.6.2 Overflow Handling   104
   3.6.3 Scaling of Signals   105
3.7   Implementation Procedure for Real-Time Applications   107

3.8   Experiments of Fixed-Point Implementations   108
      3.8.1 Experiment 3A – Quantization of Sinusoidal Signals   109
      3.8.2 Experiment 3B – Quantization of Speech Signals   111
      3.8.3 Experiment 3C – Overflow and Saturation Arithmetic   112
      3.8.4 Experiment 3D – Quantization of Coefficients   115
      3.8.5 Experiment 3E – Synthesizing Sine Function   117
   References   121
   Exercises   122

4   Frequency Analysis   127

4.1   Fourier Series and Transform   127
      4.1.1 Fourier Series   127
      4.1.2 Fourier Transform   130
4.2   The z-Transforms   133
      4.2.1 Definitions and Basic Properties   133
      4.2.2 Inverse z-Transform   136
4.3   System Concepts   141
      4.3.1 Transfer Functions   141
      4.3.2 Digital Filters   143
      4.3.3 Poles and Zeros   144
      4.3.4 Frequency Responses   148
4.4   Discrete Fourier Transform   152
      4.4.1 Discrete-Time Fourier Series and Transform   152
      4.4.2 Aliasing and Folding   154
      4.4.3 Discrete Fourier Transform   157
      4.4.4 Fast Fourier Transform   159
4.5   Applications   160
      4.5.1 Design of Simple Notch Filters   160
      4.5.2 Analysis of Room Acoustics   162
4.6   Experiments Using the TMS320C55x   165
      4.6.1 Experiment 4A – Twiddle Factor Generation   167
      4.6.2 Experiment 4B – Complex Data Operation   169
      4.6.3 Experiment 4C – Implementation of DFT   171
      4.6.4 Experiment 4D – Experiment Using Assembly Routines   173
   References   176
   Exercises   176

5   Design and Implementation of FIR Filters   181

5.1   Introduction to Digital Filters   181
      5.1.1 Filter Characteristics   182
      5.1.2 Filter Types   183
      5.1.3 Filter Specifications   185
5.2   FIR Filtering   189
      5.2.1 Linear Convolution   189
      5.2.2 Some Simple FIR Filters   192
      5.2.3 Linear Phase FIR Filters   194
      5.2.4 Realization of FIR Filters   198
5.3   Design of FIR Filters   201
      5.3.1 Filter Design Procedure   201
      5.3.2 Fourier Series Method   202
      5.3.3 Gibbs Phenomenon   205

      5.3.4 Window Functions   208
      5.3.5 Frequency Sampling Method   214
  5.4   Design of FIR Filters Using MATLAB   219
  5.5   Implementation Considerations   221
      5.5.1 Software Implementations   221
      5.5.2 Quantization Effects in FIR Filters   223
  5.6   Experiments Using the TMS320C55x   225
      5.6.1 Experiment 5A – Implementation of Block FIR Filter   227
      5.6.2 Experiment 5B – Implementation of Symmetric FIR Filter   230
      5.6.3 Experiment 5C – Implementation of FIR Filter Using Dual-MAC   233
  References   235
  Exercises   236

6 Design and Implementation of IIR Filters   241

  6.1   Laplace Transform   241
      6.1.1 Introduction to the Laplace Transform   241
      6.1.2 Relationships between the Laplace and z-Transforms   245
      6.1.3 Mapping Properties   246
  6.2   Analog Filters   247
      6.2.1 Introduction to Analog Filters   248
      6.2.2 Characteristics of Analog Filters   249
      6.2.3 Frequency Transforms   253
  6.3   Design of IIR Filters   255
      6.3.1 Review of IIR Filters   255
      6.3.2 Impulse-Invariant Method   256
      6.3.3 Bilinear Transform   259
      6.3.4 Filter Design Using Bilinear Transform   261
  6.4   Realization of IIR Filters   263
      6.4.1 Direct Forms   263
      6.4.2 Cascade Form   266
      6.4.3 Parallel Form   268
      6.4.4 Realization Using MATLAB   269
  6.5   Design of IIR Filters Using MATLAB   271
  6.6   Implementation Considerations   273
      6.6.1 Stability   274
      6.6.2 Finite-Precision Effects and Solutions   275
      6.6.3 Software Implementations   279
      6.6.4 Practical Applications   280
  6.7   Software Developments and Experiments Using the TMS320C55x   284
      6.7.1 Design of IIR Filter   285
      6.7.2 Experiment 6A – Floating-Point C Implementation   286
      6.7.3 Experiment 6B – Fixed-Point C Implementation Using Intrinsics   289
      6.7.4 Experiment 6C – Fixed-Point C Programming Considerations 292
      6.7.5 Experiment 6D – Assembly Language Implementations   295
  References   297
  Exercises   297

7 Fast Fourier Transform and Its Applications   303

  7.1   Discrete Fourier Transform   303
      7.1.1 Definitions   304
      7.1.2 Important Properties of DFT   308
      7.1.3 Circular Convolution   311

7.2 Fast Fourier Transforms 314
7.2.1 Decimation-in-Time 315
7.2.2 Decimation-in-Frequency 319
7.2.3 Inverse Fast Fourier Transform 320
7.2.4 MATLAB Implementations 321
7.3 Applications 322
7.3.1 Spectrum Estimation and Analysis 322
7.3.2 Spectral Leakage and Resolution 324
7.3.3 Power Density Spectrum 328
7.3.4 Fast Convolution 330
7.3.5 Spectrogram 332
7.4 Implementation Considerations 333
7.4.1 Computational Issues 334
7.4.2 Finite-Precision Effects 334
7.5 Experiments Using the TMS320C55x 336
7.5.1 Experiment 7A – Radix-2 Complex FFT 336
7.5.2 Experiment 7B – Radix-2 Complex FFT Using Assembly Language 341
7.5.3 Experiment 7C – FFT and IFFT 344
7.5.4 Experiment 7D – Fast Convolution 344
References 346
Exercises 347

8 Adaptive Filtering   **351**
8.1 Introduction to Random Processes 351
8.1.1 Correlation Functions 352
8.1.2 Frequency-Domain Representations 356
8.2 Adaptive Filters 359
8.2.1 Introduction to Adaptive Filtering 359
8.2.2 Performance Function 361
8.2.3 Method of Steepest Descent 365
8.2.4 The LMS Algorithm 366
8.3 Performance Analysis 367
8.3.1 Stability Constraint 367
8.3.2 Convergence Speed 368
8.3.3 Excess Mean-Square Error 369
8.4 Modified LMS Algorithms 370
8.4.1 Normalized LMS Algorithm 370
8.4.2 Leaky LMS Algorithm 371
8.5 Applications 372
8.5.1 Adaptive System Identification 372
8.5.2 Adaptive Linear Prediction 373
8.5.3 Adaptive Noise Cancellation 375
8.5.4 Adaptive Notch Filters 377
8.5.5 Adaptive Channel Equalization 379
8.6 Implementation Considerations 381
8.6.1 Computational Issues 381
8.6.2 Finite-Precision Effects 382
8.7 Experiments Using the TMS320C55x 385
8.7.1 Experiment 8A – Adaptive System Identification 385
8.7.2 Experiment 8B – Adaptive Predictor Using the Leaky LMS Algorithm 390
References 396
Exercises 396

9   Practical DSP Applications in Communications   399

9.1   Sinewave Generators and Applications   399
       9.1.1 Lookup-Table Method   400
       9.1.2 Linear Chirp Signal   402
       9.1.3 DTMF Tone Generator   403
9.2   Noise Generators and Applications   404
       9.2.1 Linear Congruential Sequence Generator   404
       9.2.2 Pseudo-Random Binary Sequence Generator   406
       9.2.3 Comfort Noise in Communication Systems   408
       9.2.4 Off-Line System Modeling   409
9.3   DTMF Tone Detection   410
       9.3.1 Specifications   410
       9.3.2 Goertzel Algorithm   411
       9.3.3 Implementation Considerations   414
9.4   Adaptive Echo Cancellation   417
       9.4.1 Line Echoes   417
       9.4.2 Adaptive Echo Canceler   418
       9.4.3 Practical Considerations   422
       9.4.4 Double-Talk Effects and Solutions   423
       9.4.5 Residual Echo Suppressor   425
9.5   Acoustic Echo Cancellation   426
       9.5.1 Introduction   426
       9.5.2 Acoustic Echo Canceler   427
       9.5.3 Implementation Considerations   428
9.6   Speech Enhancement Techniques   429
       9.6.1 Noise Reduction Techniques   429
       9.6.2 Spectral Subtraction Techniques   431
       9.6.3 Implementation Considerations   433
9.7   Projects Using the TMS320C55x   435
       9.7.1 Project Suggestions   435
       9.7.2 A Project Example – Wireless Application   437
References   442

Appendix A   Some Useful Formulas   445

A.1   Trigonometric Identities   445
A.2   Geometric Series   446
A.3   Complex Variables   447
A.4   Impulse Functions   449
A.5   Vector Concepts   449
A.6   Units of Power   450
Reference   451

Appendix B   Introduction of MATLAB for DSP
                Applications   453

B.1   Elementary Operations   453
       B.1.1 Initializing Variables and Vectors   453
       B.1.2 Graphics   455
       B.1.3 Basic Operators   457
       B.1.4 Files   459
B.2   Generation and Processing of Digital Signals   460
B.3   DSP Applications   463
B.4   User-Written Functions   465

B.5   Summary of Useful MATLAB Functions   466
References   467

Appendix C   Introduction of C Programming for DSP Applications   469

C.1   A Simple C Program   470
   C.1.1   Variables and Assignment Operators   472
   C.1.2   Numeric Data Types and Conversion   473
   C.1.3   Arrays   474
C.2   Arithmetic and Bitwise Operators   475
   C.2.1   Arithmetic Operators   475
   C.2.2   Bitwise Operators   476
C.3   An FIR Filter Program   476
   C.3.1   Command-Line Arguments   477
   C.3.2   Pointers   477
   C.3.3   C Functions   478
   C.3.4   Files and I/O Operations   480
C.4   Control Structures and Loops   481
   C.4.1   Control Structures   481
   C.4.2   Logical Operators   483
   C.4.3   Loops   484
C.5   Data Types Used by the TMS320C55x   485
References   486

Appendix D   About the Software   487

Index   489

Preface

Real-time digital signal processing (DSP) using general-purpose DSP processors is very challenging work in today's engineering fields. It promises an effective way to design, experiment, and implement a variety of signal processing algorithms for real-world applications. With DSP penetrating into various applications, the demand for high-performance digital signal processors has expanded rapidly in recent years. Many industrial companies are currently engaged in real-time DSP research and development. It becomes increasingly important for today's students and practicing engineers to master not only the theory of DSP, but equally important, the skill of real-time DSP system design and implementation techniques.

This book offers readers a hands-on approach to understanding real-time DSP principles, system design and implementation considerations, real-world applications, as well as many DSP experiments using MATLAB, C/C++, and the TMS320C55x. This is a practical book about DSP and using digital signal processors for DSP applications. This book is intended as a text for senior/graduate level college students with emphasis on real-time DSP implementations and applications. This book can also serve as a desktop reference for practicing engineer and embedded system programmer to learn DSP concepts and to develop real-time DSP applications at work. We use a practical approach that avoids a lot of theoretical derivations. Many useful DSP textbooks with solid mathematical proofs are listed at the end of each chapter. To efficiently develop a DSP system, the reader must understand DSP algorithms as well as basic DSP chip architecture and programming. It is helpful to have several manuals and application notes on the TMS320C55x from Texas Instruments at *http://www.ti.com*.

The DSP processor we will use as an example in this book is the TMS320C55x, the newest 16-bit fixed-point DSP processor from Texas Instruments. To effectively illustrate real-time DSP concepts and applications, MATLAB will be introduced for analysis and filter design, C will be used for implementing DSP algorithms, and Code Composer Studio (CCS) of the TMS320C55x are integrated into lab experiments, projects, and applications. To efficiently utilize the advanced DSP architecture for fast software development and maintenance, the mixing of C and assembly programs are emphasized.

Chapter 1 reviews the fundamentals of real-time DSP functional blocks, DSP hardware options, fixed- and floating-point DSP devices, real-time constraints, algorithm development, selection of DSP chips, and software development. In Chapter 2, we introduce the architecture and assembly programming of the TMS320C55x. Chapter 3 presents some fundamental DSP concepts in time domain and practical considerations for the implementation of digital filters and algorithms on DSP hardware. Readers who are familiar with these DSP fundamentals should be able to skip through some of these sections. However, most notations used throughout the book will be defined in this chapter. In Chapter 4, the Fourier series, the Fourier transform, the z-transform, and the discrete Fourier transforms are introduced. Frequency analysis is extremely helpful in understanding the characteristics of both signals and systems. Chapter 5 is focused on the design, implementation, and application of FIR filters; digital IIR filters are covered in Chapter 6, and adaptive filters are presented in Chapter 8. The development, implementation, and application of FFT algorithms are introduced in Chapter 7. In Chapter 9, we introduce some selected DSP applications in communications that have played an important role in the realization of the systems.

As with any book attempting to capture the state of the art at a given time, there will necessarily be omissions that are necessitated by the rapidly evolving developments in this dynamic field of exciting practical interest. We hope, at least, that this book will serve as a guide for what has already come and as an inspiration for what will follow. To aid teaching of the course a Solution Manual that presents detailed solutions to most of the problems in the book is available from the publisher.

Availability of Software

The MATLAB, C, and assembly programs that implement many DSP examples and applications are listed in the book. These programs along with many other programs for DSP implementations and lab experiments are available in the software package at *http://www.ceet.niu.edu/faculty/kuo/books/rtdsp.html* and *http://pages.prodigy.net/ sunheel/web/dspweb.htm*. Several real-world data files for some applications introduced in the book also are included in the software package. The list of files in the software package is given in Appendix D. It is not critical you have this software as you read the book, but it will help you to gain insight into the implementation of DSP algorithms, and it will be required for doing experiments at the last section of each chapter. Some of these experiments involve minor modification of the example code. By examining, studying and modifying the example code, the software can also be used as a prototype for other practical applications. Every attempt has been made to ensure the correctness of the code. We would appreciate readers bringing to our attention (kuo@ceet.niu.edu) any coding errors so that we can correct and update the codes available in the software package on the web.

Acknowledgments

We are grateful to Maria Ho and Christina Peterson at Texas Instruments, and Naomi Fernandes at Math Works, who provided the necessary support to write the book in a short period. The first author thanks many of his students who have taken his DSP courses, Senior Design Projects, and Master Thesis courses. He is indebted to Gene Frentz, Dr. Qun S. Lin, and Dr. Panos Papamichalis of Texas Instruments, John Kronenburger of Tellabs, and Santo LaMantia of Shure Brothers, for their support of DSP activities at Northern Illinois University. He also thanks Jennifer Y. Kuo for the proofreading of the book. The second author wishes to thank Robert DeNardo, David Baughman, and Chuck Brokish of Texas Instruments, for their valuable inputs, help, and encouragement during the course of writing this book. We would like to thank Peter Mitchell, editor at Wiley, for his support of this project. We also like to thank the staff at Wiley for the final preparation of the book. Finally, we thank our parents and families for their endless love, encouragement, and the understanding they have shown during the whole time.

Sen M. Kuo and **Bob H. Lee**

1

Introduction to Real-Time Digital Signal Processing

Signals can be divided into three categories – continuous-time (analog) signals, discrete-time signals, and digital signals. The signals that we encounter daily are mostly analog signals. These signals are defined continuously in time, have an infinite range of amplitude values, and can be processed using electrical devices containing both active and passive circuit elements. Discrete-time signals are defined only at a particular set of time instances. Therefore they can be represented as a sequence of numbers that have a continuous range of values. On the other hand, digital signals have discrete values in both time and amplitude. In this book, we design and implement digital systems for processing digital signals using digital hardware. However, the analysis of such signals and systems usually uses discrete-time signals and systems for mathematical convenience. Therefore we use the term 'discrete-time' and 'digital' interchangeably.

Digital signal processing (DSP) is concerned with the digital representation of signals and the use of digital hardware to analyze, modify, or extract information from these signals. The rapid advancement in digital technology in recent years has created the implementation of sophisticated DSP algorithms that make real-time tasks feasible. A great deal of research has been conducted to develop DSP algorithms and applications. DSP is now used not only in areas where analog methods were used previously, but also in areas where applying analog techniques is difficult or impossible.

There are many advantages in using digital techniques for signal processing rather than traditional analog devices (such as amplifiers, modulators, and filters). Some of the advantages of a DSP system over analog circuitry are summarized as follows:

1. *Flexibility*. Functions of a DSP system can be easily modified and upgraded with software that has implemented the specific algorithm for using the same hardware. One can design a DSP system that can be programmed to perform a wide variety of tasks by executing different software modules. For example, a digital camera may be easily updated (reprogrammed) from using JPEG (joint photographic experts group) image processing to a higher quality JPEG2000 image without actually changing the hardware. In an analog system, however, the whole circuit design would need to be changed.

2. *Reproducibility*. The performance of a DSP system can be repeated precisely from one unit to another. This is because the signal processing of DSP systems work directly with binary sequences. Analog circuits will not perform as well from each circuit, even if they are built following identical specifications, due to component tolerances in analog components. In addition, by using DSP techniques, a digital signal can be transferred or reproduced many times without degrading its signal quality.

3. *Reliability*. The memory and logic of DSP hardware does not deteriorate with age. Therefore the field performance of DSP systems will not drift with changing environmental conditions or aged electronic components as their analog counterparts do. However, the data size (wordlength) determines the accuracy of a DSP system. Thus the system performance might be different from the theoretical expectation.

4. *Complexity*. Using DSP allows sophisticated applications such as speech or image recognition to be implemented for lightweight and low power portable devices. This is impractical using traditional analog techniques. Furthermore, there are some important signal processing algorithms that rely on DSP, such as error correcting codes, data transmission and storage, data compression, perfect linear phase filters, etc., which can barely be performed by analog systems.

With the rapid evolution in semiconductor technology in the past several years, DSP systems have a lower overall cost compared to analog systems. DSP algorithms can be developed, analyzed, and simulated using high-level language and software tools such as C/C++ and MATLAB (matrix laboratory). The performance of the algorithms can be verified using a low-cost general-purpose computer such as a personal computer (PC). Therefore a DSP system is relatively easy to develop, analyze, simulate, and test.

There are limitations, however. For example, the bandwidth of a DSP system is limited by the sampling rate and hardware peripherals. The initial design cost of a DSP system may be expensive, especially when large bandwidth signals are involved. For real-time applications, DSP algorithms are implemented using a fixed number of bits, which results in a limited dynamic range and produces quantization and arithmetic errors.

1.1 Basic Elements of Real-Time DSP Systems

There are two types of DSP applications – non-real-time and real time. Non-real-time signal processing involves manipulating signals that have already been collected and digitized. This may or may not represent a current action and the need for the result is not a function of real time. Real-time signal processing places stringent demands on DSP hardware and software design to complete predefined tasks within a certain time frame. This chapter reviews the fundamental functional blocks of real-time DSP systems.

The basic functional blocks of DSP systems are illustrated in Figure 1.1, where a real-world analog signal is converted to a digital signal, processed by DSP hardware in

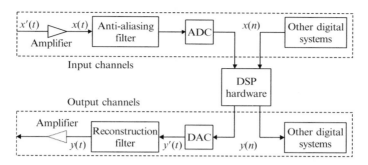

Figure 1.1 Basic functional blocks of real-time DSP system

digital form, and converted back into an analog signal. Each of the functional blocks in Figure 1.1 will be introduced in the subsequent sections. For some real-time applications, the input data may already be in digital form and/or the output data may not need to be converted to an analog signal. For example, the processed digital information may be stored in computer memory for later use, or it may be displayed graphically. In other applications, the DSP system may be required to generate signals digitally, such as speech synthesis used for cellular phones or pseudo-random number generators for CDMA (code division multiple access) systems.

1.2 Input and Output Channels

In this book, a time-domain signal is denoted with a lowercase letter. For example, $x(t)$ in Figure 1.1 is used to name an analog signal of x with a relationship to time t. The time variable t takes on a continuum of values between $-\infty$ and ∞. For this reason we say $x(t)$ is a continuous-time signal. In this section, we first discuss how to convert analog signals into digital signals so that they can be processed using DSP hardware. The process of changing an analog signal to a xdigital signal is called analog-to-digital (A/D) conversion. An A/D converter (ADC) is usually used to perform the signal conversion.

Once the input digital signal has been processed by the DSP device, the result, $y(n)$, is still in digital form, as shown in Figure 1.1. In many DSP applications, we need to reconstruct the analog signal after the digital processing stage. In other words, we must convert the digital signal $y(n)$ back to the analog signal $y(t)$ before it is passed to an appropriate device. This process is called the digital-to-analog (D/A) conversion, typically performed by a D/A converter (DAC). One example would be CD (compact disk) players, for which the music is in a digital form. The CD players reconstruct the analog waveform that we listen to. Because of the complexity of sampling and synchronization processes, the cost of an ADC is usually considerably higher than that of a DAC.

1.2.1 Input Signal Conditioning

As shown in Figure 1.1, the analog signal, $x'(t)$, is picked up by an appropriate electronic sensor that converts pressure, temperature, or sound into electrical signals.

For example, a microphone can be used to pick up sound signals. The sensor output, $x'(t)$, is amplified by an amplifier with gain value g. The amplified signal is

$$x(t) = gx'(t). \tag{1.2.1}$$

The gain value g is determined such that $x(t)$ has a dynamic range that matches the ADC. For example, if the peak-to-peak range of the ADC is ± 5 volts (V), then g may be set so that the amplitude of signal $x(t)$ to the ADC is scaled between ± 5V. In practice, it is very difficult to set an appropriate fixed gain because the level of $x'(t)$ may be unknown and changing with time, especially for signals with a larger dynamic range such as speech. Therefore an automatic gain controller (AGC) with time-varying gain determined by DSP hardware can be used to effectively solve this problem.

1.2.2 A/D Conversion

As shown in Figure 1.1, the ADC converts the analog signal $x(t)$ into the digital signal sequence $x(n)$. Analog-to-digital conversion, commonly referred as digitization, consists of the sampling and quantization processes as illustrated in Figure 1.2. The sampling process depicts a continuously varying analog signal as a sequence of values. The basic sampling function can be done with a 'sample and hold' circuit, which maintains the sampled level until the next sample is taken. Quantization process approximates a waveform by assigning an actual number for each sample. Therefore an ADC consists of two functional blocks – an ideal sampler (sample and hold) and a quantizer (including an encoder). Analog-to-digital conversion carries out the following steps:

1. The bandlimited signal $x(t)$ is sampled at uniformly spaced instants of time, nT, where n is a positive integer, and T is the sampling period in seconds. This sampling process converts an analog signal into a discrete-time signal, $x(nT)$, with continuous amplitude value.

2. The amplitude of each discrete-time sample is quantized into one of the 2^B levels, where B is the number of bits the ADC has to represent for each sample. The discrete amplitude levels are represented (or encoded) into distinct binary words $x(n)$ with a fixed wordlength B. This binary sequence, $x(n)$, is the digital signal for DSP hardware.

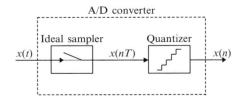

Figure 1.2 Block diagram of A/D converter

The reason for making this distinction is that each process introduces different distortions. The sampling process brings in aliasing or folding distortions, while the encoding process results in quantization noise.

1.2.3 Sampling

An ideal sampler can be considered as a switch that is periodically open and closed every T seconds and

$$T = \frac{1}{f_s},$$

$$(1.2.2)$$

where f_s is the sampling frequency (or sampling rate) in hertz (Hz, or cycles per second). The intermediate signal, $x(nT)$, is a discrete-time signal with a continuous-value (a number has infinite precision) at discrete time nT, $n = 0, 1, \ldots, \infty$ as illustrated in Figure 1.3. The signal $x(nT)$ is an impulse train with values equal to the amplitude of $x(t)$ at time nT. The analog input signal $x(t)$ is continuous in both time and amplitude. The sampled signal $x(nT)$ is continuous in amplitude, but it is defined only at discrete points in time. Thus the signal is zero except at the sampling instants $t = nT$.

In order to represent an analog signal $x(t)$ by a discrete-time signal $x(nT)$ accurately, two conditions must be met:

1. The analog signal, $x(t)$, must be bandlimited by the bandwidth of the signal f_M.

2. The sampling frequency, f_s, must be at least twice the maximum frequency component f_M in the analog signal $x(t)$. That is,

$$f_s \geq 2f_M.$$

$$(1.2.3)$$

This is Shannon's sampling theorem. It states that when the sampling frequency is greater than twice the highest frequency component contained in the analog signal, the original signal $x(t)$ can be perfectly reconstructed from the discrete-time sample $x(nT)$. The sampling theorem provides a basis for relating a continuous-time signal $x(t)$ with

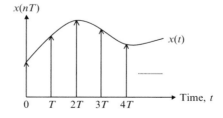

Figure 1.3 Example of analog signal $x(t)$ and discrete-time signal $x(nT)$

the discrete-time signal $x(nT)$ obtained from the values of $x(t)$ taken T seconds apart. It also provides the underlying theory for relating operations performed on the sequence to equivalent operations on the signal $x(t)$ directly.

The minimum sampling frequency $f_s = 2f_M$ is the Nyquist rate, while $f_N = f_s/2$ is the Nyquist frequency (or folding frequency). The frequency interval $[-f_s/2, f_s/2]$ is called the Nyquist interval. When an analog signal is sampled at sampling frequency, f_s, frequency components higher than $f_s/2$ fold back into the frequency range $[0, f_s/2]$. This undesired effect is known as aliasing. That is, when a signal is sampled perversely to the sampling theorem, image frequencies are folded back into the desired frequency band. Therefore the original analog signal cannot be recovered from the sampled data. This undesired distortion could be clearly explained in the frequency domain, which will be discussed in Chapter 4. Another potential degradation is due to timing jitters on the sampling pulses for the ADC. This can be negligible if a higher precision clock is used.

For most practical applications, the incoming analog signal $x(t)$ may not be band-limited. Thus the signal has significant energies outside the highest frequency of interest, and may contain noise with a wide bandwidth. In other cases, the sampling rate may be pre-determined for a given application. For example, most voice communication systems use an 8 kHz (kilohertz) sampling rate. Unfortunately, the maximum frequency component in a speech signal is much higher than 4 kHz. Out-of-band signal components at the input of an ADC can become in-band signals after conversion because of the folding over of the spectrum of signals and distortions in the discrete domain. To guarantee that the sampling theorem defined in Equation (1.2.3) can be fulfilled, an anti-aliasing filter is used to band-limit the input signal. The anti-aliasing filter is an analog lowpass filter with the cut-off frequency of

$$f_c \leq \frac{f_s}{2}. \tag{1.2.4}$$

Ideally, an anti-aliasing filter should remove all frequency components above the Nyquist frequency. In many practical systems, a bandpass filter is preferred in order to prevent undesired DC offset, 60 Hz hum, or other low frequency noises. For example, a bandpass filter with passband from 300 Hz to 3200 Hz is used in most telecommunication systems.

Since anti-aliasing filters used in real applications are not ideal filters, they cannot completely remove all frequency components outside the Nyquist interval. Any frequency components and noises beyond half of the sampling rate will alias into the desired band. In addition, since the phase response of the filter may not be linear, the components of the desired signal will be shifted in phase by amounts not proportional to their frequencies. In general, the steeper the roll-off, the worse the phase distortion introduced by a filter. To accommodate practical specifications for anti-aliasing filters, the sampling rate must be higher than the minimum Nyquist rate. This technique is known as oversampling. When a higher sampling rate is used, a simple low-cost anti-aliasing filter with minimum phase distortion can be used.

Example 1.1: Given a sampling rate for a specific application, the sampling period can be determined by (1.2.2).

(a) In narrowband telecommunication systems, the sampling rate $f_s = 8\,\text{kHz}$, thus the sampling period $T = 1/8\,000$ seconds $= 125\,\mu\text{s}$ (microseconds). Note that $1\,\mu\text{s} = 10^{-6}$ seconds.

(b) In wideband telecommunication systems, the sampling is given as $f_s = 16\,\text{kHz}$, thus $T = 1/16\,000$ seconds $= 62.5\,\mu\text{s}$.

(c) In audio CDs, the sampling rate is $f_s = 44.1\,\text{kHz}$, thus $T = 1/44\,100$ seconds $= 22.676\,\mu\text{s}$.

(d) In professional audio systems, the sampling rate $f_s = 48\,\text{kHz}$, thus $T = 1/48\,000$ seconds $= 20.833\,\mu\text{s}$.

1.2.4 Quantizing and Encoding

In the previous sections, we assumed that the sample values $x(nT)$ are represented exactly with infinite precision. An obvious constraint of physically realizable digital systems is that sample values can only be represented by a finite number of bits. The fundamental distinction between discrete-time signal processing and DSP is the wordlength. The former assumes that discrete-time signal values $x(nT)$ have infinite wordlength, while the latter assumes that digital signal values $x(n)$ only have a limited B-bit.

We now discuss a method of representing the sampled discrete-time signal $x(nT)$ as a binary number that can be processed with DSP hardware. This is the quantizing and encoding process. As shown in Figure 1.3, the discrete-time signal $x(nT)$ has an analog amplitude (infinite precision) at time $t = nT$. To process or store this signal with DSP hardware, the discrete-time signal must be quantized to a digital signal $x(n)$ with a finite number of bits. If the wordlength of an ADC is B bits, there are 2^B different values (levels) that can be used to represent a sample. The entire continuous amplitude range is divided into 2^B subranges. Amplitudes of waveform that are in the same subrange are assigned the same amplitude values. Therefore quantization is a process that represents an analog-valued sample $x(nT)$ with its nearest level that corresponds to the digital signal $x(n)$. The discrete-time signal $x(nT)$ is a sequence of real numbers using infinite bits, while the digital signal $x(n)$ represents each sample value by a finite number of bits which can be stored and processed using DSP hardware.

The quantization process introduces errors that cannot be removed. For example, we can use two bits to define four equally spaced levels (00, 01, 10, and 11) to classify the signal into the four subranges as illustrated in Figure 1.4. In this figure, the symbol 'o' represents the discrete-time signal $x(nT)$, and the symbol '•' represents the digital signal $x(n)$.

In Figure 1.4, the difference between the quantized number and the original value is defined as the quantization error, which appears as noise in the output. It is also called quantization noise. The quantization noise is assumed to be random variables that are uniformly distributed in the intervals of quantization levels. If a B-bit quantizer is used, the signal-to-quantization-noise ratio (SNR) is approximated by (will be derived in Chapter 3)

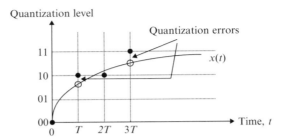

Figure 1.4   Digital samples using a 2-bit quantizer

$$\text{SNR} \approx 6B\,\text{dB}. \tag{1.2.5}$$

This is a theoretical maximum. When real input signals and converters are used, the achievable SNR will be less than this value due to imperfections in the fabrication of A/D converters. As a result, the effective number of bits may be less than the number of bits in the ADC. However, Equation (1.2.5) provides a simple guideline for determining the required bits for a given application. For each additional bit, a digital signal has about a 6-dB gain in SNR. For example, a 16-bit ADC provides about 96 dB SNR. The more bits used to represent a waveform sample, the smaller the quantization noise will be. If we had an input signal that varied between 0 and 5 V, using a 12-bit ADC, which has 4096 (2^{12}) levels, the least significant bit (LSB) would correspond to 1.22 mV resolution. An 8-bit ADC with 256 levels can only provide up to 19.5 mV resolution. Obviously with more quantization levels, one can represent the analog signal more accurately. The problems of quantization and their solutions will be further discussed in Chapter 3.

If the uniform quantization scheme shown in Figure 1.4 can adequately represent loud sounds, most of the softer sounds may be pushed into the same small value. This means soft sounds may not be distinguishable. To solve this problem, a quantizer whose quantization step size varies according to the signal amplitude can be used. In practice, the non-uniform quantizer uses a uniform step size, but the input signal is compressed first. The overall effect is identical to the non-uniform quantization. For example, the logarithm-scaled input signal, rather than the input signal itself, will be quantized. After processing, the signal is reconstructed at the output by expanding it. The process of compression and expansion is called companding (compressing and expanding). For example, the μ-law (used in North America and parts of Northeast Asia) and A-law (used in Europe and most of the rest of the world) companding schemes are used in most digital communications.

As shown in Figure 1.1, the input signal to DSP hardware may be a digital signal from other DSP systems. In this case, the sampling rate of digital signals from other digital systems must be known. The signal processing techniques called interpolation or decimation can be used to increase or decrease the existing digital signals' sampling rates. Sampling rate changes are useful in many applications such as interconnecting DSP systems operating at different rates. A multirate DSP system uses more than one sampling frequency to perform its tasks.

1.2.5 D/A Conversion

Most commercial DACs are zero-order-hold, which means they convert the binary input to the analog level and then simply hold that value for T seconds until the next sampling instant. Therefore the DAC produces a staircase shape analog waveform $y'(t)$, which is shown as a solid line in Figure 1.5. The reconstruction (anti-imaging and smoothing) filter shown in Figure 1.1 smoothes the staircase-like output signal generated by the DAC. This analog lowpass filter may be the same as the anti-aliasing filter with cut-off frequency $f_c \leq f_s/2$, which has the effect of rounding off the corners of the staircase signal and making it smoother, which is shown as a dotted line in Figure 1.5. High quality DSP applications, such as professional digital audio, require the use of reconstruction filters with very stringent specifications.

From the frequency-domain viewpoint (will be presented in Chapter 4), the output of the DAC contains unwanted high frequency or image components centered at multiples of the sampling frequency. Depending on the application, these high-frequency components may cause undesired side effects. Take an audio CD player for example. Although the image frequencies may not be audible, they could overload the amplifier and cause inter-modulation with the desired baseband frequency components. The result is an unacceptable degradation in audio signal quality.

The ideal reconstruction filter has a flat magnitude response and linear phase in the passband extending from the DC to its cut-off frequency and infinite attenuation in the stopband. The roll-off requirements of the reconstruction filter are similar to those of the anti-aliasing filter. In practice, switched capacitor filters are preferred because of their programmable cut-off frequency and physical compactness.

1.2.6 Input/Output Devices

There are two basic ways of connecting A/D and D/A converters to DSP devices: serial and parallel. A parallel converter receives or transmits all the B bits in one pass, while the serial converters receive or transmit B bits in a serial data stream. Converters with parallel input and output ports must be attached to the DSP's address and data buses,

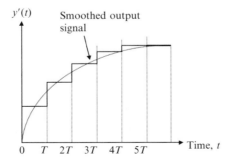

Figure 1.5 Staircase waveform generated by a DAC

which are also attached to many different types of devices. With different memory devices (RAM, EPROM, EEPROM, or flash memory) at different speeds hanging on DSP's data bus, driving the bus may become a problem. Serial converters can be connected directly to the built-in serial ports of DSP devices. This is why many practical DSP systems use serial ADCs and DACs.

Many applications use a single-chip device called an analog interface chip (AIC) or coder/decoder (CODEC), which integrates an anti-aliasing filter, an ADC, a DAC, and a reconstruction filter all on a single piece of silicon. Typical applications include modems, speech systems, and industrial controllers. Many standards that specify the nature of the CODEC have evolved for the purposes of switching and transmission. These devices usually use a logarithmic quantizer, i.e., A-law or μ-law, which must be converted into a linear format for processing. The availability of inexpensive companded CODEC justifies their use as front-end devices for DSP systems. DSP chips implement this format conversion in hardware or in software by using a table lookup or calculation.

The most popular commercially available ADCs are successive approximation, dual slope, flash, and sigma-delta. The successive-approximation ADC produces a B-bit output in B cycles of its clock by comparing the input waveform with the output of a digital-to-analog converter. This device uses a successive-approximation register to split the voltage range in half in order to determine where the input signal lies. According to the comparator result, one bit will be set or reset each time. This process proceeds from the most significant bit (MSB) to the LSB. The successive-approximation type of ADC is generally accurate and fast at a relatively low cost. However, its ability to follow changes in the input signal is limited by its internal clock rate, so that it may be slow to respond to sudden changes in the input signal.

The dual-slope ADC uses an integrator connected to the input voltage and a reference voltage. The integrator starts at zero condition, and it is charged for a limited time. The integrator is then switched to a known negative reference voltage and charged in the opposite direction until it reaches zero volts again. At the same time, a digital counter starts to record the clock cycles. The number of counts required for the integrator output voltage to get back to zero is directly proportional to the input voltage. This technique is very precise and can produce ADCs with high resolution. Since the integrator is used for input and reference voltages, any small variations in temperature and aging of components have little or no effect on these types of converters. However, they are very slow and generally cost more than successive-approximation ADCs.

A voltage divider made by resistors is used to set reference voltages at the flash ADC inputs. The major advantage of a flash ADC is its speed of conversion, which is simply the propagation delay time of the comparators. Unfortunately, a B-bit ADC needs $(2^B - 1)$ comparators and laser-trimmed resistors. Therefore commercially available flash ADCs usually have lower bits.

The block diagram of a sigma–delta ADC is illustrated in Figure 1.6. Sigma–delta ADCs use a 1-bit quantizer with a very high sampling rate. Thus the requirements for an anti-aliasing filter are significantly relaxed (i.e., the lower roll-off rate and smaller flat response in passband). In the process of quantization, the resulting noise power is spread evenly over the entire spectrum. As a result, the noise power within the band of interest is lower. In order to match the output frequency with the system and increase its resolution, a decimator is used. The advantages of the sigma–delta ADCs are high resolution and good noise characteristics at a competitive price because they use digital filters.

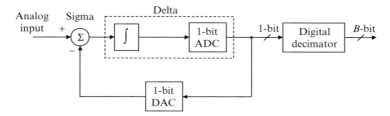

Figure 1.6 A block of a sigma-delta ADC

1.3 DSP Hardware

DSP systems require intensive arithmetic operations, especially multiplication and addition. In this section, different digital hardware architectures for DSP applications will be discussed.

1.3.1 DSP Hardware Options

As shown in Figure 1.1, the processing of the digital signal $x(n)$ is carried out using the DSP hardware. Although it is possible to implement DSP algorithms on any digital computer, the throughput (processing rate) determines the optimum hardware platform. Four DSP hardware platforms are widely used for DSP applications:

1. general-purpose microprocessors and microcontrollers (μP),

2. general-purpose digital signal processors (DSP chips),

3. digital building blocks (DBB) such as multiplier, adder, program controller, and

4. special-purpose (custom) devices such as application specific integrated circuits (ASIC).

The hardware characteristics are summarized in Table 1.1.

ASIC devices are usually designed for specific tasks that require a lot of DSP MIPS (million instructions per second), such as fast Fourier transform (FFT) devices and Reed–Solomon coders used by digital subscriber loop (xDSL) modems. These devices are able to perform their limited functions much faster than general-purpose DSP chips because of their dedicated architecture. These application-specific products enable the use of high-speed functions optimized in hardware, but they lack the programmability to modify the algorithm, so they are suitable for implementing well-defined and well-tested DSP algorithms. Therefore applications demanding high speeds typically employ ASICs, which allow critical DSP functions to be implemented in the hardware. The availability of core modules for common DSP functions has simplified the ASIC design tasks, but the cost of prototyping an ASIC device, a longer design cycle, insufficient

Table 1.1 Summary of DSP hardware implementations

	ASIC	DBB	μP	DSP chips
Chip count	1	> 1	1	1
Flexibility	none	limited	programmable	programmable
Design time	long	medium	short	short
Power consumption	low	medium–high	medium	low–medium
Processing speed	high	high	low–medium	medium–high
Reliability	high	low–medium	high	high
Development cost	high	medium	low	low
Production cost	low	high	low–medium	low–medium

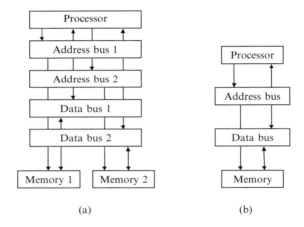

(a)   (b)

Figure 1.7 Different memory architectures: (a) Harvard architecture, and (b) von Newmann architecture

standard development tools support, and the lack of reprogramming flexibility sometimes outweigh their benefits.

Digital building blocks offer a more general-purpose approach to high-speed DSP design. These components, including multipliers, arithmetic logic units (ALUs), sequencers, etc., are joined together to build a custom DSP architecture for a specific application. Performance can be significantly higher than general-purpose DSP devices. However, the disadvantages are similar to those of the special-purpose DSP devices – lack of standard design tools, extended design cycles, and high component cost.

General architectures for computers and microprocessors fall into two categories: Harvard architecture and von Neumann architecture. Harvard architecture has a separate memory space for the program and the data, so that both memories can be accessed simultaneously, see Figure 1.7(a). The von Neumann architecture assumes that there is no intrinsic difference between the instructions and the data, and that the instructions can be partitioned into two major fields containing the operation command and the address of the operand. Figure 1.7(b) shows the memory architecture of the von Neumann model. Most general-purpose microprocessors use the von Neumann architecture. Operations such as add, move, and subtract are easy to perform. However, complex instructions such as multiplication and division are slow since they need a series of shift, addition, or subtraction operations. These devices do not have the architecture or the on-chip facilities required for efficient DSP operations. They may be used when a small amount of signal processing work is required in a much larger system. Their real-time DSP performance does not compare well with even the cheaper general-purpose DSP devices, and they would not be a cost-effective solution for many DSP tasks.

A DSP chip (digital signal processor) is basically a microprocessor whose architecture is optimized for processing specific operations at high rates. DSP chips with architectures and instruction sets specifically designed for DSP applications have been launched by Texas Instruments, Motorola, Lucent Technologies, Analog Devices, and many other companies. The rapid growth and the exploitation of DSP semiconductor technology are not a surprise, considering the commercial advantages in terms of the fast, flexible, and potentially low-cost design capabilities offered by these devices. General-purpose-programmable DSP chip developments are supported by software development tools such as C compilers, assemblers, optimizers, linkers, debuggers, simulators, and emulators. Texas Instruments' TMS320C55x, a programmable, high efficiency, and ultra low-power DSP chip, will be discussed in the next chapter.

1.3.2 Fixed- and Floating-Point Devices

A basic distinction between DSP chips is their fixed-point or floating-point architectures. The fixed-point representation of signals and arithmetic will be discussed in Chapter 3. Fixed-point processors are either 16-bit or 24-bit devices, while floating-point processors are usually 32-bit devices. A typical 16-bit fixed-point processor, such as the TMS320C55x, stores numbers in a 16-bit integer format. Although coefficients and signals are only stored with 16-bit precision, intermediate values (products) may be kept at 32-bit precision within the internal accumulators in order to reduce cumulative rounding errors. Fixed-point DSP devices are usually cheaper and faster than their floating-point counterparts because they use less silicon and have fewer external pins.

A typical 32-bit floating-point DSP device, such as the TMS320C3x, stores a 24-bit mantissa and an 8-bit exponent. A 32-bit floating-point format gives a large dynamic range. However, the resolution is still only 24 bits. Dynamic range limitations may be virtually ignored in a design using floating-point DSP chips. This is in contrast to fixed-point designs, where the designer has to apply scaling factors to prevent arithmetic overflow, which is a very difficult and time-consuming process.

Floating-point devices may be needed in applications where coefficients vary in time, signals and coefficients have a large dynamic range, or where large memory structures are required, such as in image processing. Other cases where floating-point devices can be justified are where development costs are high and production volumes are low. The faster development cycle for a floating-point device may easily outweigh the extra cost of the DSP device itself. Floating-point DSP chips also allow the efficient use of the high-level C compilers and reduce the need to identify the system's dynamic range.

1.3.3   Real-Time Constraints

A limitation of DSP systems for real-time applications is that the bandwidth of the system is limited by the sampling rate. The processing speed determines the rate at which the analog signal can be sampled. For example, a real-time DSP system demands that the signal processing time, t_p, must be less than the sampling period, T, in order to complete the processing task before the new sample comes in. That is,

$$t_p < T. \tag{1.3.1}$$

This real-time constraint limits the highest frequency signal that can be processed by a DSP system. This is given as

$$f_M \leq \frac{f_s}{2} < \frac{1}{2t_p}. \tag{1.3.2}$$

It is clear that the longer the processing time t_p, the lower the signal bandwidth f_M.

   Although new and faster DSP devices are introduced, there is still a limit to the processing that can be done in real time. This limit becomes even more apparent when system cost is taken into consideration. Generally, the real-time bandwidth can be increased by using faster DSP chips, simplified DSP algorithms, optimized DSP programs, and parallel processing using multiple DSP chips, etc. However, there is still a trade-off between costs and system performances, with many applications simply not being economical at present.

1.4   DSP System Design

A generalized DSP system design is illustrated in Figure 1.8. For a given application, the theoretical aspects of DSP system specifications such as system requirements, signal analysis, resource analysis, and configuration analysis are first performed to define the system requirements.

1.4.1   Algorithm Development

The algorithm for a given application is initially described using difference equations or signal-flow block diagrams with symbolic names for the inputs and outputs. In documenting the algorithm, it is sometimes helpful to further clarify which inputs and outputs are involved by means of a data flow diagram. The next stage of the development process is to provide more details on the sequence of operations that must be performed in order to derive the output from the input. There are two methods for characterizing the sequence of steps in a program: flowcharts or structured descriptions.

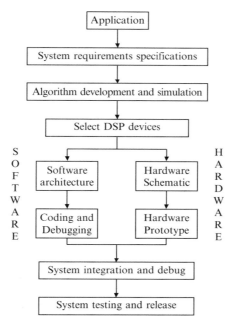

Figure 1.8 Simplified DSP system design flow

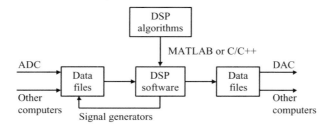

Figure 1.9 DSP software development using a general-purpose computer

At the algorithm development stage, we most likely work with high-level DSP tools (such as MATLAB or C/C++) that enable algorithmic-level system simulations. We then migrate the algorithm to software, hardware, or both, depending on our specific needs. A DSP application or algorithm can be first simulated using a general-purpose computer, such as a PC, so that it can be analyzed and tested off-line using simulated input data. A block diagram of general-purpose computer implementation is illustrated in Figure 1.9. The test signals may be internally generated by signal generators or digitized from an experimental setup based on the given application. The program uses the stored signal samples in data file(s) as input(s) to produce output signals that will be saved in data file(s).

Advantages of developing DSP software on a general-purpose computer are:

1. Using the high-level languages such as MATLAB, C/C++, or other DSP software packages can significantly save algorithm and software development time. In addition, C programs are portable to different DSP hardware platforms.

2. It is easy to debug and modify programs.

3. Input/output operations based on disk files are simple to implement and the behaviors of the system are easy to analyze.

4. Using the floating-point data format can achieve higher precision.

5. With fixed-point simulation, bit-true verification of an algorithm against fixed-point DSP implementation can easily be compared.

1.4.2   Selection of DSP Chips

A choice of DSP chip from many available devices requires a full understanding of the processing requirements of the DSP system under design. The objective is to select the device that meets the project's time-scales and provides the most cost-effective solution. Some decisions can be made at an early stage based on computational power, resolution, cost, etc. In real-time DSP, the efficient flow of data into and out of the processor is also critical. However, these criteria will probably still leave a number of candidate devices for further analysis. For high-volume applications, the cheapest device that can do the job should be chosen. For low- to medium-volume applications, there will be a trade-off between development time, development tool cost, and the cost of the DSP device itself. The likelihood of having higher-performance devices with upwards-compatible software in the future is also an important factor.

When processing speed is at a premium, the only valid comparison between devices is on an algorithm-implementation basis. Optimum code must be written for both devices and then the execution time must be compared. Other important factors are memory size and peripheral devices, such as serial and parallel interfaces, which are available on-chip.

In addition, a full set of development tools and supports are important for DSP chip selection, including:

1. Software development tools, such as assemblers, linkers, simulators, and C compilers.

2. Commercially available DSP boards for software development and testing before the target DSP hardware is available.

3. Hardware testing tools, such as in-circuit emulators and logic analyzers.

4. Development assistance, such as application notes, application libraries, data books, real-time debugging hardware, low-cost prototyping, etc.

1.4.3 Software Development

The four common measures of good DSP software are reliability, maintainability, extensibility, and efficiency. A reliable program is one that seldom (or never) fails. Since most programs will occasionally fail, a maintainable program is one that is easy to fix. A truly maintainable program is one that can be fixed by someone other than the original programmer. In order for a program to be truly maintainable, it must be portable on more than one type of hardware. An extensible program is one that can be easily modified when the requirements change, new functions need to be added, or new hardware features need to be exploited. An efficient DSP program will use the processing capabilities of the target hardware to minimize execution time.

A program is usually tested in a finite number of ways much smaller than the number of input data conditions. This means that a program can be considered reliable only after years of bug-free use in many different environments. A good DSP program often contains many small functions with only one purpose, which can be easily reused by other programs for different purposes. Programming tricks should be avoided at all costs as they will often not be reliable and will almost always be difficult for someone else to understand even with lots of comments. In addition, use variable names that are meaningful in the context of the program.

As shown in Figure 1.8, the hardware and software design can be conducted at the same time for a given DSP application. Since there is a lot of interdependence factors between hardware and software, the ideal DSP designer will be a true 'system' engineer, capable of understanding issues with both hardware and software. The cost of hardware has gone down dramatically in recent years. The majority of the cost of a DSP solution now resides in software development. This section discussed some issues regarding software development.

The software life cycle involves the completion of a software project: the project definition, the detailed specification, coding and modular testing, integration, and maintenance. Software maintenance is a significant part of the cost of a software system. Maintenance includes enhancing the software, fixing errors identified as the software is used, and modifying the software to work with new hardware and software. It is essential to document programs thoroughly with titles and comment statements because this greatly simplifies the task of software maintenance.

As discussed earlier, good programming technique plays an essential part in successful DSP application. A structured and well-documented approach to programming should be initiated from the beginning. It is important to develop an overall specification for signal processing tasks prior to writing any program. The specification includes the basic algorithm/task description, memory requirements, constraints on the program size, execution time, etc. Specification review is an important component of the software development process. A thoroughly reviewed specification can catch mistakes before code is written and reduce potential code rework risk at system integration stage. The potential use of subroutines for repetitive processes should also be noted. A flow diagram will be a very helpful design tool to adopt at this stage. Program and data blocks should be allocated to specific tasks that optimize data access time and addressing functions.

A software simulator or a hardware platform can be used for testing DSP code. Software simulators run on a host computer to mimic the behavior of a DSP chip. The simulator is able to show memory contents, all the internal registers, I/O, etc., and the effect on these after each instruction is performed. Input/output operations are simulated using disk files, which require some format conversion. This approach reduces the development process for software design only. Full real-time emulators are normally used when the software is to be tested on prototype target hardware.

Writing and testing DSP code is a highly iterative process. With the use of a simulator or an evaluation board, code may be tested regularly as it is written. Writing code in modules or sections can help this process, as each module can be tested individually, with a greater chance of the whole system working at the system integration stage.

There are two commonly used methods in developing software for DSP devices: an assembly program or a C/C++ program. Assembly language is one step removed from the machine code actually used by the processor. Programming in assembly language gives the engineers full control of processor functions, thus resulting in the most efficient program for mapping the algorithm by hand. However, this is a very time-consuming and laborious task, especially for today's highly paralleled DSP architectures. A C program is easier for software upgrades and maintenance. However, the machine code generated by a C compiler is inefficient in both processing speed and memory usage. Recently, DSP manufactures have improved C compiler efficiency dramatically.

Often the ideal solution is to work with a mixture of C and assembly code. The overall program is controlled by C code and the run-time critical loops are written in assembly language. In a mixed programming environment, an assembly routine may be either called as a function, or in-line coded into the C program. A library of hand-optimized functions may be built up and brought into the code when required. The fundamentals of C language for DSP applications will be introduced in Appendix C, while the assembly programming for the TMS320C55x will be discussed in Chapter 2. Mixed C and assembly programming will be introduced in Chapter 3. Alternatively, there are many high-level system design tools that can automatically generate an implementation in software, such as C and assembly language.

1.4.4 High-Level Software Development Tools

Software tools are computer programs that have been written to perform specific operations. Most DSP operations can be categorized as being either analysis tasks or filtering tasks. Signal analysis deals with the measurement of signal properties. MATLAB is a powerful environment for signal analysis and visualization, which are critical components in understanding and developing a DSP system. Signal filtering, such as removal of unwanted background noise and interference, is usually a time-domain operation. C programming is an efficient tool for performing signal filtering and is portable over different DSP platforms.

In general, there are two different types of data files: binary files and ASCII (text) files. A binary file contains data stored in a memory-efficient binary format, whereas an ASCII file contains information stored in ASCII characters. A binary file may be viewed as a sequence of characters, each addressable as an offset from the first position in the file. The system does not add any special characters to the data except null characters appended at the end of the file. Binary files are preferable for data that is going to be generated and used by application programs. ASCII files are necessary if the

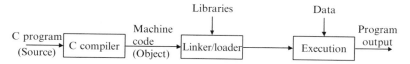

Figure 1.10   Program compilation, linking, and execution

data is to be shared by programs using different languages and different computer platforms, especially for data transfer over computer networks. In addition, an ASCII file can be generated using a word processor program or an editor.

MATLAB is an interactive, technical computing environment for scientific and engineering numerical analysis, computation, and visualization. Its strength lies in the fact that complex numerical problems can be solved easily in a fraction of the time required with a programming language such as C. By using its relatively simple programming capability, MATLAB can be easily extended to create new functions, and is further enhanced by numerous toolboxes such as the *Signal Processing Toolbox*. MATLAB is available on most commonly used computers such as PCs, workstations, Macintosh, and others. The version we use in this book is based on MATLAB for Windows, version 5.1. The brief introduction of using MATLAB for DSP is given in Appendix B.

The purpose of a programming language is to solve a problem involving the manipulation of information. The purpose of a DSP program is to manipulate signals in order to solve a specific signal-processing problem. High-level languages are computer languages that have English-like commands and instructions. They include languages such as C/C++, FORTRAN, Basic, and Pascal. High-level language programs are usually portable, so they can be recompiled and run on many different computers. Although C is categorized as a high-level language, it also allows access to low-level routines. In addition, a C compiler is available for most modern DSP devices such as the TMS320C55x. Thus C programming is the most commonly used high-level language for DSP applications.

C has become the language of choice for many DSP software development engineers not only because it has powerful commands and data structures, but also because it can easily be ported on different DSP platforms and devices. The processes of compilation, linking/loading, and execution are outlined in Figure 1.10. A C compiler translates a high-level C program into machine language that can be executed by the computer. C compilers are available for a wide range of computer platforms and DSP chips, thus making the C program the most portable software for DSP applications. Many C programming environments include debugger programs, which are useful in identifying errors in a source program. Debugger programs allow us to see values stored in variables at different points in a program, and to step through the program line by line.

1.5   Experiments Using Code Composer Studio

The code composer studio (CCS) is a useful utility that allows users to create, edit, build, debug, and analyze DSP programs. The CCS development environment supports several Texas Instruments DSP processors, including the TMS320C55x. For building applications, the CCS provides a project manager to handle the programming tasks. For debugging purposes, it provides breakpoint, variable watch, memory/register/stack viewing, probe point to stream data to and from the target, graphical analysis, execution profiling, and the capability to display mixed disassembled and C instructions. One important feature of the CCS is its ability to create and manage large projects from a graphic-user-interface environment. In this section, we will use a simple sinewave example to introduce the basic built-in editing features, major CCS components, and the use of the C55x development tools. We will also demonstrate simple approaches to software development and debugging process using the TMS320C55x simulator. The CCS version 1.8 was used in this book.

Installation of the CCS on a PC or a workstation is detailed in the Code Composer Studio Quick Start Guide [8]. If the C55x simulator has not been installed, use the CCS setup program to configure and set up the TMS320C55x simulator. We can start the CCS setup utility, either from the Windows start menu, or by clicking the Code Composer Studio Setup icon. When the setup dialogue box is displayed as shown in Figure 1.11(a), follow these steps to set up the simulator:

– Choose Install a Device Driver and select the C55x simulator device driver, `tisimc55.dvr` for the TMS320C55x simulator. The C55x simulator will appear in the middle window named as Available Board/Simulator Types if the installation is successful, as shown in Figure 1.11(b).

– Drag the C55x simulator from Available Board/Simulator Types window to the System Configuration window and save the change. When the system configuration is completed, the window label will be changed to Available Processor Types as shown in Figure 1.11(c).

1.5.1 Experiment 1A – Using the CCS and the TMS320C55x Simulator

This experiment introduces the basic features to build a project with the CCS. The purposes of the experiment are to:

(a) create projects,

(b) create source files,

(c) create linker command file for mapping the program to DSP memory space,

(d) set paths for C compiler and linker to search include files and libraries, and

(e) build and load program for simulation.

Let us begin with the simple sinewave example to get familiar with the TMS320C55x simulator. In this book, we assume all the experiment files are stored on a disk in the computer's A drive to make them portable for users, especially for students who may share the laboratory equipment.

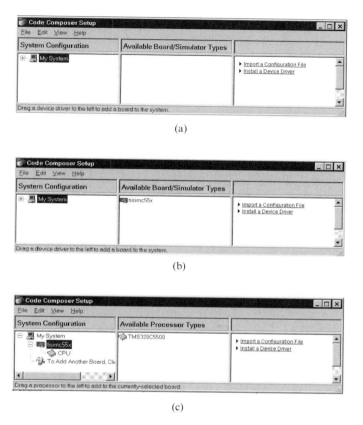

Figure 1.11 CCS setup dialogue boxes: (a) install the C55x simulator driver, (b) drag the C55x simulator to system configuration window, and (c) save the configuration

The best way to learn a new software tool is by using it. This experiment is partitioned into following six steps:

1. Start the CCS and simulator:
   - Invoke the CCS from the Start menu or by clicking on the Code Composer Studio icon on the PC. The CCS with the C55x simulator will appear on the computer screen as shown in Figure 1.12.

2. Create a project for the CCS:
   - Choose Project→New to create a new project file and save it as `exp1` to `A:\Experiment1`. The CCS uses the project to operate its built-in utilities to create a full build application.

3. Create a C program file using the CCS:
   - Choose File→New to create a new file, then type in the example C code listed in Table 1.2, and save it as `exp1.c` to `A:\Experiment1`. This example reads

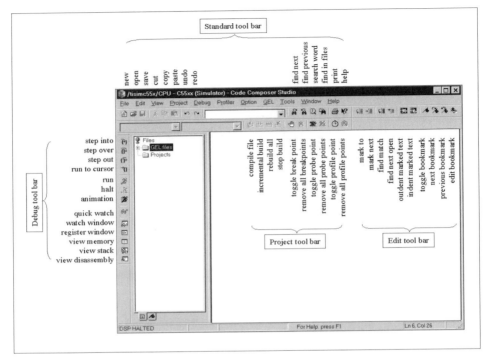

Figure 1.12   CCS integrated development environment

Table 1.2   List of sinewave example code, `exp1.c`

```
#define BUF_SIZE 40
const int sineTable[BUF_SIZE] =
    {0x0000,0x000f,0x001e,0x002d,0x003a,0x0046,0x0050,0x0059,
     0x005f,0x0062,0x0063,0x0062,0x005f,0x0059,0x0050,0x0046,
     0x003a,0x002d,0x001e,0x000f,0x0000,0xfff1,0xffe2,0xffd3,
     0xffc6,0xffba,0xffb0,0xffa7,0xffa1,0xff9e,0xff9d,0xff9e,
     0xffa1,0xffa7,0xffb0,0xffba,0xffc6,0xffd3,0xffe2,0xfff1};
int in_buffer[BUF_SIZE];
int out_buffer[BUF_SIZE];
int Gain;

void main()
{
    int i,j;
    Gain = 0x20;
```

Table 1.2   (*continued*)

```
while (1)
{          /* <- set profile point on this line */
    for (i = BUF_SIZE − 1; i >= 0; i−−)
    {
        j = BUF_SIZE − 1 − i;
        out_buffer[j] = 0;
        in_buffer[j] = 0;
    }
    for (i = BUF_SIZE − 1; i >= 0; i−−)
    {
        j = BUF_SIZE − 1 − i;
        in_buffer[i] = sineTable[i] ; /* <- set breakpoint */
        in_buffer[i] = 0 − in_buffer[i] ;
        out_buffer[j] = Gain* in_buffer[i] ;
    }
}          /* <- set probe and profile points on this line */
}
```

pre-calculated sinewave values from a table, negates, and stores the values in a reversed order to an output buffer. Note that the program exp1.c is included in the experimental software package.

However, it is recommended that we create this program with the editor to get familiar with the CCS editing functions.

4. Create a linker command file for the simulator:
   - Choose File→New to create another new file and type in the linker command file listed in Table 1.3 (or copy the file exp1.cmd from the experimental software package). Save this file as exp1.cmd to A:\Experiment1. Linker uses a command file to map different program segments into a pre-partitioned system memory space. A detailed description on how to define and use the linker command file will be presented in Chapter 2.

5. Setting up the project:
   - After exp1.c and exp1.cmd are created, add them to the project by choosing Project→Add Files, and select files exp1.c and exp1.cmd from A:\Experiment1.
   - Before building a project, the search paths should be set up for the C compiler, assembler, and linker. To set up options for the C compiler, assembler, and linker, choose Project→Options. The paths for the C55x tools should be set up during the CCS installation process. We will need to add search paths in order to include files and libraries that are not included in the C55x tools directories, such as the libraries and included files we have created in

Table 1.3 Linker command file

```
/* Specify the system memory map */

MEMORY
{
  RAM  (RWIX) : origin = 000100h, length = 01feffh /* Data memory    */
  RAM2 (RWIX) : origin = 040100h, length = 040000h /* Program memory */
  ROM  (RIX)  : origin = 020100h, length = 020000h /* Program memory */
  VECS (RIX)  : origin = 0ffff00h, length = 00100h /* Reset vector   */
}

/* Specify the sections allocation into memory */

SECTIONS
{
  vectors > VECS  /* Interrupt vector table          */
  .text   > ROM   /* Code                            */
  .switch > RAM   /* Switch table information        */
  .const  > RAM   /* Constant data                   */
  .cinit  > RAM2  /* Initialization tables           */
  .data   > RAM   /* Initialized data                */
  .bss    > RAM   /* Global & static variables       */
  .sysmem > RAM   /* Dynamic memory allocation area  */
  .stack  > RAM   /* Primary system stack            */
}
```

the working directory. For programs written in C language, it requires using the run-time support library, `rts55.lib` for DSP system initialization. This can be done by selecting Libraries under Category in the Linker dialogue box, and enter the C55x run-time support library, `rts55.lib`. We can also specify different directories to store the output executable file and map file. Figure 1.13 shows an example of how to set the search paths for compiler, assembler, or linker.

6. Build and run the program:
    – Once all the options are set, use Project→Rebuild All command to build the project. If there are no errors, the CCS will generate the executable output file, `exp1.out`. Before we can run the program, we need to load the executable output file to the simulator from File→Load Program menu. Pick the file `exp1.out` in A:\Experiment1 and open it.
    – Execute this program by choosing Debug→Run. DSP status at the bottom left-hand corner of the simulator will be changed from DSP HALTED to DSP RUNNING. The simulation process can be stopped withthe Debug→Halt command. We can continue the program by reissuing the run command or exiting the simulator by choosing File→Exit menu.

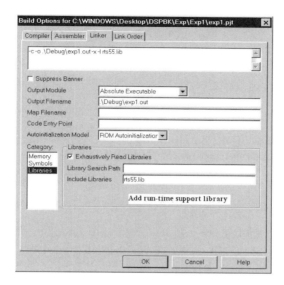

Figure 1.13 Setup search paths for C compiler, assembler, or linker

1.5.2 Experiment 1B – Debugging Program on the CCS

The CCS has extended traditional DSP code generation tools by integrating a set of editing, emulating, debugging, and analyzing capabilities in one entity. In this section of the experiment, we will introduce some DSP program building steps and software debugging capabilities including:

(a)   the CCS standard tools,

(b)   the advanced editing features,

(c)   the CCS project environment, and

(d)   the CCS debugging settings.

For a more detailed description of the CCS features and sophisticated configuration settings, please refer to Code Composer Studio User's Guide [7].

   Like most editors, the standard tool bar in Figure 1.12 allows users to create and open files, cut, copy, and paste texts within and between files. It also has undo and re-do capabilities to aid file editing. Finding or replacing texts can be done within one file or in different files. The CCS built-in context-sensitive help menu is also located in the standard toolbar menu. More advanced editing features are in the edit toolbar menu, refer to Figure 1.12. It includes mark to, mark next, find match, and find next open parenthesis capabilities for C programs. The features of out-indent and in-indent can be used to move a selected block of text horizontally. There are four bookmarks that allow users to create, remove, edit, and search bookmarks.

The project environment contains a C compiler, assembler, and linker for users to build projects. The project toolbar menu (see Figure 1.12) gives users different choices while working on projects. The compile only, incremental build, and build all functions allow users to build the program more efficiently. Breakpoints permit users to set stop points in the program and halt the DSP whenever the program executes at those breakpoint locations. Probe points are used to transfer data files in and out of programs. The profiler can be used to measure the execution time of the program. It provides program execution information, which can be used to analyze and identify critical run-time blocks of the program. Both the probe point and profile will be discussed in detail in the next section.

The debug toolbar menu illustrated in Figure 1.12 contains several step operations: single step, step into a function, step over a function, and step out from a function back to its caller function. It can also perform the run-to-cursor operation, which is a very convenient feature that allows users to step through the code. The next three hot buttons in the debug tool bar are run, halt, and animate. They allow users to execute, stop, and animate the program at anytime. The watch-windows are used to monitor variable contents. DSP CPU registers, data memory, and stack viewing windows provide additional information for debugging programs. More custom options are available from the pull-down menus, such as graphing data directly from memory locations.

When we are developing and testing programs, we often need to check the values of variables during program execution. In this experiment, we will apply debugging settings such as breakpoints, step commands, and watch-window to understand the CCS. The experiment can be divided into the following four steps.

1. Add and remove breakpoints:
   - Start with Project→Open, select exp1 in the A:\Experiment1 directory. Build and load the experiment exp1.out. Double-click on the C file exp1.c in the project-viewing window to open it from the source folder.
   - Adding and removing a breakpoint to a specific line is quite simple. To add a breakpoint, move the cursor to the line where we want to set a breakpoint. The command to enable a breakpoint can be given from the Toggle Breakpoint hot button on the project toolbar or by clicking the right mouse button and choosing toggle breakpoint. The function key <F9> is a shortcut key that also enables breakpoints. Once a breakpoint is enabled, a red dot will appear on the left to indicate where the breakpoint is set. The program will run up to that line without exceeding it. To remove breakpoints, we can either toggle breakpoints one by one, or we can select the Delete All tap from the debug tool bar to clear all the breakpoints at once. Now put the cursor on the following line:

     ```
     in_buffer[i] = sineTable[i];   /* <- set breakpoint */
     ```

     and click the Toggle Breakpoint toolbar button (or press <F9>).

2. Set up viewing windows:
   - On the standard tool menu bar click View→CPU Registers→CPU Registers to open the CPU registers window. We can edit the contents of any

CPU register by double clicking on it. Right click on the CPU Register window and select Allow Docking. We can now move and resize the window. Try to change the temporary register T0 and accumulator AC0 to T0 = 0x1234 and AC0 = 0x56789ABC.

– On the CCS menu bar click Tools→Command Window to add the Command Window. We can resize and dock it as in the previous step. The command window will appear each time we rebuild the project.

– We can customize the CCS display and settings using the workspace feature. To save a workspace, click File→Workspace→Save Workspace and give the workspace a name. When we restart CCS, we can reload that workspace by clicking File→Workspace→Load Workspace and select the proper workspace filename.

– Click View→Dis-Assembly on the menu bar to see the disassembly window. Every time we reload an executable file, the disassembly window will appear automatically.

3. Using the single step features:
   – When using C programs, the C55x system uses a function called `boot` from the run-time support library `rts55.lib` to initialize the system. After we load the `exp1.out`, the program counter (PC) should be at the start of the boot function and the assembly code, `boot.asm`, should be displayed in the disassembly window. For a project starting with C programs, there must be a function called `main()` from which the C functions logically begin to execute. We can issue the command, Go Main, from the Debug menu to start the C program.
   – After the Go Main command, the DSP will be halted at the location where the function `main()` is. Hit the <F8> key or click the single step button on the debug toolbar repeatedly and single-step through the program `exp1.c`, watching the values of the CPU registers change. Move the cursor to a different location in the code and try the run-to-cursor command (hold down the <Ctrl> and <F10> keys simultaneously).

4. Resource monitoring:
   – From View→Watch Window, open the Watch Window area. At run time, this area shows the values of watched variables. Right-click on the Watch Window area and choose Insert New Expression from the pop-up list. Type the output buffer name, `out_buffer`, into the expression box and click OK, expand the `out_buffer` to view each individual element of the buffer.
   – From View→Memory, open a memory window and enter the starting address of the `in_buffer` in data page to view the data in the input and output buffers. Since global variables are defined globally, we can use the variable name as the address for memory viewing.
   – From View→Graph→Time/Frequency, open the Graphic Property dialogue. Set the display parameters as shown in Figure 1.14. The CCS allows the user to plot data directly from memory by specifying the memory location and its length.
   – Set a breakpoint on the line of the following C statement:

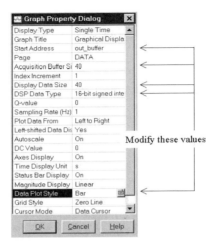

Figure 1.14   Graphics display settings

```
in_buffer[i] = sineTable[i];   /* <- set breakpoint  */
```

Start animation execution, and view CPU registers, `in_buffer` and `out_buffer` data in both the watch-window and the memory window. Figure 1.15 shows one instant snapshot of the animation. The yellow arrow represents the current program counter's location, and the red dot shows where the breakpoint is set. The data and register values in red color are the ones that have just been updated.

1.5.3   Experiment 1C – File Input and Output

Probe point is a useful tool for algorithm development, such as simulating real-time input and output operations. When a probe point is reached, the CCS can either read a selected amount of data samples from a file on the host PC to DSP memory on the target, or write processed data samples to the host PC. In the following experiment, we will learn how to set up a probe point to transfer data from the example program to a file on the host computer.

– Set the probe point at the end of the `while{}` loop at the line of the close bracket as:

```
while(1)
{
    ... ...
}          /* <- set probe point on this line */
```

where the data in the output buffer is ready to be transferred out. Put the cursor on the line and click Toggle Probe Point. A blue dot on the left indicates the probe point is set (refer to Figure 1.15).

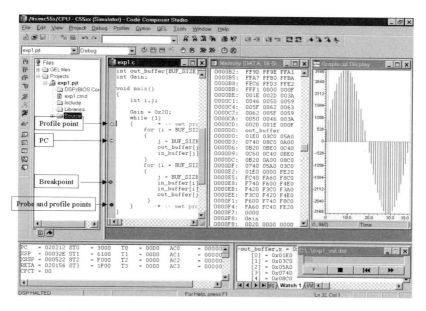

Figure 1.15   CCS screen snapshot of the Experiment 1B

– From File→File I/O, open the file I/O dialogue box and select File Output tab. From the Add File tab, enter `exp1_out.dat` as file name, then select Open. Using the output variable name, `out_buffer`, as the address and 40 (`BUF_SIZE`) as the length of the data block for transferring 40 data samples to the host computer from the buffer every time the probe point is reached. Now select Add Probe Point tab to connect the probe point with the output file `exp1_out.dat` as shown in Figure 1.16.

– Restart the program. After execution, we can view the data file `exp1_out.dat` using the built-in editor by issue File→Open command. If we want to view or edit the data file using other editors/viewers, we need to exit the CCS or disconnect the file from the File I/O.

An example data file is shown in Table 1.4. The first line contains the header information in TI Hexadecimal format, which uses the syntax illustrated in Figure 1.17.

For the example given in Table 1.4, the data stored is in hexadecimal format with the address of `out_buffer` at 0xa8 on data page and each block containing 40 (0x28) data values. If we want to use probe to connect an input data file to the program, we will need to use the same hex format to include a header in the input data file.

1.5.4   Experiment 1D – Code Efficiency Analysis

The profiler can be used to measure the system execution status of specific segments of the code. This feature gives users an immediate result about the program's performance.

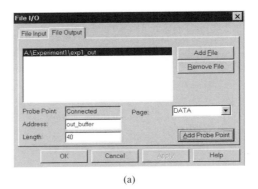

(a)

(b)

Figure 1.16   Connect probe point to a file: (a) set up probe point address and length, and (b) connect probe point with a file

Table 1.4   Data file saved by CCS

```
1651 1 a8 1 28
0x01E0
0x03C0
0x05A0
0x0740
0x08C0
0x0A00
0x0B20
0x0BE0
. . .
```

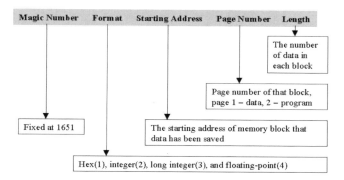

Figure 1.17 CCS File header format

It is a very useful tool for analyzing and optimizing DSP code for large complex projects. In the following experiment, we use the profiling features of the CCS to obtain statistics about code execution time.

– Open the project `exp1` and load the file `exp1.out`. Open the source file `exp1.c` and identify the line numbers on the source code where we like to set profile marks. For a demonstration purpose, we will profile the entire code within the `while {}` loop in the experiment. The profile points are set at line 32 and 46 as shown below:

```
while (1)
{       /* <- set profile point here */
    ... ...
}       /* <- set profile point here */
```

– From Profiler menu, select Start New Session to open the profile window. Click the Create Profile Area hot button, and in the Manual Profile Area Creation dialogue box (see Figure 1.18), enter the number for starting and ending lines. In the meantime, make sure the Source File Lines and the Generic type are selected. Finally, click on the Ranges tab to switch the window that displays the range of the code segments we just selected.

– The profiler is based on the system clock. We need to select Profile→ Enable Clock to enable the system clock before starting the profiler. This clock counts instruction cycles. The clock setting can also be adjusted. Since the C55x simulator does not have the connection to real hardware, the profiler of the simulator can only display CPU cycles in the count field (refer to the example given in Figure 1.18). More information can be obtained by using real DSP hardware such as the C55x EVMs.

– Run the program and record the cycle counts shown in the profile status window.

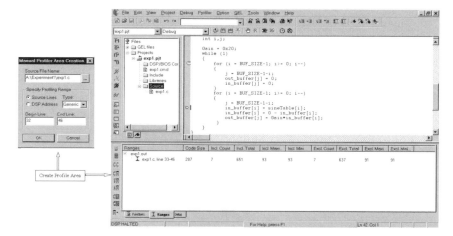

Figure 1.18   Profile window displaying DSP run-time status

Table 1.5   Gain control GEL function

```
Menuitem "Gain Control"
slider Gain(1, 0x20, 1, 1, gainParam)
{
    Gain = gainParam;
}
```

Figure 1.19   GEL slide bar

1.5.5   Experiment 1E – General Extension Language

The CCS uses General Extension Language (GEL) to extend its functions. GEL is a useful tool for automated testing and design of workspace customization. The Code Composer Studio User's Guide [7] provides a detailed description of GEL functions. In this experiment, we will use a simple example to introduce it.

Create a file called `Gain.gel` and type the simple GEL code listed in Table 1.5. From the CCS, load this GEL file from File→Load GEL and bring the Gain control slide bar shown in Figure 1.19 out from GEL→Gain Control. While animating the program using the CCS, we can change the gain by moving the slider up and down.

References

[1] A. V. Oppenheim and R. W. Schafer, *Discrete-Time Signal Processing*, Englewood Cliffs, NJ: Prentice-Hall, 1989.
[2] S. J. Orfanidis, *Introduction to Signal Processing*, Englewood Cliffs, NJ: Prentice-Hall, 1996.
[3] J. G. Proakis and D. G. Manolakis, *Digital Signal Processing – Principles, Algorithms, and Applications*, 3rd Ed., Englewood Cliffs, NJ: Prentice-Hall, 1996.
[4] A Bateman and W. Yates, *Digital Signal Processing Design*, New York: Computer Science Press, 1989.
[5] S. M. Kuo and D. R. Morgan, *Active Noise Control Systems – Algorithms and DSP Implementations*, New York: Wiley, 1996.
[6] J. H. McClellan, R. W. Schafer, and M. A. Yoder, *DSP First: A Multimedia Approach*, 2nd Ed., Englewood Cliffs, NJ: Prentice-Hall, 1998.
[7] Texas Instruments, Inc., *Code Composer Studio User's Guide*, Literature no. SPRU32, 1999.
[8] Texas Instruments, Inc., *Code Composer Studio Quick Start Guide*, Literature no. SPRU368A, 1999.

Exercises

Part A

1. Given an analog audio signal with frequencies up to 10 kHz.

   (a) What is the minimum required sampling frequency that allows a perfect reconstruction of the signal from its samples?

   (b) What will happen if a sampling frequency of 8 kHz is used?

   (c) What will happen if the sampling frequency is 50 kHz?

   (d) When sampled at 50 kHz, if only taking every other samples (this is a decimation by 2), what is the frequency of the new signal? Is this causing aliasing?

2. Refer to Example 1.1, assuming that we have to store 50 ms (milliseconds, 1 ms $= 10^{-3}$ seconds) of digitized signals. How many samples are needed for (a) narrowband telecommunication systems with $f_s = 8$ kHz, (b) wideband telecommunication systems with $f_s = 16$ kHz, (c) audio CDs with $f_s = 44.1$ kHz, and (d) professional audio systems with $f_s = 48$ kHz.

3. Given a discrete time sinusoidal signal of $x(n) = 5\sin(n\pi/100)$ V.

   (a) Find its peak-to-peak range?

   (b) What is the quantization resolution of an 8-bit ADC for this signal?

   (c) In order to obtain the quantization resolution of below 1 mV, how many bits are required in the ADC?

Part B

4. From the Option menu, set the CCS for automatically loading the program after the project has been built.

5. To reduce the number of mouse click, many pull-down menu items have been mapped to the hot buttons for the standard, advanced edit, project management, and debug tools bar. There are still some functions; however, do not associate with any hot buttons. Using the Option menu to create shortcut keys for the following menu items:

    (a) map Go Main in the debug menu to Alt+M (Alt key and M key),

    (b) map Reset in the debug menu to Alt+R,

    (c) map Restart in the debug menu to Alt+S, and

    (d) map File reload in the file menu to Ctrl+R.

6. After having loaded a program into the simulator and enabled Source/ASM mixed display mode from View→Mixed Source/ASM, what is showing in the CCS source display window besides the C source code?

7. How to change the format of displayed data in the watch-window to hex, long, and floating-point format from integer format?

8. What does File→Workspace do? Try the save and reload workspace commands.

9. Besides using file I/O with the probe point, data values in a block of memory space can also be stored to a file. Try the File→Data→Store and File→Data→Load commands.

10. Use Edit→Memory command we can manipulate (edit, copy, and fill) system memory:

    (a) open memory window to view `out_buffer`,

    (b) fill `out_buffer` with data 0x5555, and

    (c) copy the constant `sineTable[]` to `out_buffer`.

11. Using CCS context-sensitive on-line help menu to find the TMS320C55x CUP diagram, and name all the buses and processing units.

2

Introduction to TMS320C55x Digital Signal Processor

Digital signal processors with architecture and instructions specifically designed for DSP applications have been launched by Texas Instruments, Motorola, Lucent Technologies, Analog Devices, and many other companies. DSP processors are widely used in areas such as communications, speech processing, image processing, biomedical devices and equipment, power electronics, automotive, industrial electronics, digital instruments, consumer electronics, multimedia systems, and home appliances.

To efficiently design and implement DSP systems, we must have a solid knowledge of DSP algorithms as well as a basic concept of processor architecture. In this chapter, we will introduce the architecture and assembly programming of the Texas Instruments TMS320C55x fixed-point processor.

2.1 Introduction

Wireless communications, telecommunications, medical, and multimedia applications are developing rapidly. Increasingly traditional analog devices are being replaced with digital systems. The fast growth of DSP applications is not a surprise when considering the commercial advantages of DSP in terms of the potentially fast time to market, flexibility for upgrades to new technologies and standards, and low design cost offered by various DSP devices. The rising demand from the digital handheld devices in the consumer market to the digital networks and communication infrastructures coupled with the emerging internet applications are the driving forces for DSP applications.

In 1982, Texas Instruments introduced its first general-purpose fixed-point DSP device, the TMS32010, to the consumer market. Since then, the TMS320 family has extended into two major classes: the fixed-point and floating-point processors. The TMS320 fixed-point family consists of C1x, C2x, C5x, C2xx, C54x, C55x, C62x, and C64x. The TMS320 floating-point family includes C3x, C4x, and C67x. Each generation of the TMS320 series has a unique central processing unit (CPU) with a variety of memory and peripheral configurations. In this book, we chose the TMS320C55x as an example for real-time DSP implementations, applications, and experiments.

The C55x processor is designed for low power consumption, optimum performance, and high code density. Its dual multiply–accumulate (MAC) architecture provides twice the cycle efficiency computing vector products – the fundamental operation of digital signal processing, and its scaleable instruction length significantly improves the code density. In addition, the C55x is source code compatible with the C54x. This greatly reduces the migration cost from the popular C54x based systems to the C55x systems.

Some essential features of the C55x device are listed below:

- Upward source-code compatible with all TMS320C54x devices.

- 64-byte instruction buffer queue that works as a program cache and efficiently implements block repeat operations.

- Two 17-bit by 17-bit MAC units can execute dual multiply-and-accumulate operations in a single cycle.

- A 40-bit arithmetic and logic unit (ALU) performs high precision arithmetic and logic operations with an additional 16-bit ALU performing simple arithmetic operations parallel to the main ALU.

- Four 40-bit accumulators for storing computational results in order to reduce memory access.

- Eight extended auxiliary registers for data addressing plus four temporary data registers to ease data processing requirements.

- Circular addressing mode supports up to five circular buffers.

- Single-instruction repeat and block repeat operations of program for supporting zero-overhead looping.

Detailed information about the TMS320C55x can be found in the manufacturer's manuals listed in references [1–6].

2.2   TMS320C55x Architecture

The C55x CPU consists of four processing units: an instruction buffer unit (IU), a program flow unit (PU), an address-data flow unit (AU), and a data computation unit (DU). These units are connected to 12 different address and data buses as shown in Figure 2.1.

2.2.1   TMS320C55x Architecture Overview

Instruction buffer unit (IU): This unit fetches instructions from the memory into the CPU. The C55x is designed for optimum execution time and code density. The instruction set of the C55x varies in length. Simple instructions are encoded using eight bits

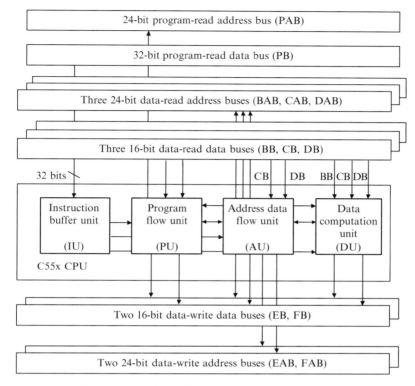

Figure 2.1   Block diagram of TMS320C55x CPU

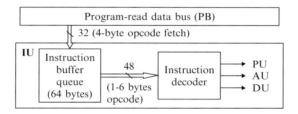

Figure 2.2   Simplified block diagram of the C55x instruction buffer unit

(one byte), while more complicated instructions may contain as many as 48 bits (six bytes). For each clock cycle, the IU can fetch four bytes of program code via its 32-bit program-read data bus. At the same time, the IU can decode up to six bytes of program. After four program bytes are fetched, the IU places them into the 64-byte instruction buffer. At the same time, the decoding logic decodes an instruction of one to six bytes previously placed in the instruction decoder as shown in Figure 2.2. The decoded instruction is passed to the PU, the AU, or the DU.

The IU improves the efficiency of the program execution by maintaining a constant stream of instruction flow between the four units within the CPU. If the IU is able to hold a segment of the code within a loop, the program execution can be repeated many times without fetching additional code. Such a capability not only improves the loop execution time, but also saves the power consumption by reducing program accesses from the memory. Another advantage is that the instruction buffer can hold multiple instructions that are used in conjunction with conditional program flow control. This can minimize the overhead caused by program flow discontinuities such as conditional calls and branches.

Program flow unit (PU): This unit controls DSP program execution flow. As illustrated in Figure 2.3, the PU consists of a program counter (PC), four status registers, a program address generator, and a pipeline protection unit. The PC tracks the C55x program execution every clock cycle. The program address generator produces a 24-bit address that covers 16 Mbytes of program space. Since most instructions will be executed sequentially, the C55x utilizes pipeline structure to improve its execution efficiency. However, instructions such as branches, call, return, conditional execution, and interrupt will cause a non-sequential program address switch. The PU uses a dedicated pipeline protection unit to prevent program flow from any pipeline vulnerabilities caused by a non-sequential execution.

Address-data flow unit (AU): The address-data flow unit serves as the data access manager for the data read and data write buses. The block diagram illustrated in Figure 2.4 shows that the AU generates the data-space addresses for data read and data write. It also shows that the AU consists of eight 23-bit extended auxiliary registers (XAR0–XAR7), four 16-bit temporary registers (T0–T3), a 23-bit extended coefficient data pointer (XCDP), and a 23-bit extended stack pointer (XSP). It has an additional 16-bit ALU that can be used for simple arithmetic operations. The temporary registers may be utilized to expand compiler efficiency by minimizing the need for memory access. The AU allows two address registers and a coefficient pointer to be used together for processing dual-data and one coefficient in a single clock cycle. The AU also supports up to five circular buffers, which will be discussed later.

Data computation unit (DU): The DU handles data processing for most C55x applications. As illustrated in Figure 2.5, the DU consists of a pair of MAC units, a 40-bit ALU, four 40-bit accumulators (AC0, AC1, AC2, and AC3), a barrel shifter, rounding and saturation control logic. There are three data-read data buses that allow two data paths and a coefficient path to be connected to the dual-MAC units simultaneously. In a single cycle, each MAC unit can perform a 17-bit multiplication

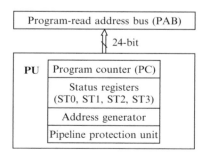

Figure 2.3 Simplified block diagram of the C55x program flow unit

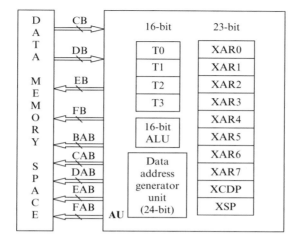

Figure 2.4 Simplified block diagram of the C55x address-data flow unit

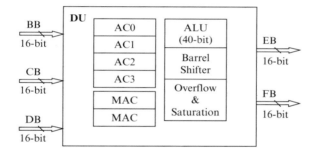

Figure 2.5 Simplified block diagram of the C55x data computation unit

and a 40-bit addition or subtraction operation with a saturation option. The ALU can perform 40-bit arithmetic, logic, rounding, and saturation operations using the four accumulators. It can also be used to achieve two 16-bit arithmetic operations in both the upper and lower portions of an accumulator at the same time. The ALU can accept immediate values from the IU as data and communicate with other AU and PU registers. The barrel shifter may be used to perform a data shift in the range of 2^{-32} (shift right 32-bit) to 2^{31} (shift left 31-bit).

2.2.2 TMS320C55x Buses

As illustrated in Figure 2.1, the TMS320C55x has one 32-bit program data bus, five 16-bit data buses, and six 24-bit address buses. The program buses include a 32-bit program-read data bus (PB) and a 24-bit program-read address bus (PAB). The PAB carries the program memory address to read the code from the program space. The unit of program address is in bytes. Thus the addressable program space is in the range of

0x000000–0xFFFFFF (the prefix 0x indicates the following number is in hexadecimal format). The PB transfers four bytes of program code to the IU each clock cycle. The data buses consist of three 16-bit data-read data buses (BB, CB, and DB) and three 24-bit data-read addresses buses (BAB, CAB, and DAB). This architecture supports three simultaneous data reads from data memory or I/O space. The C bus and D buses (CB and DB) can send data to the PU, AU, and DU; while the B bus (BB) can only work with the DU. The primary function of the BB is to connect memory to a dual-MAC; so some specific operations can access all three data buses, such as fetching two data and one coefficient. The data-write operations are carried out using two 16-bit data-write data buses (EB and FB) and two 24-bit data-write address buses (EAB and FAB). For a single 16-bit data write, only the EB is used. A 32-bit data write will use both the EB and FB in one cycle. The data-write address buses (EAB and FAB) have the same 24-bit addressing range. Since the data access uses a word unit (2-byte), the data memory space becomes 23-bit word addressable from address 0x000000 to 0x7FFFFF.

The C55x architecture is built around these 12 buses. The program buses carry the instruction code and immediate operands from program memory, while the data buses connect various units. This architecture maximizes the processing power by maintaining separate memory bus structures for full-speed execution.

2.2.3   TMS320C55x Memory Map

The C55x uses a unified program, data, and I/O memory configurations. All 16 Mbytes of memory are available as program or data space. The program space is used for instructions and the data space is used for general-purpose storage and CPU memory mapped registers. The I/O space is separated from the program/data space, and is used for duplex communication with peripherals. When the CPU fetches instructions from the program space, the C55x address generator uses the 24-bit program-read address bus. The program code is stored in byte units. When the CPU accesses data space, the C55x address generator masks the least-significant-bit (LSB) of the data address since data stored in memory is in word units. The 16 Mbytes memory map is shown in Figure 2.6. Data space is divided into 128 data pages (0–127). Each page has 64 K words. The memory block from address 0 to 0x5F in page 0 is reserved for memory mapped registers (MMRs).

2.3   Software Development Tools

The manufacturers of DSP processors typically provide a set of software tools for the user to develop efficient DSP software. The basic software tools include an assembler, linker, C compiler, and simulator. As discussed in Section 1.4, DSP programs can be written in either C or assembly language. Developing C programs for DSP applications requires less time and effort than those applications using assembly programs. However, the run-time efficiency and the program code density of the C programs are generally worse than those of the assembly programs. In practice, high-level language tools such

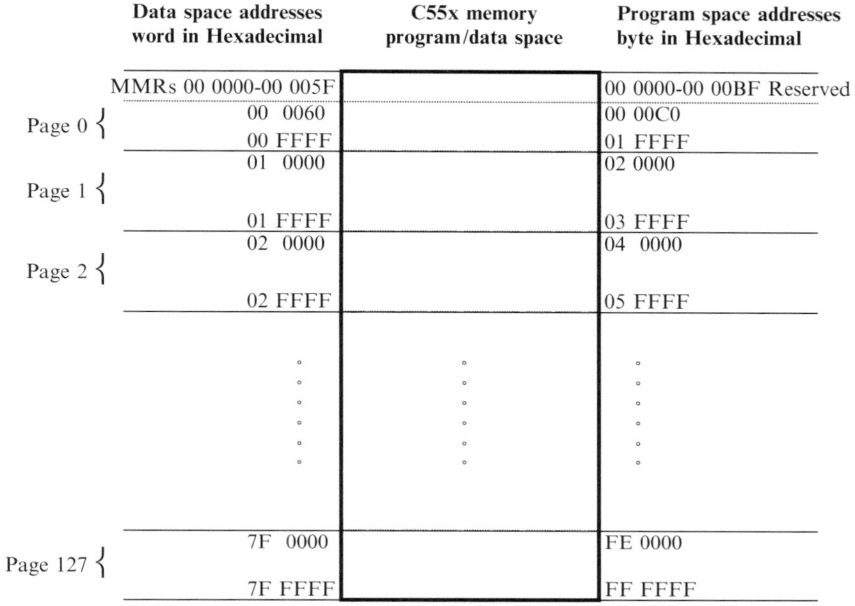

Data space addresses word in Hexadecimal	C55x memory program/data space	Program space addresses byte in Hexadecimal
MMRs 00 0000-00 005F		00 0000-00 00BF Reserved
00 0060		00 00C0
00 FFFF		01 FFFF
01 0000		02 0000
01 FFFF		03 FFFF
02 0000		04 0000
02 FFFF		05 FFFF
7F 0000		FE 0000
7F FFFF		FF FFFF

Page 0 { (00 0060 – 00 FFFF)
Page 1 { (01 0000 – 01 FFFF)
Page 2 { (02 0000 – 02 FFFF)
Page 127 { (7F 0000 – 7F FFFF)

Figure 2.6 TMS320C55x program space and data space memory map

as MATLAB and C are used in early development stages to verify and analyze the functionality of the algorithms. Due to real-time constraints and/or memory limitations, part (or all) of the C functions have to be replaced with assembly programs.

In order to execute the designed DSP algorithms on the target system, the C or assembly programs must first be translated into binary machine code and then linked together to form an executable code for the target DSP hardware. This code conversion process is carried out using the software development tools illustrated in Figure 2.7.

The TMS320C55x software development tools include a C compiler, an assembler, a linker, an archiver, a hex conversion utility, a cross-reference utility, and an absolute lister. The debugging tools can either be a simulator or an emulator. The C55x C compiler generates assembly code from the C source files. The assembler translates assembly source files; either hand-coded by the engineers or generated by the C compiler, into machine language object files. The assembly tools use the common object file format (COFF) to facilitate modular programming. Using COFF allows the programmer to define the system's memory map at link time. This maximizes performance by enabling the programmer to link the code and data objects into specific memory locations. The archiver allows users to collect a group of files into a single archived file. The linker combines object files and libraries into a single executable COFF object module. The hex conversion utility converts a COFF object file into a format that can be downloaded to an EPROM programmer.

In this section, we will briefly describe the C compiler, assembler, and linker. A full description of these tools can be found in the user's guides [2,3].

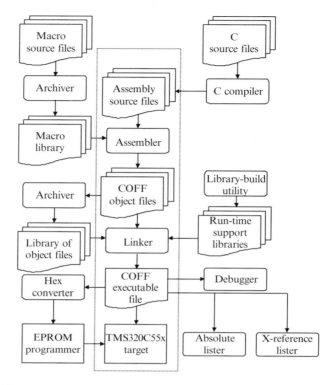

Figure 2.7   TMS320C55x software development flow and tools

2.3.1   C Compiler

As mentioned in Chapter 1, C language is the most popular high-level tool for evaluating DSP algorithms and developing real-time software for practical applications. The TMS320C55x C compiler translates the C source code into the TMS320C55x assembly source code first. The assembly code is then given to the assembler for generating machine code. The C compiler can generate either a mnemonic assembly code or algebraic assembly code. Table 2.1 gives an example of the mnemonic and algebraic assembly code generated by the C55x compiler. In this book, we will introduce only the widely used mnemonic assembly language. The C compiler package includes a shell program, code optimizer, and C-to-ASM interlister. The shell program supports automatic compile, assemble, and link modules. The optimizer improves run-time and code density efficiency of the C source files. The C-to-ASM interlister inserts the original comments in C source code into the compiler's output assembly code; so the user can view the corresponding assembly instructions generated by the compiler for each C statement.

The C55x compiler supports American National Standards Institute (ANSI) C and its run-time-support library. The run-time support library, rts55.lib, includes functions to support string operation, memory allocation, data conversion, trigonometry, and exponential manipulations. The CCS introduced in Section 1.5 has made using DSP development tools (compiler, assembly, and linker) easier by providing default setting

Table 2.1 An example of C code and the C55x compiler generated assembly code

Code	Mnemonic assembly code	Algebraic assembly code
	mov * SP (#0) , AR2	AR2 = * SP (#0)
	add #_sineTable, AR2	AR2 = AR2 + #_sineTable
in_buffer[i] = sineTable[i];	mov * SP (#0) , AR3	AR3 = * SP (#0)
	add #_in_buffer, AR3	AR3 = AR3 + #_in_buffer
	mov * AR2, * AR3	* AR3 = * AR2

parameters and prompting the options. It is still beneficial for the user to understand how to use these tools individually, and set parameters and options from the command line correctly.

We can invoke the C compiler from a PC or workstation shell by entering the following command:

```
c155 [-options] [filenames] [-z[link_options] [object_files]]
```

The `filenames` can be one or more C program source files, assembly source files, object files, or a combination of these files. If we do not supply an extension, the compiler assumes the default extension as `.c`, `.asm`, or `.obj`. The `-z` option enables the linker, while the `-c` option disables the linker. The `link_options` set up the way the linker processes the object files at link time. The `object_files` are additional objective files for the linker to add to the target file at link time. The compiler `options` have the following categories:

1. The options that control the compiler shell, such as the `-g` option that generates symbolic debug information for debugging code.

2. The options that control the parser, such as the `-ps` option that sets the strict ANSI C mode for C.

3. The options that are C55x specific, such as the `-ml` option that sets the large memory model.

4. The options that control the optimization, such as the `-o0` option that sets the register optimization.

5. The options that change the file naming conventions and specify the directories, such as the `-eo` option that sets the default object file extension.

6. The options that control the assembler, such as the `-al` option that creates assembly language listing files.

7. The options that control the linker, such as the `-ar` option that generates a relocatable output module.

There are a number of options in each of the above categories. Refer to the TMS320C55x Optimizing C Compiler User's Guide [3] for detailed information on how to use these options.

The options are preceded by a hyphen and are not case sensitive. All the single letter options can be combined together, i.e., the options of -g, -k, and -s, are the same as setting the compiler options as -gks. The two-letter operations can also be combined if they have the same first letter. For example, setting -pl, -pk, and -pi three options are the same as setting the options as -plki.

C language lacks specific DSP features, especially those of fixed-point data operations that are necessary for many DSP algorithms. To improve compiler efficiency for real-time DSP applications, the C55x compiler provides a method to add in-line assembly language routines directly into the C program. This allows the programmer to write highly efficient assembly code for the time-critical sections of a program. Intrinsic is another improvement for users to substitute DSP arithmetic operation with assembly intrinsic operators. We will introduce more compiler features in Section 2.7 when we present the mixing of C and assembly programs. In this chapter, we emphasize assembly language programming.

2.3.2   Assembler

The assembler translates processor-specific assembly language source files (in ASCII text) into binary COFF object files for specific DSP processors. Source files can contain assembler directives, macro directives, and instructions. Assembler directives are used to control various aspects of the assembly process such as the source file listing format, data alignment, section content, etc. Binary object files contain separate blocks (called sections) of code or data that can be loaded into memory space.

Assembler directives are used to control the assembly process and to enter data into the program. Assembly directives can be used to initialize memory, define global variables, set conditional assembly blocks, and reserve memory space for code and data. Some of the most important C55x assembler directives are described below:

.BSS directive: The .bss directive reserves space in the uninitialized .bss section for data variables. It is usually used to allocate data into RAM for run-time variables such as I/O buffers. For example,

```
.bss xn_buffer, size_in_words
```

where the xn_buffer points to the first location of the reserved memory space, and the size_in_words specifies the number of words to be reserved in the .bss section. If we do not specify uninitialized data sections, the assembler will put all the uninitialized data into the .bss section.

.DATA directive: The .data directive tells the assembler to begin assembling the source code into the .data section, which usually contains data tables or pre-initialized variables such as sinewave tables. The data sections are word addressable.

.SECT directive: The .sect directive defines a section and tells the assembler to begin assembling source code or data into that section. It is often used to separate long programs into logical partitions. It can separate the subroutines from the main program, or separate constants that belong to different tasks. For example,

```
.sect "section_name"
```

assigns the code into the user defined memory section called `section_name`. Code from different source files with the same section names are placed together.

.USECT directive: The `.usect` reserves space in an uninitialized section. It is similar to the `.bss` directive. It allows the placement of data into user defined sections instead of `.bss` sections. It is often used to separate large data sections into logical partitions, such as separating the transmitter data variables from the receiver data variables. The syntax of `.usect` directive is

```
symbol .usect "section_name", size_in_words
```

where `symbol` is the variable, or the starting address of a data array, which will be placed into the section named `section_name`. In the latter case, the `size_in_words` defines the number of words in the array.

.TEXT directive: The `.text` directive tells the assembler to begin assembling source code into the `.text` section, which normally contains executable code. This is the default section for program code. If we do not specify a program code section, the assembler will put all the programs into the `.text` section.

The directives, `.bss`, `.sect`, `.usect`, and `.text` are used to define the memory sections. The following directives are used to initialize constants.

.INT (.WORD) directive: The `.int` (or `.word`) directive places one or more 16-bit integer values into consecutive words in the current section. This allows users to initialize memory with constants. For example,

```
data1 .word 0x1234
data2 .int  1010111b
```

In these examples, `data1` is initialized to the hexadecimal number 0x1234 (decimal number 4660), while `data2` is initialized to the binary number of 1010111b (decimal 87). The suffix 'b' indicates the data 1010111 is in binary format.

.SET (.EQU) directive: The `.set` (or `.equ`) directive assigns values to symbols. This type of symbol is known as an assembly-time constant. It can then be used in source statements in the same manner as a numeric constant. The `.set` directive has the form:

```
symbol .set value
```

where the `symbol` must appear in the first column. This example equates the constant `value` to the `symbol`. The symbolic name used in the program will be replaced with the constant by the assembler during assembly time, thus allowing programmers to write more readable programs. The `.set` and `.equ` directives can be used interchangeably, and do not produce object code.

The assembler is used to convert assembly language source code to COFF format object files for the C55x processor. The following command invokes the C55x mnemonic assembler:

```
masm55 [input_file [object_file [list_file] ] ] [-options]
```

The `input_file` is the name of the assembly source program. If no extension is supplied, the assembler assumes that the `input_file` has the default extension

.asm. The object_file is the name of the object file that the assembler creates. The assembler uses the source file's name with the default extension .obj for the object file unless specified otherwise. The list_file is the name of the list file that the assembler creates. The assembler will use the source file's name and .lst as the default extension for the list file. The assembler will not generate list files unless the option -l is set.

The options identify the assembler options. Some commonly used assembler options are:

- The -l option tells the assembler to create a listing file showing where the program and the variables are allocated.

- The -s option puts all symbols defined in the source code into the symbol table so the debugger may access them.

- The -c option makes the case insignificant in symbolic names. For example, -c makes the symbols ABC and abc equivalent.

- The -i option specifies a directory where the assembler can find included files such as those following the .copy and .include directives.

2.3.3 Linker

The linker is used to combine multiple object files into a single executable program for the target DSP hardware. It resolves external references and performs code relocation to create the executable code. The C55x linker handles various requirements of different object files and libraries as well as target system memory configurations. For a specific hardware configuration, the system designers need to provide the memory mapping specifications for the linker. This task can be accomplished by using a linker command file. The Texas Instruments' visual linker is also a very useful tool that provides memory usage directly.

The linker commands support expression assignment and evaluation, and provides the MEMORY and SECTION directives. Using these directives, we can define the memory configuration for the given target system. We can also combine object file sections, allocate sections into specific memory areas, and define or redefine global symbols at link time.

We can use the following command to invoke the C55x linker from the host system:

```
lnk55 [-options] filename_1, . . ., filename_n
```

The filename list (filename_1, . . ., filename_n) consists of object files created by the assembler, linker command files, or achieve libraries. The default extension for object files is .obj; any other extension must be explicitly specified. The options can be placed anywhere on the command line to control different linking operations. For example, the -o filename option can be used to specify the output executable file name. If we do not provide the output file name, the default executable file name is a.out. Some of the most common linker options are:

- The `-ar` option produces a re-locatable executable object file. The linker generates an absolute executable code by default.

- The `-e entry_point` option defines the entry point for the executable module. This will be the address of the first operation code in the program after power up or reset.

- The `-stack size` option sets the system stack size.

We can put the filenames and options inside the linker command file, and then invoke the linker from the command line by specifying the command file name as follows:

```
lnk55 command_file.cmd
```

The linker command file is especially useful when we frequently invoke the linker with the same information. Another important feature of the linker command file is that it allows users to apply the MEMORY and SECTION directives to customize the program for different hardware configurations. A linker command file is an ASCII text file and may contain one or more of the following items:

- Input files (object files, libraries, etc.).

- Output files (map file and executable file).

- Linker options to control the linker as given from the command line of the shell program.

- The MEMORY and SECTION directives define the target memory configuration and information on how to map the code sections into different memory spaces.

The linker command file we used for the experiments in Chapter 1 is listed in Table 2.2. The first portion of the command file uses the MEMORY directive to identify the range of memory blocks that physically exist in the target hardware. Each memory block has a name, starting address, and block length. The address and length are given in bytes. For example, the data memory is given a name called RAM, and it starts at the byte address of hexadecimal 0x100, with a size of hexadecimal 0x1FEFF bytes.

The SECTIONS directive provides different code section names for the linker to allocate the program and data into each memory block. For example, the program in the `.text` section can be loaded into the memory block ROM. The attributes inside the parenthesis are optional to set memory access restrictions. These attributes are:

    R – the memory space can be read.
    W – the memory space can be written.
    X – the memory space contains executable code.
    I – the memory space can be initialized.

There are several additional options that can be used to initialize the memory using linker command files [2].

Table 2.2   Example of a linker command file used for the C55x simulator

```
/* Specify the system memory map */

MEMORY
{
   RAM (RWIX)    : origin = 0100h,    length = 01FEFFh /* Data memory    */
   RAM2 (RWIX)   : origin = 040100h,  length = 040000h /* Program memory */
   ROM (RIX)     : origin = 020100h,  length = 020000h /* Program memory */
   VECS (RIX)    : origin = 0FFFF00h, length = 00100h  /* Reset vector   */
}

/* Specify the sections allocation into memory */

SECTIONS
{
   vectors > VECS         /* Interrupt vector table    */
   .text   > ROM          /* Code                      */
   .switch > RAM          /* Switch table info         */
   .const  > RAM          /* Constant data             */
   .cinit  > RAM2         /* Initialization tables     */
   .data   > RAM          /* Initialized data          */
   .bss    > RAM          /* Global & static variables */
   .stack  > RAM          /* Primary system stack      */
}
```

2.3.4   Code Composer Studio

As illustrated in Figure 2.8, the code composer studio (CCS) provides interface with the C55x simulator (SIM), DSP starter kit (DSK), evaluation module (EVM), or in-circuit emulator (XDS). The CCS supports both C and assembly programs.

The C55x simulator is available for PC and workstations, making it easy and inexpensive to develop DSP software and to evaluate the performance of the processor before designing any hardware. It accepts the COFF files and simulates the instructions of the program such as the code running on the target DSP hardware. The C55x simulator enables the users to single-step through the program, and observe the contents of the CPU registers, data and I/O memory locations, and the current DSP states of the status registers. The C55x simulator also provides profiling capabilities that tell users the amount of time spent in one portion of the program relative to another. Since all the functions of the TMS320C55x are performed on the host computer, the simulation may be slow, especially for complicated DSP applications. Real world signals can only be digitized and then later fed into a simulator as test data. In addition, the timing of the algorithm under all possible input conditions cannot be tested using a simulator.

As introduced in Section 1.5, the various display windows and the commands of the CCS provide most debugging needs. Through the CCS, we can load the executable object code, display a disassembled version of the code along with the original source code, and

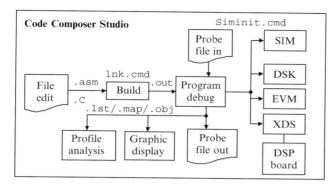

Figure 2.8 TMS320C55x software development using CCS

view the contents of the registers and the memory locations. The data in the registers and the memory locations can be modified manually. The data can be displayed in hexadecimal, decimal integer, or floating-point formats. The execution of the program can be single-stepped through the code, run-to-cursor, or controlled by applying breakpoints.

DSK and EVM are development boards with the C55x processor. They can be used for real-time analysis of DSP algorithms, code logic verification, and simple application tests. The XDS allows breakpoints to be set at a particular point in a program to examine the registers and the memory locations in order to evaluate the real-time results using a DSP board. Emulators allow the DSP software to run at full-speed in a real-time environment.

2.3.5 Assembly Statement Syntax

The TMS320C55x assembly program statements may be separated into four ordered fields. The basic syntax expression for a C55x assembly statement is

```
[label] [:] mnemonic [operand list] [;comment]
```

The elements inside the brackets are optional. Statements must begin with a label, blank, asterisk, or semicolon. Each field must be separated by at least one blank. For ease of reading and maintenance, it is strongly recommended that we use meaningful mnemonics for labels, variables, and subroutine names, etc. An example of a C55x assembly statement is shown in Figure 2.9. In this example, the auxiliary register, AR1, is initialized to a constant value of 2.

Label field: A label can contain up to 32 alphanumeric characters (A–Z, a–z, 0–9, _, and $). It associates a symbolic address with a unique program location. The line that is labeled in the assembly program can then be referenced by the defined symbolic name. This is useful for modular programming and branch instructions. Labels are optional, but if used, they must begin in column 1. Labels are case sensitive and must start with an alphabetic letter. In the example depicted in Figure 2.9, the symbol `start` is a label and is placed in the first column.

Mnemonic field: The mnemonic field can contain a mnemonic instruction, an assembler directive, macro directive, or macro call. The C55x instruction set supports both

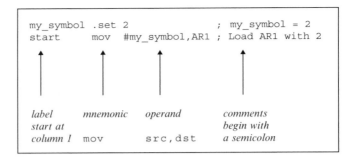

Figure 2.9   An example of TMS320C55x assembly statement

DSP-specific operations and general-purpose applications (see the TMS320C55x DSP Mnemonic Instruction Set Reference Guide [4] for details). Note that the mnemonic field cannot start in column 1 otherwise it would be interpreted as a label. The mnemonic instruction `mov` (used in Figure 2.9) copies the constant, `my_symbol` (which is set to be 2 by `.set` directive) into the auxiliary register AR1.

Operand field: The operand field is a list of operands. An operand can be a constant, a symbol, or a combination of constants and symbols in an expression. An operand can also be an assembly-time expression that refers to memory, I/O ports, or pointers. Another category of the operands can be the registers and accumulators. Constants can be expressed in binary, decimal, or hexadecimal formats. For example, a binary constant is a string of binary digits (0s and 1s) followed by the suffix B (or b) and a hexadecimal constant is a string of hexadecimal digits (0, 1, ..., 9, A, B, C, D, E, and F) followed by the suffix H (or h). A hexadecimal number can also use a 0x prefix similar to those used by C language. The prefix # is used to indicate an immediate constant. For example, #123 indicates that the operand is a constant of decimal number 123, while #0x53CD is the hexadecimal number of 53CD (equal to a decimal number of 21 453). Symbols defined in an assembly program with assembler directives may be labels, register names, constants, etc. For example, we use the `.set` directive to assign a value to `my_symbol` in the example given by Figure 2.9. Thus the symbol `my_symbol` becomes a constant of value during assembly time.

Comment field: Comments are notes about the program that are significant to the programmer. A comment can begin with an asterisk or a semicolon in column one. Comments that begin in any other column must begin with a semicolon.

2.4   TMS320C55x Addressing Modes

The TMS320C55x can address a total of 16 Mbytes of memory space. The C55x supports the following addressing modes:

- Direct addressing mode
- Indirect addressing mode

- Absolute addressing mode

- Memory-mapped register addressing mode

- Register bits addressing mode

- Circular addressing mode

To explain the different addressing modes of the C55x, Table 2.3 lists the move instruction (mov) with different syntax.

As illustrated in Table 2.3, each addressing mode uses one or more operands. Some of the operand types are explained as follows:

- Smem means a data word (16-bit) from data memory, I/O memory, or MMRs.

- Lmem means a long data word (32-bit) from either data memory space or MMRs.

- Xmem and Ymem are used by an instruction to perform two 16-bit data memory accesses simultaneously.

- src and dst are source and destination registers, respectively.

- #k is a signed immediate 16-bit constant ranging from −32 768 to 32 767.

- dbl is a memory containing a long data word.

- xdst is an extended register (23-bit).

Table 2.3   C55x mov instruction with different operand forms

Instruction	Description
1. mov #k, dst	Load the 16-bit signed constant k to the destination register dst
2. mov src, dst	Load the content of source register src to the destination register dst
3. mov Smem, dst	Load the content of memory location Smem to the destination register dst
4. mov Xmem, Ymem, ACx	The content of Xmem is loaded into the lower part of ACx while the content of Ymem is sign extended and loaded into upper part of ACx
5. mov dbl(Lmem), pair(TAx)	Load upper 16-bit data and lower 16-bit data from Lmem to the TAx and TA(x+1), respectively
6. amov #k23, xdst	Load the effective address of k23 (23-bit constant) into extended destination register (xdst)

2.4.1   Direct Addressing Mode

There are four types of direct addressing modes: data-page pointer (DP) direct, stack pointer (SP) direct, register-bit direct, and peripheral data-page pointer (PDP) direct.

The DP direct mode uses the main data page specified by the 23-bit extended data-page pointer (XDP). Figure 2.10 shows a generation of DP direct address. The upper seven bits of the XDP (DPH) determine the main data page (0–127). The lower 16 bits of the XDP (DP) define the starting address in the data page selected by the DPH. The instruction contains the seven-bit offset in the data page (@x) that directly points to the variable x (Smem). The data-page registers DPH, DP, and XDP can be loaded by the mov instruction as

```
mov #k7, DPH   ; Load DPH with a 7-bit constant k7
mov #k16, DP   ; Load DP with a 16-bit constant k16
```

These instructions initialize the data pointer DPH and DP, respectively, using the assembly code syntax, mov #k,dst, given in Table 2.3. The first instruction loads the high portion of the extended data-page pointer, DPH, with a 7-bit constant k7 to set up the main data page. The second instruction initializes the starting address of the data-page pointer. The following is an example that initializes the DPH and DP pointers:

Example 2.1:   Instruction

```
mov #0x3, DPH
mov #0x0100, DP
```

DPH	0
DP	0000

DPH	03
DP	0100

Before instruction   After instruction

The data-page pointer also can be initialized using a 23-bit constant as

```
amov #k23, XDP   ; Load XDP with a 23-bit constant
```

This instruction initializes the XDP in one instruction. The syntax used in the assembly code is given in Table 2.3, amov #k23, xdst, where #k23 is a 23-bit address and the destination xdst is an extended register. The following example initializes the data-page pointer XDP to data page 1 with starting address 0x4000:

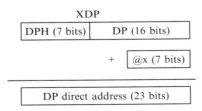

Figure 2.10   The DP direct addressing mode to variable x

Example 2.2: Instruction

amov #0x14000, XDP

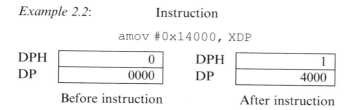

DPH	0
DP	0000

Before instruction

DPH	1
DP	4000

After instruction

The following code details how to use DP direct addressing mode:

```
X       .set 0x1FFEF
        mov  #0x1, DPH        ; Load DPH with 1
        mov  #0x0FFEF, DP     ; Load DP with starting address
        .dp  X
        mov  #0x5555, @X      ; Store 0x5555 to memory location X
        mov  #0xFFFF, @(X+5)  ; Store 0xFFFF to memory location X+5
```

In this example, the symbol @ tells the assembler that this access is using the direct address mode. The directive .dp does not use memory space. It is used to indicate the base address of the variable X.

The stack pointer (SP) direct addressing mode is similar to the DP direct addressing mode. The 23-bit address can be formed with the extended stack pointer (XSP) in the same way as the direct address that uses XDP. The upper seven bits (SPH) select the main data page and the lower 16 bits (SP) determine the starting address of the stack pointer. The 7-bit stack offset is contained in the instruction. When SPH = 0 (main page 0), the stack must not use the reserved memory space for MMRs from address 0 to 0x5F.

The I/O space address mode only has a 16-bit address range. The 512 peripheral data pages are selected by the upper 9 bits of the PDP register. The 7-bit offset determines the location inside the selected peripheral data page as illustrated in Figure 2.11.

2.4.2 Indirect Addressing Mode

Indirect addressing modes using index and displacement are the most powerful and commonly used addressing modes. There are four types of indirect addressing modes. The AR indirect mode uses one of the eight auxiliary registers as a pointer to data memory, I/O space, and MMRs. The dual-AR indirect mode uses two

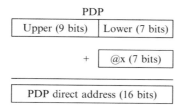

PDP

Upper (9 bits)	Lower (7 bits)

+ | @x (7 bits) |

| PDP direct address (16 bits) |

Figure 2.11   The PDP direct addressing mode to variable x

auxiliary registers for dual data memory access. The coefficient data pointer (CDP) indirect mode uses the CDP to point to data memory space. The coefficient-dual-AR indirect mode uses the CDP and the dual-AR indirect modes for generating three addresses. The coefficient-dual-AR indirect mode will be discussed later along with pipeline parallelism.

The indirect addressing is the most frequently used addressing mode because it provides powerful pointer update/modification schemes. Several pointer modification schemes are listed in Table 2.4.

The AR indirect addressing mode uses an auxiliary register (AR0–AR7) to point to data memory space. The upper seven-bit of the extended auxiliary register (XAR) points to the main data page, while the lower 16-bit points to a data location on that page. Since the I/O space address is limited to a 16-bit range, the upper portion of the XAR must be set to zero when accessing I/O space. The next example uses indirect addressing mode, where AR0 is used as the address pointer, and the instruction loads the data content stored in data memory pointed by AR0 to the destination register AC0.

Example 2.3:   Instruction

mov * AR0, AC0

AC0	00 0FAB 8678		AC0	00 0000 12AB
AR0	0100		AR0	0100

Data memory   Data memory

0x100	12AB		0x100	12AB

Before instruction   After instruction

Table 2.4   The AR and CDP indirect addressing pointer modifications

Operand	ARn/CDP pointer modifications
* ARn or * CDP	ARn (or CDP) is not modified.
* ARn± or * CDP±	ARn (or CDP) is modified after the operation by: ±1 for 16-bit operation (ARn=ARn±1) ±2 for 32-bit operation (ARn=ARn±2)
* ARn (#k16) or * CDP (#k16)	ARn (or CDP) is not modified. The signed 16-bit constant k16 is used as the offset for the base pointer ARn (or CDP).
* +ARn (#k16) or * +CDP (#k16)	ARn (or CDP) is modified before the operation. The signed 16-bit constant k16 is added as the offset to the base pointer ARn (or CDP) before generating new address.
* (ARn±T0/T1)	ARn is modified after the operation by ±16-bit content in T0 or T1, (ARn = ARn±T0/T1)
* ARn (T0/T1)	ARn is not modified. T0 or T1 is used as the offset for the base pointer ARn.

The dual-AR indirect addressing mode allows two data-memory accesses through the auxiliary registers AR0–AR7. It can access two 16-bit data in memory using the syntax, mov Xmem, Ymem, ACx given in Table 2.3. The next example performs dual 16-bit data load with AR2 and AR3 as the data pointers to Xmem and Ymem, respectively. The data pointed at by AR3 is sign-extended to 24-bit, loaded into the upper portion of the destination register AC0(39:16), and the data pointed at by AR2 is loaded into the lower portion of AC0(15:0). The data pointers AR2 and AR3 are also updated.

Example 2.4:   Instruction

mov *AR2+, *AR3−, AC0

AC0	FF FFAB 8678		AC0	00 3333 5555
AR2	0100		AR2	0101
AR3	0300		AR3	02FF

Data memory   Data memory

0x100	5555		0x100	5555

0x300	3333		0x300	3333

            Before instruction   After instruction

The extended coefficient data pointer (XCDP) is the concatenation of the CDPH (the upper 7-bit) and the CDP (the lower 16-bit). The CDP indirect addressing mode uses the upper 7-bit to define the main data page and the lower 16-bit to point to the data memory location within the specified data page. For the I/O space, only the 16-bit address is used. An example of using the CDP indirect addressing mode is given as follows:

Example 2.5:   Instruction

mov *+CDP (#2), AC3

AC3	00 0FAB EF45		AC3	00 0000 5631
CDP	0400		CDP	0402

Data memory   Data memory

0x402	5631		0x402	5631

            Before instruction   After instruction

In this example, CDP is the pointer that contains the address of the coefficient in data memory with an offset. This instruction increments the CDP pointer by 2 first, then loads a coefficient pointed by the updated coefficient pointer to the destination register AC3.

2.4.3 Absolute Addressing Mode

The memory can also be addressed using absolute addressing modes in either k16 or k23 absolute addressing modes. The k23 absolute mode specifies an address as a 23-bit unsigned constant. The following example loads the data content at address 0x1234 on main data page 1 into the temporary register, T2, where the symbol, *( ), represents the absolute address mode.

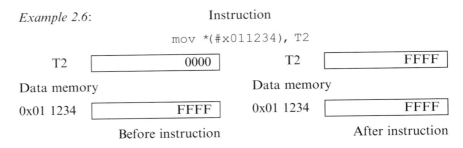

Example 2.6:   Instruction

mov *(#x011234), T2

| T2 | 0000 | | T2 | FFFF |

Data memory   Data memory

| 0x01 1234 | FFFF | | 0x01 1234 | FFFF |

Before instruction   After instruction

The k16 absolute addressing mode uses the operand *abs(#k16), where k16 is a 16-bit unsigned constant. The DPH (7-bit) is forced to 0 and concatenated with the unsigned constant k16 to form a 23-bit data-space memory address. The I/O absolute addressing mode uses the operand port(#k16). The absolute address can also be the variable name such as the variable, x, in the following example:

mov *(x), AC0

This instruction loads the accumulator AC0 with a content of variable x. When using absolute addressing mode, we do not need to worry about what is loaded into the data-page pointer. The drawback of the absolute address is that it uses more code space to represent the 23-bit address.

2.4.4 Memory-Mapped Register Addressing Mode

The absolute, direct, and indirect addressing modes introduced above can be used to address MMRs. The MMRs are located in the data memory from address 0x0 to 0x5F on the main data page 0 as shown in Figure 2.6. To access the MMRs using the k16 absolute operand, the DPH must be set to zero. The following example uses the absolute addressing mode to load the 16-bit content of the AR2 into the temporary register T2:

Example 2.7:   Instruction

mov *abs16(#AR2), T2

| AR2 | 1357 | | AR2 | 1357 |
| T2 | 0000 | | T2 | 1357 |

Before instruction   After instruction

For the MMR direct addressing mode, the DP addressing mode must be selected. The example given next uses direct addressing mode to load the content of the lower portion of the accumulator AC0(15:0), into the temporary register T0. When the mmap() qualifier for the MMR direct addressing mode is used, it forces the data address generator to act as if the access is made to the main data page 0. That is, XDP = 0.

Example 2.8:   Instruction

```
mov mmap(@AC0L), T0
```

AC0	00 12DF 0202		AC0	00 12DF 0202
T0	0000		T0	0202

Before instruction   After instruction

Accessing the MMRs using indirect addressing mode is the same as addressing the data memory space. The address pointer can be either an auxiliary register or a CDP. Since the MMRs are all located on data page 0, the XAR and XCDP must be initialized to page 0 by setting all upper 7-bit to zero. The following instructions load the content of AC0 into T1 and T2 temporary registers:

```
amov #AC0H, XAR6
mov  *AR6-, T2
mov  *AR6+, T1
```

In this example, the first instruction loads the effective address of the upper portion of the accumulator AC0 (AC0H, located at address 0x9 of page 0) to the extended auxiliary register XAR6. That is, XAR6 = 0x000009. The second instruction uses AR6 as a pointer to copy the content of AC0H into the T2 register, and then the pointer decrements by 1 to point to the lower portion of AC0 (AC0L, located at address 0x8 of page 0). The third instruction copies the content of AC0L into the register T1 and modifies AR6 to point to AC0H again.

2.4.5 Register Bits Addressing Mode

Both direct and indirect addressing modes can be used to address one bit or a pair of bits of a specific register. The direct addressing mode uses a bit offset to access a particular register's bit. The offset is the number of bits counting from the least significant bit (LSB), i.e., bit 0. The bit test instruction will update the test condition bits, TC1 and TC2, of the status register ST0. The instruction of register-bit direct addressing mode is shown in the next example.

Example 2.9:   Instruction

```
btstp @30, AC1
```

AC1	00 7ADF 3D05		AC1	00 7ADF 3D05
TC1	0		TC1	1
TC2	0		TC2	0

Before instruction   After instruction

Using the indirect addressing modes to specify register bit(s) can be done as follows:

```
mov   #2, AR4    ; AR4 contains the bit offset 2
bset  *AR4, AC3  ; Set the AC3 bit pointed by AR4 to 1
btstp *AR4, AC1  ; Test AC1 bit-pair pointed by AR4
```

The register bit-addressing mode supports only the bit test, bit set, bit clear, and bit complement instructions in conjunction with the accumulators (AC0–AC3), auxiliary registers (AR0–AR7), and temporary registers (T0–T3).

2.4.6   Circular Addressing Mode

Circular addressing mode provides an efficient method for accessing data buffers continuously without having to reset the data pointers. After accessing data, the data buffer pointer is updated in a modulo fashion. That is, when the pointer reaches the end of the buffer, it will wrap back to the beginning of the buffer for the next iteration. Auxiliary registers (AR0–AR7) and the CDP can be used as circular pointers in indirect addressing mode. The following steps are commonly used to set up circular buffers:

1. Initialize the most significant 7-bit extended auxiliary register (ARnH or CDPH) to select the main data page for a circular buffer. For example, mov #k7, AR2H.

2. Initialize the 16-bit circular pointer (ARn or CDP). The pointer can point to any memory location within the buffer. For example, mov #k16, AR2 (the initialization of the address pointer in the example of steps 1 and 2 can also be done using the amov #k23, XAR2 instruction).

3. Initialize the 16-bit circular buffer starting address register (BSA01, BSA23, BSA45, BSA67, or BSAC) associated with the auxiliary registers. For example, mov #k16, BSA23, if AR2 (or AR3) is used as the circular addressing pointer register. The main data page concatenated with the content of this register defines the 23-bit starting address of the circular buffer.

4. Initialize the data buffer size register (BK03, BK47, or BKC). When using AR0–AR3 (or AR4–AR7) as the circular pointer, BK03 (or BK47) should be initialized. The instruction, mov #16, BK03, sets up a circular buffer of 16 elements for the auxiliary registers, AR0–AR3.

5. Enable the circular buffer configuration by setting the appropriate bit in the status register ST2. For example, the instruction bset AR2LC enables AR2 for circular addressing.

Refer to the TMS320C55x DSP CPU Reference Guide [1] for details on circular addressing mode. The following example demonstrates how to initialize a four integer circular buffer, COEFF[4], and how the circular addressing mode accesses data in the buffer:

```
amov #COEFF, XAR2      ; Main data page for COEFF[4]
mov  #COEFF, BSA23     ; Buffer base address is COEFF[0]
mov  #0x4, BK03        ; Set buffer size of 4-word
mov  #2, AR2           ; AR2 points to COEFF[2]
bset AR2LC             ; AR2 is configured as circular pointer
mov  *AR2+, T0         ; T0 is loaded with COEFF[2]
mov  *AR2+, T1         ; T1 is loaded with COEFF[3]
mov  *AR2+, T2         ; T2 is loaded with COEFF[0]
mov  *AR2+, T3         ; T3 is loaded with COEFF[1]
```

Since the circular addressing uses the indirect addressing modes, the circular pointers can be updated using the modifications listed in Table 2.4. The use of circular buffers for FIR filtering will be introduced in Chapter 5 in details.

2.5 Pipeline and Parallelism

The pipeline technique has been widely used by many DSP manufacturers to improve processor performance. The pipeline execution breaks a sequence of operations into smaller segments and executes these smaller pieces in parallel. The TMS320C55x uses the pipelining mechanism to efficiently execute its instructions to reduce the overall execution time.

2.5.1 TMS320C55x Pipeline

Separated by the instruction buffer unit, the pipeline operation is divided into two independent pipelines – the program fetch pipeline and the program execution pipeline (see Figure 2.12). The program fetch pipeline consists of the following three stages (it uses three clock cycles):

PA (program address): The C55x instruction unit places the program address on the program-read address bus (PAB).

PM (program memory address stable): The C55x requires one clock cycle for its program memory address bus to be stabilized before that memory can be read.

PB (program fetch from program data bus): In this stage, four bytes of the program code are fetched from the program memory via the 32-bit program data-read bus (PB).

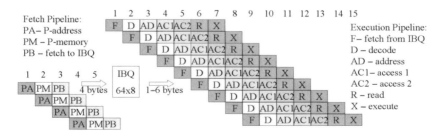

Figure 2.12 The C55x pipeline execution diagram

The code is placed into the instruction buffer queue (IBQ). For every clock cycle, the IU will fetch four bytes to the IBQ. The numbers on the top of the diagram represent the CPU clock cycle.

At the same time, the seven-stage execution pipeline performs the fetch, decode, address, access, read, and execution sequence independent of the program fetch pipeline. The C55x program execution pipeline stages are summarized as follows:

F (fetch): In the fetch stage, an instruction is fetched from the IBQ. The size of the instruction can be one byte for simple operations, or up to six bytes for more complex operations.

D (decode): During the decoding process, decode logic gets one to six bytes from the IBQ and decodes these bytes into an instruction or an instruction pair under the parallel operation. The decode logic will dispatch the instruction to the program flow unit (PU), address flow unit (AU), or data computation unit (DU).

AD (address): In this stage, the AU calculates data memory addresses using its data-address generation unit (DAGEN), modifies pointers if required, and computes the program-space address for PC-relative branching instructions.

AC (access cycles 1 and 2): The first cycle is used for the C55x CPU to send the address for read operations to the data-read address buses (BAB, CAB, and DAB), or transfer an operand to the CPU via the C-bus (CB). The second access cycle is inserted to allow the address lines to be stabilized before the memory is read.

R (read): In the read stage, the data and operands are transferred to the CPU via the CB for the Ymem operand, the B-bus (BB) for the Cmem operand, and the D-bus (DB) for the Smem or the Xmem operands. For the Lmem operand read, both the CB and the DB will be used. The AU will generate the address for the operand write and send the address to the data-write address buses (EAB and FAB).

X (execute): Most data processing work is done in this stage. The ALU inside the AU and the ALU inside the DU performs data processing execution, stores an operand via the F-bus (FB), or stores a long operand via the E-bus and F-bus (EB and FB).

The C55x pipeline diagram illustrated in Figure 2.12 explains how the C55x pipeline works. It is clear that the execution pipeline is full after seven cycles and every execution cycle that follows will complete an instruction. If the pipeline is always full, this technique increases the processing speed seven times. However, the pipeline flow efficiency is based on the sequential execution of instruction. When a disturbing execution such as a branch instruction occurs, the sudden change of the program flow breaks the pipeline sequence. Under such circumstances, the pipeline will be flushed and will need to be refilled. This is called pipeline break down. The use of IBQ can minimize the impact of the pipeline break down. Proper use of conditional execution instructions to replace branch instructions can also reduce the pipeline break down.

2.5.2 Parallel Execution

The parallelism of the TMS320C55x uses the processor's multiple-bus architecture, dual MAC units, and separated PU, AU, and DU. The C55x supports two parallel processing types – implied and explicit. The implied parallel instructions are the built-in instructions. They use the symbol of parallel columns, ': :', to separate the pair of instructions that will be processed in parallel. The explicit parallel instructions are the user-built instructions. They use the symbol of parallel bar, '||', to indicate the pair of parallel instructions. These two types of parallel instructions can be used together to form a combined parallel instruction. The following examples show the user-built, built-in, and combined parallel instructions. Each example is carried out in just one clock cycle.

User-built:

```
    mpym * AR1+, * AR2+, AC0   ; User-built parallel instruction
||and   AR4, T1                ; Using DU and AU
```

Built-in:

```
    mac * AR0-, * CDP-, AC0   ; Built-in parallel instruction
::mac * AR1-, * CDP-, AC1     ; Using dual-MAC units
```

Built-in and User-built Combination:

```
    mpy * AR2+, * CDP+, AC0   ; Combined parallel instruction
::mpy * AR3+, * CDP+, AC1     ; Using dual-MAC units and PU
||rpt #15
```

Some of the restrictions when using parallel instructions are summarized as follows:

- For either the user-built or the built-in parallelism, only two instructions can be executed in parallel, and these two instructions must not exceed six bytes.

- Not all instructions can be used for parallel operations.

- When addressing memory space, only the indirect addressing mode is allowed.

- Parallelism is allowed between and within execution units, but there cannot be any hardware resources conflicts between units, buses, or within the unit itself.

There are several restrictions that define the parallelism within each unit when applying parallelism to assembly coding. The detailed descriptions are given in the TMS320C55x DSP Mnemonic Instruction Set Reference Guide [4].

The PU, AU, and DU can all be involved in parallel operations. Understanding the register files in each of these units will help to be aware of the potential conflicts when using the parallel instructions. Table 2.5 lists some of the registers in PU, AU, and DU.

The parallel instructions used in the following example are incorrect because the second instruction uses the direct addressing mode:

```
    mov * AR2, AC0
||mov T1, @x
```

We can correct the problem by replacing the direct addressing mode, @x, with an indirect addressing mode, * AR1, so both memory accesses are using indirect addressing mode as follows:

Table 2.5 Partial list of the C55x registers and buses

PU Registers/Buses	AU Registers/Buses	DU Registers/Buses
RPTC	T0, T1, T2, T3	AC0, AC1, AC2, AC3
BRC0, BRC1	AR0, AR1, AR2, AR3,	TRN0, TRN1
RSA0, RSA1	AR4, AR5, AR6, AR7	
REA0, REA1	CDP	
	BSA01, BSA23, BSA45, BSA67	
	BK01, BK23, BK45, BK67	
Read Buses: C, D	Read Buses: C, D	Read Buses: B, C, D
Write Buses: E, F	Write Buses: E, F	Write Buses: E, F

```
   mov *AR2, AC0
| |mov T1, *AR1
```

Consider the following example where the first instruction loads the content of AC0 that resides inside the DU to the auxiliary register AR2 inside the AU. The second instruction attempts to use the content of AC3 as the program address for a function call. Because there is only one link between AU and DU, when both instructions try to access the accumulators in the DU via the single link, it creates a conflict.

```
   mov AC0, AR2
| |call AC3
```

To solve the problem, we can change the subroutine call from call by accumulator to call by address as follows:

```
   mov AC0, AR2
| |call my_func
```

This is because the instruction, `call my_func`, only needs the PU.

The coefficient-dual-AR indirect addressing mode is used to perform operations with dual-AR indirect addressing mode. The coefficient indirect addressing mode supports three simultaneous memory-accesses (Xmem, Ymem, and Cmem). The finite impulse response (FIR) filter (will be introduced in Chapter 3) is an application that can effectively use coefficient indirect addressing mode. The following code segment is an example of using the coefficient indirect addressing mode:

```
    mpy *AR1+, *CDP+, AC2   ; AR1 pointer to data X1
 : :mpy *AR2+, *CDP+, AC3   ; AR2 pointer to data X2
 | |rpt #6                   ; Repeat the following 7 times
    mac *AR1+, *CDP+, AC2   ; AC2 has accumulated result
 : :mac *AR2+, *CDP+, AC3   ; AC3 has another result
```

In this example, the memory buffers (Xmem and Ymem) are pointed at by AR2 and AR3, respectively, while the coefficient array is pointed at by CDP. The multiplication results are added with the contents in the accumulators AC2 and AC3, and the final results are stored back to AC2 and AC3.

2.6 TMS320C55x Instruction Set

We briefly introduced the TMS320C55x instructions and assembly syntax expression in Section 2.3.5. In this section, we will introduce more useful instructions for DSP applications. In general, we can divide the C55x instruction set into four categories: arithmetic instructions, logic and bit manipulation instructions, load and store (move) instructions, and program flow control instructions.

2.6.1 Arithmetic Instructions

Instructions used to perform addition (ADD), subtraction (SUB), and multiplication (MPY) are arithmetic instructions. Most arithmetic operations can be executed conditionally. The combination of these basic arithmetic operations produces another powerful subset of instructions such as the multiply–accumulation (MAC) and multiply–subtraction (MAS) instructions. The C55x also supports extended precision arithmetic such as add-with-carry, subtract-with-borrow, signed/signed, signed/unsigned, and unsigned/unsigned arithmetic instructions. In the following example, the multiplication instruction, mpym, multiplies the data pointed by AR1 and CDP, and the multiplication product is stored in the accumulator AC0. After the multiplication, both pointers (AR1 and CDP) are updated.

Example 2.10:   Instruction

mpym * AR1+, * CDP−, AC0

AC0	FF FFFF FF00
FRC	0
AR1	02E0
CDP	0400

AC0	00 0000 0020
FRC	0
AR1	02E1
CDP	03FF

Data memory

0x2E0	0002
0x400	0010

Data memory

0x2E0	0002
0x400	0010

Before instruction   After instruction

In the next example, the macmr40 instruction uses AR1 and AR2 as data pointers and performs multiplication–accumulation. At the same time, the instruction also carries out the following operations:

1. The key word 'r' produces a rounded result in the high portion of the accumulator AC3. After rounding, the lower portion of AC3(15:0) is cleared.

2. 40-bit overflow detection is enabled by the key word '40'. If overflow is detected, the result in accumulator AC3 will be saturated to its 40-bit maximum value.

3. The option 'T3 = *AR1+' loads the data pointed at by AR1 into the temporary register T3 for later use.

4. Finally, AR1 and AR2 are incremented by one to point to the next data location in memory space.

Example 2.11:　　　　Instruction

macmr40 T3 = *AR1+, *AR2+, AC3

AC3	00 0000 0020		AC3	00 235B 0000
FRC	1		FRC	1
T3	FFF0		T3	3456
AR1	0200		AR1	0201
AR2	0380		AR2	0381

Data memory　　　　　　　　　Data memory

0x200	3456		0x200	3456
0x380	5678		0x380	5678

Before instruction　　　　　　After instruction

2.6.2　Logic and Bits Manipulation Instructions

Logic operation instructions such as AND, OR, NOT, and XOR (exclusive-OR) on data values are widely used in program decision-making and execution flow control. They are also found in many applications such as error correction coding in data communications. For example, the instruction and #0xf, AC0 clears all upper bits in the accumulator AC0 but the four least significant bits.

Example 2.12:　　　　Instruction

and #0xf, AC0

AC0	00 1234 5678	AC0	00 0000 0008

Before instruction　　　　　　After instruction

The bit manipulation instructions act on an individual bit or a pair of bits of a register or data memory. These types of instructions consist of bit clear, bit set, and bit test to a specified bit (or a pair of bits). Similar to logic operations, the bit manipulation instructions are often used with logic operations in supporting decision-making processes. In the following example, the bit clear instruction clears the carry bit (bit 11) of the status register ST0.

Example 2.13: Instruction

```
bclr #11, ST0
```

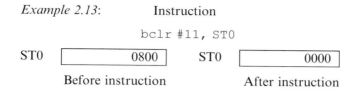

ST0 | 0800 | ST0 | 0000 |

Before instruction   After instruction

2.6.3 Move Instruction

The move instruction is used to copy data values between registers, memory locations, register to memory, or memory to register. For example, to initialize the upper portion of the 32-bit accumulator AC1 with a constant and zero out the lower portion of the AC1, we can use the instruction mov #k≪16, AC1, where the constant k is first shifted left by 16-bit and then loaded into the upper portion of the accumulator AC1(31:16) and the lower portion of the accumulator AC1(15:0) is zero filled. The 16-bit constant that follows the # can be any signed number.

Example 2.14: Instruction

```
mov #5≪16, AC1
```

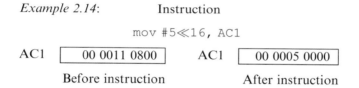

AC1 | 00 0011 0800 | AC1 | 00 0005 0000 |

Before instruction   After instruction

A more complicated instruction completes the following several operations in one clock cycle:

Example 2.15: Instruction

```
mov uns(rnd(HI(satuate(AC0≪T2)))), *AR1+
```

ACO | 00 0FAB 8678 | ACO | 00 0FAB 8678 |
AR1 | 0x100 | AR1 | 0x101 |
T2 | 0x2 | T2 | 0x2 |

Data memory   Data memory

0x100 | 1234 | 0x100 | 3EAE |

Before instruction   After instruction

1. The unsigned data content in AC0 is shifted to the left according to the content in the temporary register T2.

2. The upper portion of the AC0(31:16) is rounded.

3. The data value in AC0 may be saturated if the left-shift or the rounding process causes the result in AC0 to overflow.

4. The final result after left shifting, rounding, and maybe saturation, is stored into the data memory pointed at by the pointer AR1.

5. Pointer AR1 is automatically incremented by 1.

2.6.4 Program Flow Control Instructions

The program flow control instructions are used to control the execution flow of the program, including branching (B), subroutine call (CALL), loop operation (RPTB), return to caller (RET), etc. All these instructions can be either conditionally or unconditionally executed. For example,

```
callcc my_routine, TC1
```

is the conditional instruction that will call the subroutine my_routine only if the test control bit TC1 of the status register ST0 is set. Conditional branch (BCC) and conditional return (RETCC) can be used to control the program flow according to certain conditions.

The conditional execution instruction, xcc, can be implemented in either conditional execution or partial conditional execution. In the following example, the conditional execution instruction tests the TC1 bit. If TC1 is set, the instruction, mov *AR1+, AC0, will be executed, and both AC0 and AR1 are updated. If the condition is false, AC0 and AR1 will not be changed. Conditional execution instruction xcc allows for the conditional execution of one instruction or two paralleled instructions. The label is used for readability, especially when two parallel instructions are used.

Example 2.16:   Instruction

```
xcc label, TC1
mov *AR1+, AC0
label
```

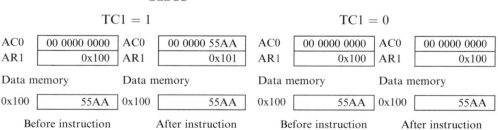

In addition to conditional execution, the C55x also provides the capability of partially conditional execution of an instruction. An example of partial conditional execution is given as follows:

Example 2.17:   Instruction

```
                    xccpart label, TC1
                    mov *AR1+, AC0
              label
```

	TC1 = 1				TC1 = 0		
AC0	00 0000 0000	AC0	00 0000 55AA	AC0	00 0000 0000	AC0	00 0000 0000
AR1	0x100	AR1	0x101	AR1	0x100	AR1	0x101

Data memory		Data memory		Data memory		Data memory	
0x100	55AA	0x100	55AA	0x100	55AA	0x100	55AA

Before instruction	After instruction	Before instruction	After instruction

When the condition is true, both AR1 and AC0 will be updated. However, if the condition is false, the execution phase of the pipeline will not be carried out. Since the first operand (the address pointer AR1) is updated in the read phase of the pipeline, AR1 will be updated whether or not the condition is true, while the accumulator AC0 will remain unchanged at the execution phase. That is, the instruction is only partially executed.

Many real-time DSP applications require repeated executions of some instructions such as filtering processes. These arithmetic operations may be located inside nested loops. If the number of data processing instructions in the inner loop is small, the percentage of overhead for loop control may be very high. The loop control instructions, such as testing and updating the loop counter(s), pointer(s), and branches back to the beginning of the loop to execute the loop again, impose a heavy overhead for the processor. To minimize the loop overhead, the C55x includes built-in hardware for zero-overhead loop operations.

The single-repeat instruction (RPT) repeats the following single-cycle instruction or two single-cycle instructions that are executed in parallel. For example,

```
    rpt #N−1          ; Repeat next instruction N times
    instruction_A
```

The number, N−1, is loaded into the single-repeat counter (RPTC) by the RPT instruction. The following instruction_A will be executed N times.

The block-repeat instruction (RPTB) forms a loop that repeats a block of instructions. It supports a nested loop with an inner loop being placed inside an outer loop. Block-repeat registers use block-repeat counters BRC0 and BRC1. For example,

```
        mov  #N−1, BRC0      ; Repeat outer loop N times
        mov  #M−1, BRC1      ; Repeat inner loop M times
        rptb outloop-1       ; Repeat outer loop up to outloop
        mpy  *AR1+, *CDP+, AC0
        mpy  *AR2+, *CDP+, AC1
        rptb inloop-1        ; Repeat inner loop up to inloop
        mac  *AR1+, *CDP+, AC0
        mac  *AR2+, *CDP+, AC1
  inloop                     ; End of inner loop
```

```
        mov  AC0, *AR3+          ; Save result in AC0
        mov  AC1, *AR4+          ; Save result in AC1
    outloop                      ; End of outer loop
```

The above example uses two repeat instructions to control a nested repetitive operation. The block-repeat structure

```
        rptb label_name-1
        (more instructions ...)
    label_name
```

executes a block of instructions between the `rptb` instruction and the end label `label_name`. The maximum number of instructions that can be used inside a block-repeat loop is limited to 64 Kbytes of code. Because of the pipeline scheme, the minimum cycles within a block-repeat loop are two. The maximum number of times that a loop can be repeated is limited to 65 536 ($= 2^{16}$) because of the 16-bit block-repeat counters.

2.7   Mixed C and Assembly Language Programming

As discussed in Chapter 1, the mixing of C and assembly programs are used for many DSP applications. C code provides the ease of maintenance and portability, while assembly code has the advantages of run-time efficiency and code density. We can develop C functions and assembly routines, and use them together. In this section, we will introduce how to interface C with assembly programs and review the guidelines of the C function calling conventions for the TMS320C55x.

The assembly routines called by a C function can have arguments and return values just like C functions. The following guidelines are used for writing the C55x assembly code that is callable by C functions.

Naming convention: Use the underscore '_' as a prefix for all variables and routine names that will be accessed by C functions. For example, use `_my_asm_func` as the name of an assembly routine called by a C function. If a variable is defined in an assembly routine, it must use the underscore prefix for C function to access it, such as `_my_var`. The prefix '_' is used by the C compiler only. When we access assembly routines or variables in C, we don't need to use the underscore as a prefix. For example, the following C program calls the assembly routine using the name `my_asm_func` without the underscore:

```
    extern int my_asm_func  /* Reference an assembly function */
    void main()
    {
        int c;                  /* Define local variable          */
        c = my_asm_func();      /* Call the assembly function     */
    }
```

This C program calls the following assembly routine:

```
    .global _my_asm_func  ; Define the assembly function
    _my_asm_func          ; Name of assembly routine
        mov #0x1234, T0
        ret               ; Return to the call function
```

Variable definition: The variables that are accessed by both C and assembly routines must be defined as global variables using the directive `.global`, `.def`, or `.ref` by the assembler.

Compiler mode: By using the C compiler, the C55x CPL (compiler mode) bit is automatically set for using stack-pointer (SP) relative addressing mode when entering an assembly routine. The indirect addressing modes are preferred under this configuration. If we need to use direct addressing modes to access data memory in a C callable assembly routine, we must change to DP (data-page) direct addressing mode. This can be done by clearing the CPL bit. However, before the assembly routine returns to its C caller function, the CPL bit must be restored. The bit clear and bit set instructions, `bclr CPL` and `bset CPL`, can be used to reset and set the CPL bit in the status register ST1, respectively. The following code can be used to check the CPL bit, turn CPL bit off if it is set, and restore the CPL bit before returning it to the caller.

```
     btstclr #14, *(ST1), TC1   ; Turn off CPL bits if it is set
     (more instructions . . . )
     xcc continue, TC1          ; TC1 is set if we turned CPL bit off
     bset CPL                   ; Turn CPL bit on
continue
     ret
```

Passing arguments: To pass arguments from a C function to an assembly routine, we must follow the strict rules of C-callable conversions set by the C compiler. When passing an argument, the C compiler assigns it to a particular data type and then places it in a register according to its data type. The C55x C compiler uses the following three classes to define the data types:

- Data pointer: `int *`, or `long *`.

- 16-bit data: `char`, `short`, or `int`.

- 32-bit data: `long`, `float`, `double`, or function pointers.

If the arguments are pointers to data memory, they are treated as data pointers. If the argument can fit into a 16-bit register such as `int` and `char`, it is considered to be 16-bit data. Otherwise, it is considered 32-bit data. The arguments can also be structures. A structure of two words (32 bits) or less is treated as a 32-bit data argument and is passed using a 32-bit register. For structures larger than two words, the arguments are passed by reference. The C compiler will pass the address of a structure as a pointer, and this pointer is treated like a data argument.

For a subroutine call, the arguments are assigned to registers in the order that the arguments are listed in the function. They are placed in the following registers according to their data type, in an order shown in Table 2.6.

Note in Table 2.6 the overlap between AR registers used for data pointers and the registers used for 16-bit data. For example, if T0 and T1 hold 16-bit data arguments, and AR0 already holds a data pointer argument, a third 16-bit data argument would be placed into AR1. See the second example in Figure 2.13. If the registers of the appropriate type are not available, the arguments are passed onto the stack. See the third example in Figure 2.13.

Table 2.6 Argument classes assigned to registers

Argument type	Register assignment order
16-bit data pointer	AR0, AR1, AR2, AR3, AR4
23-bit data pointer	XAR0, XAR1, XAR2, XAR3, XAR4
16-bit data	T0, T1, AR0, AR1, AR2, AR3, AR4
32-bit data	AC0, AC1, AC2

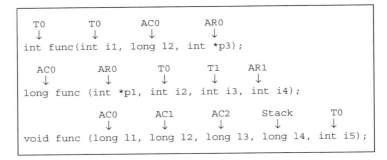

```
   T0        T0        AC0        AR0
   ↓         ↓         ↓          ↓
int func(int i1, long 12, int *p3);

  AC0       AR0       T0        T1      AR1
   ↓         ↓        ↓         ↓       ↓
long func (int *p1, int i2, int i3, int i4);

            AC0       AC1       AC2       Stack       T0
             ↓         ↓         ↓         ↓          ↓
void func (long 11, long 12, long 13, long 14, int i5);
```

Figure 2.13 Examples of arguments passing conventions

Return values: The calling function/routine collects the return value from the called function/subroutine. A 16-bit integer data is returned by the register T0 and a 32-bit data is returned in the accumulator AC0. A data pointer is returned in (X)AR0. When the called routine returns a structure, the structure is on the local stack.

Register use and preservation: When making a function call, the register assignments and preservations between the caller and called functions are strictly defined. Table 2.7 describes how the registers are preserved during a function call. The called function must save the contents of the save-on-entry registers (T2, T3, AR5, AR6, and AR7) if it will use these registers. The calling function must push the contents of any other save-on-call registers onto the stack if these register's contents are needed after the function/subroutine call. A called function can freely use any of the save-on-call registers (AC0–AC2, T0, T1, and AR0–AR4) without saving its value. More detailed descriptions can be found in the TMS320C55x Optimizing C Compiler User's Guide [3].

2.8 Experiments – Assembly Programming Basics

We have introduced the TMS320C55x assembly language and several addressing modes. Experiments given in this section will help to use different addressing modes for writing assembly code. We also introduced the C function interfacing with assembly routines and we will explore the C-assembly interface first.

Table 2.7 Register use and preservation conventions

Registers	Preserved by	Used for
AC0–AC2	Calling function Save-on-call	16, 32, or 40-bit data 24-bit code pointers
(X)AR0–(X)AR4	Calling function Save-on-call	16-bit data 16 or 23-bit pointers
T0 and T1	Calling function Save-on-call	16-bit data
AC3	Called function Save-on-entry	16, 32, or 40-bit data
(X)AR5–(X)AR7	Called function Save-on-entry	16-bit data 16 or 23-bit pointers
T2 and T3	Called function Save-on-entry	16-bit data

2.8.1   Experiment 2A – Interfacing C with Assembly Code

In this experiment, we will learn how to write C-callable assembly routines. The following example illustrates a C function `main`, which calls an assembly routine to perform a summation, and returns the result back to the C main function. The C program `exp2a.c` is listed as follows:

```
extern int sum(int *);        /* Assembly routine sum        */
int x[2]  = {0x1234, 0x4321} ;  /* Define x[ ] as global array   */
int s;                        /* Define s as global variable  */
void main ( )
{
    s = sum(x);               /* Call assembly routine _sum   */
}
```

The assembly routine `exp2_sum.asm` is listed as follows:

```
        .global _sum
_sum
        mov * AR0+, AC0   ; AC0 = x[1]
        add * AR0+, AC0   ; AC0 = x[1]+x[2]
        mov AC0, T0       ; Return value in T0
        ret               ; Return to calling function
```

where the label _sum defines the starting or entry of the assembly subroutine and directive .global defines that the assembly routine _sum as a global function.

Perform the following steps for Experiment 2A:

1. Use the CCS to create a project called `exp2a` in A:\Experiment2.

2. Write `exp2a.c` based on the C sample code given above, save it in A:\Experiment2.

3. Write `exp2_sum.asm` based on the assembly sample code given above and save it in A:\Experiment2.

4. Copy the linker command file, `exp1.cmd`, from previous experiment, rename it to `exp2.cmd` and save it to A:\Experiment2.

5. From the CCS Project-Options-Linker-Library tab, to include the run-time support library `rst55.lib`.

6. From CCS Project-Options-Compiler, sets include file search path if necessary. Under the Category-Assembly, check the Keep `.asm` files box and interlist C and ASM statements box.

7. Compile and debug the code, then load `exp2a.out` and issue the Go-Main command.

8. Watch and record the changes in the AC0, AR0, and T0 in the CPU register window. Watch memory location 's' and 'x' and record when the content of result 's' is updated. Why?

9. Single-step through the C and assembly code.

10. Examine the assembly code `exp2a.asm` that the C compiler generated. How is the return value passed to the C calling function?

11. If we define the result 's' inside the function `main( )`, from which location can we view its value? Why?

2.8.2   Experiment 2B – Addressing Mode Experiments

In Section 2.4, we introduced six C55x addressing modes. In the second part of the experiments, we will write assembly routines to exercise different addressing modes to understand how each of these addressing modes work. The assembly routines for this experiment are called by a C function.

1. Write a C function called `exp2b.c` as follows:

```
int Ai[8] ;                 /* Define array Ai[ ] */
int Xi[8] ;                 /* Define array Xi[ ] */
int result1, result2;       /* Define variables */
main ( )
{
    void exp2b_1 (void);
    void exp2b_2 (void);
```

```
    result1 = exp2b_3(Ai, Xi);
    result2 = exp2b_4(Ai, Xi);
}
```

Save the file as `exp2b.c` in `A:\Experiment2`.

2. Write assembly routine `void exp2b_1(void)` using the absolute addressing mode to initialize the array `Ai[8] = { 1, 2, 3, 4, 5, 6, 7, 8}` in data memory as in the following example:

```
.global _Ai
mov      #1, *(_Ai)
mov      #2, *(_Ai+1)
(more instructions . . . )
```

Since the array `Ai[8]` is defined in the C function, the assembly routine references it using the directive `.global` (or `.ref`).

3. Write the assembly routine `void exp2b_2(void)` using the direct addressing mode to initialize the array `X[8] = { 9, 3, 2, 0, 1, 9, 7, 1}` in data memory. As we mentioned in Section 2.4.1, there are four different direct addressing modes. We use the DP direct addressing mode for this part of the experiment. The data-page pointer, XDP, needs to be set before we can start using the direct addressing mode. The `Xi` array can be initialized using the following assembly code:

```
btstclr #14, *(ST1), TC1   ; Clear CPL bit for DP addressing
amov     #_Xi, XDP          ; Initialize XDP to point to Xi
.dp      _Xi
mov      #9, @_Xi           ; Using direct addressing mode
mov      #3, @_Xi+1
(more instructions . . . )
xcc      continue, TC1
bset     CPL                ; Reset CPL bit if it was cleared
continue
```

The instruction `btstclr #14, *(ST1), TC1` tests the CPL (compiler mode) bit (bit 14) of the status register ST1. The compiler mode will be set if this routine is called by a C function. If the test is true, the test flag bit, TC1 (bit 13 of status register ST0) will be set, and the instruction clears the CPL bit. This is necessary for using DP direct addressing mode instead of SP direct addressing mode. At the end of the code section, the conditional execution instruction, `xcc continue, TC1`, is used to set the CPL bit if TC1 was set.

4. The sum-of-product operation (dot product) is one of the most commonly used functions by DSP systems. A dot product of two vectors of length L can be expressed as

$$Y = \sum_{i=0}^{L-1} A_i X_i = A_0 X_0 + A_1 X_1 + \cdots + A_{L-1} X_{L-1},$$

where the vectors A_i and X_i are one-dimensional arrays of length L. There are many different ways to access the elements of the arrays, such as direct, indirect, and absolute addressing modes. In the following experiment, we will write a subroutine int exp2b_3(int *Ai, int *Xi) to perform the dot product using indirect addressing mode, and store the returned value in the variable result in data memory. The code example is given as follows:

```
; Assume AR0 and AR1 are pointing to Ai and Xi
mpym  *AR0+, *AR1+, AC0   ; Multiply Ai[0] and Xi[0]
mpym  *AR0+, *AR1+, AC1   ; Multiply Ai[1] and Xi[1]
add   AC1, AC0            ; Accumulate the partial result
mpym  *AR0+, *AR1+, AC1   ; Multiply Ai[2] and Xi[2]
add   AC1, AC0            ; Accumulate the partial result
(more instructions . . . )
mov   AC0, T0
```

In the program, arrays A_i and X_i are defined as global arrays in the exp2.c. The A_i and X_i arrays have the same data values as given previously. The return value is passed to the calling function by T0.

5. Write an assembly routine int exp2b_4(int *Ai, int *Xi) using the indirect addressing mode in conjunction with parallel instructions and repeat instructions to improve the code density and efficiency. The following is an example of the code:

```
  mpym * AR0+, * AR1+, AC0   ; Multiply Ai[0] and Xi[0]
||rpt #6                     ; Multiply and accumulate the rest
  macm * AR0+, * AR1+, AC0
```

The auxiliary registers, AR0 and AR1, are used as data pointers to array A_i and array X_i, respectively. The instruction macm performances multiply-and-accumulate operation. The parallel bar || indicates the parallel operation of two instructions. The repeat instruction, rpt #K will repeat the following instruction K+1 times.

6. Create a project called exp2b and save it in A:\Experiment2.

7. Use exp2.cmd, exp2b.c, exp2b_1.asm, exp2b_2.asm, exp2b_3.asm, and exp2b_4.asm to build the project.

8. Open the memory watch window to watch how the arrays A_i and X_i are initialized in data memory by the assembly routine exp2b_1.asm and exp2b_2.asm.

9. Open the CPU registers window to see how the dot product is computed by exp2b_3.asm, and exp2b_4.asm.

10. Use the profile capability learned from the experiments given in Chapter 1 to measure the run-time of the sum-of-product operations and compare the cycle difference of the routine exp2b_3.asm and exp2b_4.asm.

11. Use the map file to compare the assembly program code size of routine `exp2b_3.asm` and `exp2b_4.asm`. Note that the program size is given in bytes.

References

[1] Texas Instruments, Inc., *TMS320C55x DSP CPU Reference Guide*, Literature no. SPRU371A, 2000.
[2] Texas Instruments, Inc., *TMS320C55x Assembly Language Tools User's Guide*, Literature no. SPRU380, 2000.
[3] Texas Instruments, Inc., *TMS320C55x Optimizing C Compiler User's Guide*, Literature no. SPRU281, 2000.
[4] Texas Instruments, Inc., *TMS320C55x DSP Mnemonic Instruction Set Reference Guide*, Literature no. SPRU374A, 2000.
[5] Texas Instruments, Inc., *TMS320C55x DSP Algebraic Instruction Set Reference Guide*, Literature no. SPRU375, 2000.
[6] Texas Instruments, Inc., *TMS320C55x Programmer's Reference Guide*, Literature no. SPRU376, 2000.

Exercises

1. Check the following examples to determine if these are correct parallel instructions. If not, correct the problems.

   (a)   `mov * AR1+, AC1`
       `: : add @x, AR2`

   (b)   `mov AC0, dbl(* AR2+)`
       `: : mov dbl(*(AR1+T0)), AC2`

   (c)   `mpy * AR1+, * AR2+, AC0`
       `: : mpy * AR3+, * AR2+, AC1`
       `| | rpt #127`

2. Given a memory block and XAR0, XDP, and T0 as shown in Figure 2.14. Determine the contents of AC0, AR0, and T0 after the execution of the following instructions:

   (a) `mov *(#x+2), AC0`
   (b) `mov @(x−x+1), AC0`
   (c) `mov @(x−x+0x80), AC0`
   (d) `mov * AR0+, AC0`
   (e) `mov *(AR0+T0), AC0`
   (f) `mov * AR0(T0), AC0`
   (g) `mov * AR0(#−1), AC0`
   (h) `mov * AR0(#2), AC0`
   (i) `mov * AR0(#0x80), AC0`

3. C functions are defined as follows. Use Table 2.8 to show how the C compiler passes parameters for each of the following functions:

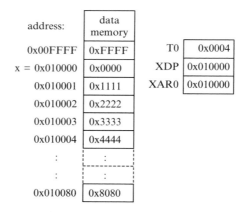

Figure 2.14 Data memory and registers

(a) int func_a(long, int, int, int, int, int,
       int *, int *, int *, int *);
   var = func_a(0xD32E0E1D, 0, 1, 2, 3, 4, pa, pb, pc, pd);

(b) int func_b(long, long, long, int, int, int,
       int *, int *, int *, int *);
   var = func_b(0x12344321, 0, 1, 2, 3, 4, pa, pb, pc, pd);

(c) long func_c(int, int *, int *, int, int, long,
       int *, int *, long, long);
   var = func_c(0x2468, pa, pb, 1, 2, 0x1001,
       pc, pd, 0x98765432, 0x0);

Table 2.8 List of parameters passed by the C55x C compiler

T0	T1	T2	T3	AC0	AC1	AC2	AC3
XAR0	XAR1	XAR2	XAR3	XAR4	XAR5	XAR6	XAR7
SP(-3)	SP(-2)	SP(-1)	SP(0)	SP(1)	SP(2)	SP(3)	var

3

DSP Fundamentals and Implementation Considerations

The derivation of discrete-time systems is based on the assumption that the signal and system parameters have infinite precision. However, most digital systems, filters, and algorithms are implemented on digital hardware with finite wordlength. Therefore DSP implementation with fixed-point hardware requires special attention because of the potential quantization and arithmetic errors, as well as the possibility of overflow. These effects must always be taken into consideration in DSP system design and implementation.

This chapter presents some fundamental DSP concepts in time domain and practical considerations for the implementation of digital filters and algorithms on DSP hardware. Sections 3.1 and 3.2 briefly review basic time-domain DSP issues. Section 3.3 introduces probability and random processes, which are useful in analyzing the finite-precision effects in the latter half of the chapter and adaptive filtering in Chapter 8. The rigorous treatment of these subjects can be found in other DSP books listed in the reference. Readers who are familiar with these DSP fundamentals should be able to skip through some of these sections. However, most notations used throughout the book will be defined in this chapter.

3.1 Digital Signals and Systems

In this section, we will define some widely used digital signals and simple DSP systems. The purpose of this section is to provide the necessary background for understanding the materials presented in subsequent sections and later chapters.

3.1.1 Elementary Digital Signals

There are several ways to describe signals. For example, signals encountered in communications are classified as deterministic or random. Deterministic signals are used for testing purposes and for mathematically describing certain phenomena. Random signals are information-bearing signals such as speech. Some deterministic signals will be introduced in this section, while random signals will be discussed in Section 3.3.

As discussed in Chapter 1, a digital signal is a sequence of numbers $\{x(n), -\infty < n < \infty\}$, where n is the time index. The unit-impulse sequence, with only one non-zero value at $n = 0$, is defined as

$$\delta(n) = \begin{cases} 1, & n = 0 \\ 0, & n \neq 0, \end{cases} \tag{3.1.1}$$

where $\delta(n)$ is also called the Kronecker delta function. This unit-impulse sequence is very useful for testing and analyzing the characteristics of DSP systems, which will be discussed in Section 3.1.3.

The unit-step sequence is defined as

$$u(n) = \begin{cases} 1, & n \geq 0 \\ 0, & n < 0. \end{cases} \tag{3.1.2}$$

This signal is very convenient for describing a causal (or right-sided) signal $x(n)$ for $n \geq 0$. Causal signals are the most commonly encountered signals in real-time DSP systems.

Sinusoidal signals (sinusoids or sinewaves) are the most important sine (or cosine) signals that can be expressed in a simple mathematical formula. They are also good models for real-world signals. The analog sinewave can be expressed as

$$x(t) = A \sin(\Omega t + \phi) = A \sin(2\pi f t + \phi), \tag{3.1.3}$$

where A is the amplitude of the sinewave,

$$\Omega = 2\pi f \tag{3.1.4}$$

is the frequency in radians per second (rad/s), f is the frequency in Hz, and ϕ is the phase-shift (initial phase at origin $t = 0$) in radians.

When the analog sinewave defined in (3.1.3) is connected to the DSP system shown in Figure 1.1, the digital signal $x(n)$ available for the DSP hardware is the causal sinusoidal signal

$$\begin{aligned} x(n) &= A \sin(\Omega n T + \phi), \quad n = 0, 1, \ldots, \infty \\ &= A \sin(\Omega n T + \phi) u(n) \\ &= A \sin(2\pi f n T + \phi) u(n), \end{aligned} \tag{3.1.5}$$

where T is the sampling period in seconds. This causal sequence can also be expressed as

$$x(n) = A \sin(\omega n + \phi) u(n) = A \sin(2\pi F n + \phi) u(n), \tag{3.1.6}$$

where

$$w = \Omega T = \frac{\Omega}{f_s} \qquad (3.1.7)$$

is the discrete-time frequency in radians per sample and

$$F = f T = \frac{f}{f_s} \qquad (3.1.8)$$

is the normalized frequency to its sampling frequency, f_s, in cycles per sample.

The fundamental difference between describing the frequency of analog and digital signals is summarized in Table 3.1. Analog signal sampling implies a mapping of an infinite range of real-world frequency variable f (or Ω) into a finite range of discrete-time frequency variable F (or w). The highest frequency in a digital signal is $F = 1/2$ (or $w = \pi$) based on Shannon's sampling theorem defined in (1.2.3). Therefore the spectrum of discrete-time (digital) signals is restricted to a limited range as shown in Table 3.1. Note that some DSP books define the normalized frequency as $F = f/(f_s/2)$ with frequency range $-1 \leq F \leq 1$.

Example 3.1: Generate 64 samples of a sine signal with $A = 2$, $f = 1000$ Hz, and $f_s = 8$ kHz using MATLAB. Since $F = f/f_s = 0.125$, we have $w = 2\pi F = 0.25\pi$. From Equation (3.1.6), we need to generate $x(n) = 2 \sin(wn)$, for $n = 0, 1, \ldots, 63$. These sinewave samples can be generated and plotted by the following MATLAB script:

```
n = [0:63];
omega = 0.25*pi;
xn = 2*sin(omega*n);
plot(n, xn);
```

3.1.2 Block Diagram Representation of Digital Systems

A DSP system (or algorithm) performs prescribed operations on digital signals. In some applications, we view a DSP system as an operation performed on an input signal, $x(n)$, in order to produce an output signal, $y(n)$, and express the general relationship between $x(n)$ and $y(n)$ as

Table 3.1 Units, relationships, and range of four frequency variables

Variables	Unit	Relationship	Range
Ω	radians per second	$\Omega = 2\pi f$	$-\infty < \Omega < \infty$
f	cycles per second (Hz)	$f = \dfrac{F}{T}$	$-\infty < f < \infty$
w	radians per sample	$w = 2\pi F$	$-\pi \leq w \leq \pi$
F	cycles per sample	$F = \dfrac{f}{f_s}$	$-\dfrac{1}{2} \leq F \leq \dfrac{1}{2}$

$$y(n) = T[x(n)],$$ (3.1.9)

where T denotes the computational process for transforming the input signal, $x(n)$, into the output signal, $y(n)$. A block diagram of the DSP system defined in (3.1.9) is illustrated in Figure 3.1.

The processing of digital signals can be described in terms of combinations of certain fundamental operations on signals. These operations include addition (or subtraction), multiplication, and time shift (or delay). A DSP system consists of the interconnection of three basic elements – adders, multipliers, and delay units.

Two signals, $x_1(n)$ and $x_2(n)$, can be added as illustrated in Figure 3.2, where

$$y(n) = x_1(n) + x_2(n)$$ (3.1.10)

is the adder output. With more than two inputs, the adder could be drawn as a multi-input adder, but the additions are typically done two inputs at a time in digital hardware. The addition operation of Equation (3.1.10) can be implemented as the following C55x code using direct addressing mode:

```
mov @x1n, AC0   ; AC0 = x1(n)
add @x2n, AC0   ; AC0 = x1(n)+x2(n) = y(n)
```

A given signal can be multiplied by a constant, α, as illustrated in Figure 3.3, where $x(n)$ is the multiplier input, α represents the multiplier coefficient, and

$$y(n) = \alpha x(n)$$ (3.1.11)

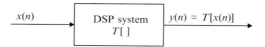

Figure 3.1   Block diagram of a DSP system

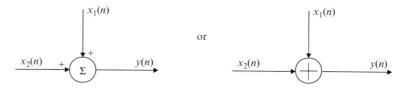

Figure 3.2   Block diagram of an adder

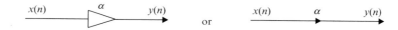

Figure 3.3   Block diagram of a multiplier

Figure 3.4 Block diagram of a unit delay

is the multiplier's output. The multiply operation of equation (3.1.11) can be implemented by the following C55x code using indirect addressing mode:

```
amov #alpha, XAR1    ; AR1 points to alpha (α)
amov #xn, XAR2       ; AR2 points to x(n)
mpy  *AR1, *AR2, AC0 ; AC0 = α*x(n) = y(n)
```

The sequence $\{x(n)\}$ can be shifted (delayed) in time by one sampling period, T, as illustrated in Figure 3.4. The box labeled z^{-1} represents the unit delay, $x(n)$ is the input signal, and

$$y(n) = x(n - 1) \tag{3.1.12}$$

is the output signal, which is the input signal delayed by one unit (a sampling period). In fact, the signal $x(n - 1)$ is actually the stored signal $x(n)$ one sampling period (T seconds) before the current time. Therefore the delay unit is very easy to implement in a digital system, but is difficult to implement in an analog system. A delay by more than one unit can be implemented by cascading several delay units in a row. Therefore an L-unit delay requires L memory locations configured as a first-in first-out buffer, which can also be implemented as a circular buffer (will be discussed in Chapter 5) in memory.

There are several ways to implement delay operations on the TMS320C55x. The following code uses a delay instruction to move the contents of the addressed data memory location into the next higher address location:

```
amov #xn, XAR1    ; AR1 points to x(n)
delay *AR1        ; Contents of x(n) is copied to x(n-1)
```

These three basic building blocks can be connected to form a block diagram representation of a DSP system. The input–output (I/O) description of a DSP system consists of mathematical expressions with addition, multiplication, and delays, which explicitly define the relationship between the input and output signals. DSP algorithms are closely related to block diagram realizations of the I/O difference equations. For example, consider a simple DSP system described by the difference equation

$$y(n) = \alpha x(n) + \alpha x(n - 1). \tag{3.1.13}$$

The block diagram of the system using the three basic building blocks is sketched in Figure 3.5(a). Note that the difference equation (3.1.13) and the block diagram show exactly how the output signal $y(n)$ is computed in the DSP system for a given input signal, $x(n)$.

The DSP algorithm shown in Figure 3.5(a) requires two multiplications and one addition to compute the output sample $y(n)$. A simple algebraic simplification may

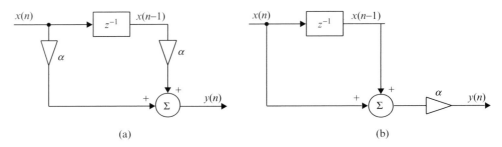

Figure 3.5 Block diagrams of DSP systems: (a) direct realization described in (3.1.13), and (b) simplified implementation given in (3.1.14)

be used to reduce computational requirements. For example, (3.1.13) can be rewritten as

$$y(n) = \alpha[x(n) + x(n-1)].$$ (3.1.14)

The block diagram implementation of this difference equation is illustrated in Figure 3.5(b), where only one multiplication is required. This example shows that with careful design (or optimization), the complexity of the system (or algorithm) can be further reduced.

The C55x implementation of (3.1.14) can be written as:

```
amov #alpha, XAR1      ; AR1 points to α
amov #temp, XAR2       ; AR2 points to temp
mov  *(x1n), AC0       ; AC0 = x1(n)
add  *(x2n), AC0       ; AC0 = x1(n)+x2(n)
mov  AC0, *AR2         ; Temp = x1(n)+x2(n), pointed by AR2
mpy  *AR1, *AR2, AC1   ; AC1 = α*[x1(n)+x2(n)]
```

Equation (3.1.14) can also be implemented as:

```
amov #x1n, XAR1        ; AR1 points to x1(n)
amov #x2n, XAR2        ; AR2 points to x2(n)
amov #alpha, XAR3      ; AR3 points to α
mpy  *AR1, *AR3, AC1   ; AC1 = α*x1(n)
mac  *AR2, *AR3, AC1   ; AC1 = α*x1(n) + α*x2(n)
```

When the multiplier coefficient α is a number with a base of 2 such as 0.25 (1/4), we can use shift operation instead of multiplication. The following example uses the absolute addressing mode:

```
mov  *(x1n)≪#-2, AC0   ; AC0 = 0.25*x1(n)
add  *(x2n)≪#-2, AC0   ; AC0 = 0.25*x1(n) + 0.25*x2(n)
```

where the right shift option, $\ll$#-2, shifts the content of x1n and x2n to the right by 2 bits (equivalent to dividing it by 4) before they are used.

3.1.3   Impulse Response of Digital Systems

If the input signal to the DSP system is the unit-impulse sequence $\delta(n)$ defined in (3.1.1), then the output signal, $h(n)$, is called the impulse response of the system. The impulse response plays a very important role in the study of DSP systems. For example, consider a digital system with the I/O equation

$$y(n) = b_0 x(n) + b_1 x(n-1) + b_2 x(n-2). \tag{3.1.15}$$

The impulse response of the system can be obtained by applying the unit-impulse sequence $\delta(n)$ to the input of the system. The outputs are the impulse response coefficients computed as follows:

$$h(0) = y(0) = b_0 \cdot 1 + b_1 \cdot 0 + b_2 \cdot 0 = b_0$$
$$h(1) = y(1) = b_0 \cdot 0 + b_1 \cdot 1 + b_2 \cdot 0 = b_1$$
$$h(2) = y(2) = b_0 \cdot 0 + b_1 \cdot 0 + b_2 \cdot 1 = b_2$$
$$h(3) = y(3) = b_0 \cdot 0 + b_1 \cdot 0 + b_2 \cdot 0 = 0$$
$$\cdots$$

Therefore the impulse response of the system defined in (3.1.15) is $\{b_0, b_1, b_2, 0, 0, \dots\}$.
The I/O equation given in (3.1.15) can be generalized as the difference equation with L parameters, expressed as

$$y(n) = b_0 x(n) + b_1 x(n-1) + \cdots + b_{L-1} x(n-L+1) = \sum_{l=0}^{L-1} b_l x(n-l). \tag{3.1.16}$$

Substituting $x(n) = \delta(n)$ into (3.1.16), the output is the impulse response expressed as

$$h(n) = \sum_{l=0}^{L-1} b_l \delta(n-l) = \begin{cases} b_n & n = 0, 1, \dots, L-1 \\ 0 & \text{otherwise.} \end{cases} \tag{3.1.17}$$

Therefore the length of the impulse response is L for the difference equation defined in (3.1.16). Such a system is called a finite impulse response (FIR) system (or filter). The impulse response coefficients, b_l, $l = 0, 1, \dots, L-1$, are called filter coefficients (weights or taps). The FIR filter coefficients are identical to the impulse response coefficients. Table 3.2 shows the relationship of the FIR filter impulse response $h(n)$ and its coefficients b_l.

3.2   Introduction to Digital Filters

As shown in (3.1.17), the system described in (3.1.16) has a finite number of non-zero impulse response coefficients b_l, $l = 0, 1, \dots, L-1$. The signal-flow diagram of the

Table 3.2 Relationship of impulse response and coefficients of an FIR filter

| b_l | b_0 | b_1 | b_2 | ... | b_{L-1} | |
n	$x(n)$	$x(n-1)$	$x(n-2)$	...	$x(n-L+1)$	$y(n) = h(n)$
0	1	0	0	...	0	$h(0) = b_0$
1	0	1	0	...	0	$h(1) = b_1$
2	0	0	1	...	0	$h(2) = b_2$
...	...	...	...	...	...	...
$L-1$	0	0	0	...	1	$h(L-1) = b_{L-1}$
L	0	0	0	...	0	0

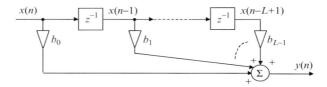

Figure 3.6 Detailed signal-flow diagram of FIR filter

system described by the I/O Equation (3.1.16) is illustrated in Figure 3.6. The string of z^{-1} functions is called a tapped-delay-line, as each z^{-1} corresponds to a delay of one sampling period. The parameter, L, is the order (length) of the FIR filter. The design and implementation of FIR filters (transversal filters) will be discussed in Chapter 5.

3.2.1 FIR Filters and Power Estimators

The moving (running) average filter is a simple example of an FIR filter. Averaging is used whenever data fluctuates and must be smoothed prior to interpretation. Consider an L-point moving-average filter defined as

$$y(n) = \frac{1}{L}[x(n) + x(n-1) + \cdots + x(n-L+1)]$$
$$= \frac{1}{L}\sum_{l=0}^{L-1} x(n-l), \tag{3.2.1}$$

where each output signal $y(n)$ is the average of L consecutive input signal samples. The summation operation that adds all samples of $x(n)$ between 1 and L can be implemented using the MATLAB statement:

```
yn = sum(xn(1:L));
```

Implementation of (3.2.1) requires $L-1$ additions and L memory locations for storing signal sequence $x(n), x(n-1), \ldots, x(n-L+1)$ in a memory buffer. As illustrated in Figure 3.7, the signal samples used to compute the output signal at time n are L samples included in the window at time n. These samples are almost the same as those samples used for the previous window at time $n-1$ to compute $y(n-1)$, except that the oldest sample $x(n-L)$ of the window at time $n-1$ is replaced by the newest sample $x(n)$ of the window at time n. Thus (3.2.1) can be computed as

$$y(n) = y(n-1) + \frac{1}{L}[x(n) - x(n-L)]. \tag{3.2.2}$$

Therefore the averaged signal, $y(n)$, can be computed recursively as expressed in (3.2.2). This recursive equation can be realized by using only two additions. However, we need $L+1$ memory locations for keeping $L+1$ signal samples $\{x(n)x(n-1)\cdots x(n-L)\}$.

The following C5xx assembly code illustrates the implementation of a moving-average filter of $L = 8$ based on Equation (3.2.2):

```
L           .set 8                      ; Order of filter
xin         .usect "indata",  1
xbuffer     .usect "indata", L          ; Length of buffer
y           .usect "outdata", 2,1,1     ; Long-word format
      amov #xbuffer+L-1, XAR3           ; AR3 points to end of x buffer
      amov #xbuffer+L-2, XAR2           ; AR2 points to next sample
      mov  dbl(*(y)), AC1               ; AC1 = y(n-1) in long format
      mov  * (xin), AC0                 ; AC0 = x(n)
      sub  *AR3, AC0                    ; AC0 = x(n) - x(n-L)
      add  AC0, #-3, AC1                ; AC1 = y(n-1) + 1/L[x(n)-x(n-L)]
      mov  AC1, dbl (*(y))              ; y(n) = AC1
      rpt  # (L-1)                      ; Update the tapped-delay-line
      mov  *AR2-, *AR3-                 ; X(n-1) = x(n)
      mov  * (xin), AC0                 ; Update the newest sample x(n)
      mov  AC0, *AR3                    ; X(n) = input xin
```

The strength of a digital signal may be expressed in terms of peak value, energy, and power. The peak value of deterministic signals is the maximum absolute value of the signal. That is,

$$M_x = \max_n\{|x(n)|\}. \tag{3.2.3}$$

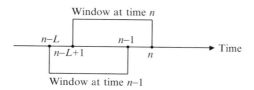

Figure 3.7   Time windows at current time n and previous time $n-1$

The maximum value of the array xn can be found using the MATLAB function

```
Mx = max(xn);
```

The energy of the signal $x(n)$ is defined as

$$E_x = \sum_n |x(n)|^2. \tag{3.2.4}$$

The energy of a real-valued $x(n)$ can be calculated by the MATLAB statement:

```
Ex = sum(abs(xn).^2);
```

Periodic signals and random processes have infinite energy. For such signals, an appropriate definition of strength is power. The power of signal $x(n)$ is defined as

$$P_x = \lim_{L \to \infty} \frac{1}{L} \sum_{n=0}^{L-1} |x(n)|^2. \tag{3.2.5}$$

If $x(n)$ is a periodic signal, we have

$$x(n) = x(n + kL), \tag{3.2.6}$$

where k is an integer and L is the period in samples. Any one period of L samples completely defines a periodic signal. From Figure 3.7, the power of $x(n)$ can be computed by

$$P_x = \frac{1}{L} \sum_{l=n-L+1}^{n} |x(l)|^2 = \frac{1}{L} \sum_{l=0}^{L-1} |x(n-l)|^2. \tag{3.2.7}$$

For example, a real-valued sinewave of amplitude A defined in (3.1.6) has the power $P_x = 0.5A^2$.

In most real-time applications, the power estimate of real-valued signals at time n can be expressed as

$$\hat{P}_x(n) = \frac{1}{L} \sum_{l=0}^{L-1} x^2(n-l). \tag{3.2.8}$$

Note that this power estimate uses L samples from the most recent sample at time n back to the oldest sample at time $n - L + 1$, as shown in Figure 3.7. Following the derivation of (3.2.2), we have the recursive power estimator

$$\hat{P}_x(n) = \hat{P}_x(n-1) + \frac{1}{L} [x^2(n) - x^2(n-L)]. \tag{3.2.9}$$

To further simplify the algorithm, we assume L is large enough so that $x^2(n - L) \approx \hat{P}_x(n - 1)$ from a statistical point of view. Thus Equation (3.2.9) can be further simplified to

$$\hat{P}_x(n) \approx \left(1 - \frac{1}{L}\right)\hat{P}_x(n - 1) + \frac{1}{L}x^2(n), \qquad (3.2.10)$$

or

$$\hat{P}_x(n) \approx (1 - \alpha)\hat{P}_x(n - 1) + \alpha x^2(n), \qquad (3.2.11a)$$

where

$$\alpha = \frac{1}{L}. \qquad (3.2.11b)$$

This is the most effective and widely used recursive algorithm for power estimation because only three multiplication operations and two memory locations are needed. For example, (3.2.11a) can be implemented by the C statement

```
pxn = (1.0-alpha)*pxn + alpha*xn*xn;
```

where `alpha` $= 1/L$ as defined in (3.2.11b). This C statement shows that we need three multiplications and only two memory locations for `xn` and `pxn`.

For stationary signals, a larger L (longer window) or smaller α can be used for obtaining a better average. However, a smaller L (shorter window) should be used for non-stationary signals for better results. In many real-time applications, the square of signal $x^2(n)$ used in (3.2.10) and (3.2.11a) can be replaced with its absolute value $|x(n)|$ in order to reduce further computation. This efficient power estimator will be further analyzed in Chapter 4 using the z-transform.

3.2.2 Response of Linear Systems

As discussed in Section 3.1.3, a digital system can be completely characterized by its impulse response $h(n)$. Consider a digital system illustrated in Figure 3.8, where $x(n)$ is the input signal and $y(n)$ is the output signal. If the impulse response of the system is $h(n)$, the output of the system can be expressed as

$$y(n) = x(n) * h(n) = \sum_{k=-\infty}^{\infty} x(k)h(n - k) = \sum_{k=-\infty}^{\infty} h(k)x(n - k), \qquad (3.2.12)$$

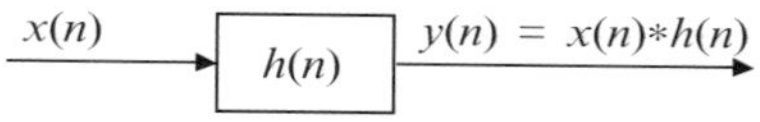

Figure 3.8   A simple linear system expressed in time domain

where * denotes the linear convolution operation and the operation defined in (3.2.12) is called the convolution sum. The input signal, $x(n)$, is convoluted with the impulse response, $h(n)$, in order to yield the output, $y(n)$. We will discuss the computation of linear convolution in detail in Chapter 5.

As shown in (3.2.12), the I/O description of a DSP system consists of mathematical expressions, which define the relationship between the input and output signals. The exact internal structure of the system is either unknown or ignored. The only way to interact with the system is by using its input and output terminals as shown in Figure 3.8. The system is assumed to be a 'black box'. This block diagram representation is a very effective way to depict complicated DSP systems.

A digital system is called the causal system if and only if

$$h(n) = 0, \quad n < 0. \tag{3.2.13}$$

A causal system is one that does not provide a response prior to input application. For a causal system, the limits on the summation of the Equation (3.2.12) can be modified to reflect this restriction as

$$y(n) = \sum_{k=0}^{\infty} h(k)x(n-k). \tag{3.2.14}$$

Thus the output signal $y(n)$ of a causal system at time n depends only on present and past input signals, and does not depend on future input signals.

Consider a causal system that has a finite impulse response of length L. That is,

$$h(n) = \begin{cases} 0, & n < 0 \\ b_n, & 0 \le n \le L - 1 \\ 0 & n \ge L . \end{cases} \tag{3.2.15}$$

Substituting this equation into (3.2.14), the output signal can be expressed identically to the Equation (3.1.16). Therefore the FIR filter output can be calculated as the input sequence convolutes with the coefficients (or impulse response) of the filter.

3.2.3 IIR Filters

A digital filter can be classified as either an FIR filter or an infinite impulse response (IIR) filter, depending on whether or not the impulse response of the filter is of finite or infinite duration. Consider the I/O difference equation of the digital system expressed as

$$y(n) = bx(n) - ay(n-1), \tag{3.2.16}$$

where each output signal $y(n)$ is dependent on the current input signal $x(n)$ and the previous output signal $y(n-1)$. Assuming that the system is causal, i.e., $y(n) = 0$ for $n < 0$ and let $x(n) = \delta(n)$. The output signals $y(n)$ are computed as

$$y(0) = bx(0) - ay(-1) = b,$$
$$y(1) = bx(1) - ay(0) = -ay(0) = -ab,$$
$$y(2) = bx(2) - ay(1) = -ay(1) = a^2b,$$
$$\cdots$$

In general, we have

$$h(n) = y(n) = (-1)^n a^n b, \quad n = 0, 1, 2, \ldots, \infty. \tag{3.2.17}$$

This system has infinite impulse response $h(n)$ if the coefficients a and b are non-zero. This system is called an IIR system (or filter). In theory, we can calculate an IIR filter output $y(n)$ using either the convolution equation (3.2.14) or the I/O difference equation (3.2.16). However, it is not computationally feasible using (3.2.14) for the impulse response $h(n)$ given in (3.2.17), because we cannot deal with an infinite number of impulse response coefficients. Therefore we must use an I/O difference equation such as the one defined in (3.2.16) for computing the IIR filter output in practical applications.

The I/O equation of the IIR system given in (3.2.16) can be generalized with the difference equation

$$y(n) = b_0 x(n) + b_1 x(n-1) + \cdots + b_{L-1} x(n-L+1) - a_1 y(n-1) - \cdots - a_M y(n-M)$$
$$= \sum_{l=0}^{L-1} b_l x(n-l) - \sum_{m=1}^{M} a_m y(n-m). \tag{3.2.18}$$

This IIR system is represented by a set of feedforward coefficients $\{b_l, \ l = 0, 1, \ldots, L-1\}$ and a set of feedback coefficients $\{a_m, \ m = 1, 2, \ldots, M\}$. Since the outputs are fed back and combined with the weighted inputs, this system is an example of the general class of feedback systems. Note that when all a_m are zero, Equation (3.2.18) is identical to (3.1.16). Therefore an FIR filter is a special case of an IIR filter without feedback coefficients. An FIR filter is also called a non-recursive filter.

The difference equation of IIR filters given in (3.2.18) can be implemented using the MATLAB function `filter` as follows:

```
yn = filter(b, a, xn);
```

where the vector b contains feedforward coefficients $\{b_l, \ l = 0, 1, \ldots, L-1\}$ and the vector a contains feedback coefficient $\{a_m, \ m = 1, 2, \ldots, M\}$. The signal vectors, xn and yn, are the input and output buffers of the system. The FIR filter defined in (3.1.16) can be implemented using MATLAB as

```
yn = filter(b, 1, xn);
```

Assuming that L is large enough so that the oldest sample $x(n-L)$ can be approximated using its average, $y(n-1)$. The moving-average filter defined in (3.2.2) can be simplified as

$$y(n) = \left(1 - \frac{1}{L}\right) y(n-1) + \frac{1}{L} x(n) = (1-\alpha) y(n-1) + \alpha x(n), \tag{3.2.19}$$

where α is defined in (3.2.11b). This is a simple first-order IIR filter. Design and implementation of IIR filters will be further discussed in Chapter 6.

3.3   Introduction to Random Variables

In Section 3.1, we treat signals as deterministic, which are known exactly and repeatable (such as a sinewave). However, the signals encountered in practice are often random signals such as speech and interference (noise). These random (stochastic) processes can be described at best by certain probability concepts. In this section, we will briefly introduce the concept of probability, followed by random variables and random signal processing.

3.3.1   Review of Probability and Random Variables

An experiment that has at least two possible outcomes is fundamental to the concept of probability. The set of all possible outcomes in any given experiment is called the sample space S. An event A is defined as a subset of the sample space S. The probability of event A is denoted by $P(A)$. Letting A be any event defined on a sample space S, we have

$$0 \leq P(A) \leq 1 \tag{3.3.1}$$

and

$$P(S) = 1. \tag{3.3.2}$$

For example, consider the experiment of rolling of a fair die N times ($N \to \infty$), we have $S = \{1 \leq A \leq 6\}$ and $P(A) = 1/6$.

A random variable, x, is defined as a function that maps all elements from the sample space S into points on the real line. A random variable is a number whose value depends on the outcome of an experiment. Given an experiment defined by a sample space with elements A, we assign to every A a real number $x = x(A)$ according to certain rules. Consider the rolling of a fair die N times and assign an integer number to each face of a die, we have a discrete random variable that can be any one of the discrete values from 1 to 6.

The cumulative probability distribution function of a random variable x is defined as

$$F(X) = P(x \leq X), \tag{3.3.3}$$

where X is a real number ranging from $-\infty$ to ∞, and $P(x \leq X)$ is the probability of $\{x \leq X\}$. Some properties of $F(X)$ are summarized as follows:

$$F(-\infty) = 0, \tag{3.3.4a}$$

$$F(\infty) = 1, \tag{3.3.4b}$$

$$0 \leq F(X) \leq 1, \tag{3.3.4c}$$

$$F(X_1) \leq F(X_2) \quad \text{if } X_1 \leq X_2, \tag{3.3.4d}$$

$$P(X_1 < x \leq X_2) = F(X_2) - F(X_1). \tag{3.3.4e}$$

The probability density function of a random variable x is defined as

$$f(X) = \frac{dF(X)}{dX} \tag{3.3.5}$$

if the derivative exists. Some properties of $f(X)$ are summarized as follows:

$$f(X) \geq 0 \quad \text{for all } X \tag{3.3.6a}$$

$$P(-\infty \leq x \leq \infty) = \int_{-\infty}^{\infty} f(X)dX = 1, \tag{3.3.6b}$$

$$F(X) = \int_{-\infty}^{X} f(\xi)d\xi, \tag{3.3.6c}$$

$$P(X_1 < x \leq X_2) = F(X_2) - F(X_1) = \int_{X_1}^{X_2} f(X)dX. \tag{3.3.6d}$$

Note that both $F(X)$ and $f(X)$ are non-negative functions. The knowledge of these two functions completely defines the random variable x.

Example 3.2: Consider a random variable x that has a probability density function

$$f(X) = \begin{cases} 0, & x < X_1 \text{ or } x > X_2 \\ a, & X_1 \leq x \leq X_2, \end{cases}$$

which is uniformly distributed between X_1 and X_2. The constant value a can be computed by using (3.3.6b). That is,

$$\int_{-\infty}^{\infty} f(X)dX = \int_{X_1}^{X_2} a \cdot dX = a[X_2 - X_1] = 1.$$

Thus we have

$$a = \frac{1}{X_2 - X_1}.$$

If a random variable x is equally likely to take on any value between the two limits X_1 and X_2 and cannot assume any value outside that range, it is said to be uniformly distributed in the range $[X_1, X_2]$. As shown in Figure 3.9, a uniform density function is defined as

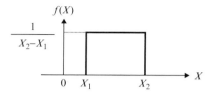

Figure 3.9   A uniform density function

$$f(X) = \begin{cases} \dfrac{1}{X_2 - X_1}, & X_1 \leq X \leq X_2 \\ 0, & \text{otherwise.} \end{cases} \tag{3.3.7}$$

This uniform density function will be used to analyze quantization noise in Section 3.4.

If x is a discrete random variable that can take on any one of the discrete values X_i, $i = 1, 2, \ldots$ as the result of an experiment, we define

$$p_i = P(x = X_i). \tag{3.3.8}$$

3.3.2   Operations on Random Variables

We can use certain statistics associated with random variables. These statistics areoften more meaningful from a physical viewpoint than the probability density function. For example, the mean and the variance are used to reveal sufficient features of random variables. The mean (expected value) of a random variable x is defined as

$$m_x = E[x] = \int_{-\infty}^{\infty} Xf(X)dX, \quad \text{continuous-time case}$$

$$= \sum_i X_i p_i, \quad \text{discrete-time case}, \tag{3.3.9}$$

where $E[\cdot]$ denotes the expectation operation (ensemble averaging).

The expectation is a linear operation. Two useful properties of the expectation operation are $E[\alpha] = \alpha$ and $E[\alpha x] = \alpha E[x]$, where α is a constant. If $E[x] = 0$, x is the zero-mean random variable. The MATLAB function `mean` calculates the mean value. For example, the statement

```
mx = mean(x);
```

computes the mean value of the elements in the vector x.

In (3.3.9), the sum is taken over all possible values of x. The mean m_x defines the level about which the random process x fluctuates. For example, consider the rolling of a die N times ($N \to \infty$), the probability of outcomes is listed in Table 3.3, as follows:

Table 3.3 Probability of rolling a die

X_i	1	2	3	4	5	6
p_i	1/6	1/6	1/6	1/6	1/6	1/6

The mean of outcomes can be computed as

$$m_x = \sum_{i=1}^{6} p_i X_i = \frac{1}{6}(1 + 2 + 3 + 4 + 5 + 6) = 3.5.$$

The variance of x, which is a measure of the spread about the mean, is defined as

$$\sigma_x^2 = E[(x - m_x)^2]$$
$$= \int_{-\infty}^{\infty} (X - m_x)^2 f(X) dX, \quad \text{continuous-time case}$$
$$= \sum_i p_i (X_i - m_x)^2, \quad \text{discrete-time case}, \tag{3.3.10}$$

where $x - m_x$ is the deviation of x from the mean value m_x. The mean of the squared deviation indicates the average dispersion of the distribution about the mean m_x. The positive square root σ_x of variance is called the standard deviation of x. The MATLAB function std calculates standard deviation. The statement

```
s = std(x);
```

computes the standard deviation of the elements in the vector x.
 The variance defined in (3.3.10) can be expressed as

$$\sigma_x^2 = E[(x - m_x)^2] = E[x^2 - 2xm_x + m_x^2] = E[x^2] - 2m_x E[x] + m_x^2$$
$$= E[x^2] - m_x^2. \tag{3.3.11}$$

We call $E[x^2]$ the mean-square value of x. Thus the variance is the difference between the mean-square value and the square of the mean value. That is, the variance is the expected value of the square of the random variable after the mean has been removed.
 The expected value of the square of a random variable is equivalent to the notation of power. If the mean value is equal to 0, then the variance is equal to the mean-square value. For a zero-mean random variable x, i.e., $m_x = 0$, we have

$$\sigma_x^2 = E[x^2] = P_x, \tag{3.3.12}$$

which is the power of x. In addition, if $y = \alpha x$ where α is a constant, it can be shown that $\sigma_y^2 = \alpha^2 \sigma_x^2$. It can also be shown (see exercise problem) that $P_x = m_x^2 + \sigma_x^2$ if $m_x \neq 0$.
 Consider a uniform density function as given in (3.3.7). The mean of the function can be computed by

$$m_x = E[x] = \int_{-\infty}^{\infty} Xf(X)dX = \frac{1}{X_2 - X_1} \int_{X_1}^{X_2} XdX$$

$$= \frac{X_2 + X_1}{2}. \tag{3.3.13}$$

The variance of the function is

$$\sigma_x^2 = E[x^2] - m_x^2 = \int_{-\infty}^{\infty} X^2 f(X)dX - m_x^2 = \frac{1}{X_2 - X_1} \int_{X_1}^{X_2} X^2 dX - m_x^2$$

$$= \frac{1}{X_2 - X_1} \cdot \frac{X_2^3 - X_1^3}{3} - m_x^2 = \frac{X_2^2 + X_1 X_2 + X_1^2}{3} - \frac{(X_2 + X_1)^2}{4}$$

$$= \frac{(X_2 - X_1)^2}{12}. \tag{3.3.14}$$

In general, if x is a uniformly distributed random variable in the interval $(-\Delta, \Delta)$, the mean is 0 ($m_x = 0$) and the variance is $\sigma_x^2 = \Delta^2/3$.

Example 3.3: The MATLAB function `rand` generates pseudo-random numbers uniformly distributed in the interval (0, 1). From Equation (3.3.13), the mean of the generated pseudo-random numbers is 0.5. From (3.3.14), the variance is computed as 1/12. To generate zero-mean random numbers, we subtract 0.5 from every generated number. The numbers are now distributed in the interval (−0.5, 0.5). To make these pseudo-random numbers with unit-variance, i.e., $\sigma_x^2 = \Delta^2/3 = 1$, the generated numbers must be equally distributed in the interval $(-\sqrt{3}, \sqrt{3})$. Therefore we have to multiply $2\sqrt{3}$ to every generated number that was subtracted by 0.5. The following MATLAB statement can be used to generate the uniformly distributed random numbers with mean 0 and variance 1:

```
xn = 2* sqrt (3)* (rand − 0.5);
```

For two random variables x and y, we have

$$E[x + y] = E[x] + E[y], \tag{3.3.15}$$

i.e., the mean value of the sum of random variables equals the sum of mean values. The correlation of x and y is denoted as $E[xy]$. In general, $E[xy] \neq E[x] \cdot E[y]$. However, if x and y are uncorrelated, then the correlation can be written in the form

$$E[xy] = E[x] \cdot E[y]. \tag{3.3.16}$$

Statistical independence of x and y is sufficient to guarantee that they are uncorrelated.

If the random variables x_i are independent with the mean m_i and variance σ_i^2, the random variable y is defined as

$$y = x_1 + x_2 + \cdots + x_N = \sum_{i=1}^{N} x_i. \tag{3.3.17}$$

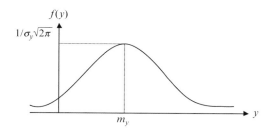

Figure 3.10 Probability density function of Gaussian random variable

The probability density function $f(Y)$ becomes a Gaussian (normal) distribution function (normal curve) as $N \to \infty$. That is,

$$f(Y) = \frac{1}{\sigma_y\sqrt{2\pi}}e^{-(y-m_y)^2/2\sigma_y^2} = \frac{1}{\sigma_y\sqrt{2\pi}}e^{-\left[\frac{1}{2}\left(\frac{y-m_y}{\sigma_y}\right)^2\right]}, \qquad (3.3.18)$$

where $m_y = \sum_{i=1}^{N} m_i$ and $\sigma_y = \sqrt{\sum_{i=1}^{N} \sigma_i^2}$. A graphical representation of the probability density function defined in (3.3.18) is illustrated in Figure 3.10.

The central limit theorem defined in (3.3.17) is useful in generating a Gaussian random variable from a uniformly distributed random variable using $N \geq 12$. The Gaussian random variable is frequently used in communication theory. The MATLAB function randn generates pseudo-random numbers normally distributed with mean 0 and variance 1.

3.4 Fixed-Point Representation and Arithmetic

The basic element in digital hardware is the two-state (binary) device that contains one bit of information. A register (or memory unit) containing B bits of information is called a B-bit word. There are several different methods of representing numbers and carrying out arithmetic operations. In fixed-point arithmetic, the binary point has a fixed location in the register. In floating-point arithmetic, it does not. In general, floating-point processors are more expensive and slower than fixed-point devices. In this book, we focus on widely used fixed-point implementations.

A B-bit fixed-point number can be interpreted as either an integer or a fractional number. It is better to limit the fixed-point representation to fractional numbers because it is difficult to reduce the number of bits representing an integer. In fixed-point fractional implementation, it is common to assume that the data is properly scaled so that their values lie between -1 and 1. When multiplying these normalized fractional numbers, the result (product) will always be less than one.

A given fractional number x has a fixed-point representation as illustrated in Figure 3.11. In the figure, M is the number of data (magnitude) bits. The most significant bit

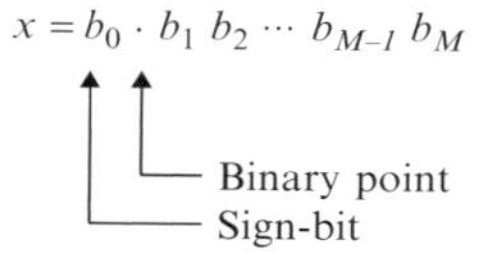

$$x = b_0 \cdot b_1 \, b_2 \, \cdots \, b_{M-1} \, b_M$$

Binary point
Sign-bit

Figure 3.11   Fixed-point representation of binary fractional numbers

$$b_0 = \begin{cases} 0, & x \geq 0 \text{ (positive number)} \\ 1, & x < 0 \text{ (negative number),} \end{cases} \qquad (3.4.1)$$

represents the sign of the number. It is called the sign-bit. The remaining M bits give the magnitude of the number. The rightmost bit, b_M, is called the least significant bit (LSB). The wordlength is $B \, (= M + 1)$ bits, i.e., each data point is represented by $B - 1$ magnitude bits and one sign-bit.

As shown in Figure 3.11, the decimal value of a positive B-bit binary fractional number x can be expressed as

$$(x)_{10} = b_1 \cdot 2^{-1} + b_2 \cdot 2^{-2} + \cdots + b_M \cdot 2^{-M} = \sum_{m=1}^{M} b_m 2^{-m}. \qquad (3.4.2)$$

For example, the largest (positive) 16-bit fractional number is $x = 0111\ 1111\ 1111\ 1111$. The decimal value of this number can be obtained as

$$x = \sum_{m=1}^{15} 2^{-m} = 2^{-1} + 2^{-2} + \cdots + 2^{-15} = \sum_{m=0}^{15} \left(\frac{1}{2}\right)^m - 1 = \frac{1 - (1/2)^{16}}{1 - (1/2)} - 1$$

$$= 1 - 2^{-15} \approx 0.999969.$$

The negative numbers can be represented using three different formats: the sign-magnitude, the 1's complement, and the 2's complement. Fixed-point DSP devices usually use the 2's complement to represent negative numbers because it allows the processor to perform addition and subtraction uses the same hardware. A positive number $(b_0 = 0)$ is represented as a simple binary value, while a negative number $(b_0 = 1)$ is represented using the 2's complement format. With the 2's complement form, a negative number is obtained by complementing all the bits of the positive binary number and then adding a 1 to the least significant bit. Table 3.4 shows an example of 3-bit binary fractional numbers using the 2's complement format and their corresponding decimal values.

In general, the decimal value of a B-bit binary fractional number can be calculated as

$$(x)_{10} = -b_0 + \sum_{m=1}^{15} b_m 2^{-m}. \qquad (3.4.3)$$

For example, the smallest (negative) 16-bit fractional number is $x = 1000\ 0000\ 0000\ 0000$. From (3.4.3), its decimal value is -1. Therefore the range of fractional binary numbers is

Table 3.4 Example of 3-bit binary fractional numbers in 2's complement format and their corresponding decimal values

Binary number	000	001	010	011	100	101	110	111
Decimal value	0.00	0.25	0.50	0.75	−1.00	−0.75	−0.50	−0.25

$$-1 \leq x \leq (1 - 2^{-M}). \tag{3.4.4}$$

For a 16-bit fractional number x, the decimal value range is $-1 \leq x \leq 1 - 2^{-15}$.

It is important to note that we use an implied binary point to represent the binary fractional number. It affects only the accuracy of the result and the location from which the result will be read. The binary point is purely a programmer's convention and has no relationship to the hardware. That is, the processor treats the 16-bit number as an integer. The programmer needs to keep track of the binary point when manipulating fractional numbers in assembly language programming. For example, if we want to initialize a data memory location x with the constant decimal value 0.625, we can use the binary form $x = 0101\,0000\,0000\,0000b$, the hexadecimal form $x = 0\text{x}5000$, or the decimal integer $x = 2^{12} + 2^{14} = 20\,480$. The easiest way to convert a normalized fractional number into the integer that can be used by the C55x assembler is to multiply the decimal value by $2^{15} = 32\,768$. For example, $0.625 * 32\,768 = 20\,480$.

Most commercially available DSP devices, such as the TMS320C55x discussed in Chapter 2, are 16-bit processors. These fixed-point DSP devices assume the binary point after the sign-bit as shown in Figure 3.11. This fractional number representation is also called the Q15 format since there are 15 magnitude bits.

Example 3.4: The following are some examples of the Q15 format data used for C55x assembly programming. The directives .set and .equ have the same functions that assign a value to a symbolic name. They do not require memory space. The directives .word and .int are used to initialize memory locations with particular data values represented in binary, hexadecimal, or integer format. Each data is treated as a 16-bit value and separated by a comma.

```
ONE           .set  32767       ; 1 - 2^-15 ≈ 0.999969 in integer
ONE_HALF      .set  0x4000       ; 0.5 in hexadecimal
ONE_EIGHTH .equ  1000h        ; 1/8 in hexadecimal
MINUS_ONE  .equ  0xffff       ; -1 in hexadecimal
COEFF         .int  0ff00h       ; COEFF of -0x100
ARRAY         .word 2048, -2048  ; ARRAY[0.0625, -0.0625]
```

Fixed-point arithmetic is often used with DSP hardware for real-time processing because it offers fast operation and relatively economical implementation. Its drawbacks include a small dynamic range (the range of numbers that can be represented) and low accuracy. Roundoff errors exist only for multiplication. However, the addition may cause an accumulator overflow. These problems will be discussed in detail in the following sections.

3.5   Quantization Errors

As discussed in Section 3.4, digital signals and system parameters are represented by a finite number of bits. There is a noticeable error between desired and actual results – the finite-precision (finite wordlength, or numerical) effects. In general, finite-precision effects can be broadly categorized into the following classes:

1. Quantization errors
    a. Input quantization
    b. Coefficient quantization

2. Arithmetic errors
    a. Roundoff (truncation) noise
    b. Overflow

The limit cycle oscillation is another phenomenon that may occur when implementing a feedback system such as an IIR filter with finite-precision arithmetic. The output of the system may continue to oscillate indefinitely while the input remains 0. This can happen because of quantization errors or overflow.

   This section briefly analyzes finite-precision effects in DSP systems using fixed-point arithmetic, and presents methods for confining these effects to acceptable levels.

3.5.1   Input Quantization Noise

The ADC shown in Figure 1.2 converts a given analog signal $x(t)$ into digital form $x(n)$. The input signal is first sampled to obtain the discrete-time signal $x(nT)$. Each $x(nT)$ value is then encoded using B-bit wordlength to obtain the digital signal $x(n)$, which consists of M magnitude bits and one sign-bit as shown in Figure 3.11. As discussed in Section 3.4, we assume that the signal $x(n)$ is scaled such that $-1 \leq x(n) < 1$. Thus the full-scale range of fractional numbers is 2. Since the quantizer employs B bits, the number of quantization levels available for representing $x(nT)$ is 2^B. Thus the spacing between two successive quantization levels is

$$\Delta = \frac{\text{full-scale range}}{\text{number of quantization levels}} = \frac{2}{2^B} = 2^{-B+1} = 2^{-M}, \qquad (3.5.1)$$

which is called the quantization step (interval, width, or resolution).

   Common methods of quantization are rounding and truncation. With rounding, the signal value is approximated using the nearest quantization level. When truncation is used, the signal value is assigned to the highest quantization level that is not greater than the signal itself. Since the truncation produces bias effect (see exercise problem), we use rounding for quantization in this book. The input value $x(nT)$ is rounded to the nearest level as illustrated in Figure 3.12. We assume there is a line between two quantization levels. The signal value above this line will be assigned to the higher quantization level, while the signal value below this line is assigned to the lower level. For example, the

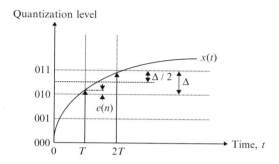

Figure 3.12 Quantization process related to ADC

discrete-time signal $x(T)$ is rounded to 010, since the real value is below the middle line between 010 and 011, while $x(2T)$ is rounded to 011 since the value is above the middle line.

The quantization error (noise), $e(n)$, is the difference between the discrete-time signal, $x(nT)$, and the quantized digital signal, $x(n)$. The error due to quantization can be expressed as

$$e(n) = x(n) - x(nT). \qquad (3.5.2)$$

Figure 3.12 clearly shows that

$$|e(n)| \leq \frac{\Delta}{2}. \qquad (3.5.3)$$

Thus the quantization noise generated by an ADC depends on the quantization interval. The presence of more bits results in a smaller quantization step, therefore it produces less quantization noise.

From (3.5.2), we can view the ADC output as being the sum of the quantizer input $x(nT)$ and the error component $e(n)$. That is,

$$x(n) = Q[x(nT)] = x(nT) + e(n), \qquad (3.5.4)$$

where $Q[\cdot]$ denotes the quantization operation. The nonlinear operation of the quantizer is modeled as a linear process that introduces an additive noise $e(n)$ to the discrete-time signal $x(nT)$ as illustrated in Figure 3.13. Note that this model is not accurate for low-amplitude slowly varying signals.

For an arbitrary signal with fine quantization (B is large), the quantization error $e(n)$ may be assumed to be uncorrelated with the digital signal $x(n)$, and can be assumed to be random noise that is uniformly distributed in the interval $\left[-\frac{\Delta}{2}, \frac{\Delta}{2}\right]$. From (3.3.13), we can show that

$$E[e(n)] = \frac{-\Delta/2 + \Delta/2}{2} = 0. \qquad (3.5.5)$$

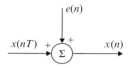

Figure 3.13   Linear model for the quantization process

That is, the quantization noise $e(n)$ has zero mean. From (3.3.14) and (3.5.1), we can show that the variance

$$\sigma_e^2 = \frac{\Delta^2}{12} = \frac{2^{-2B}}{3}. \tag{3.5.6}$$

Therefore the larger the wordlength, the smaller the input quantization error.

If the quantization error is regarded as noise, the signal-to-noise ratio (SNR) can be expressed as

$$\text{SNR} = \frac{\sigma_x^2}{\sigma_e^2} = 3 \cdot 2^{2B} \sigma_x^2, \tag{3.5.7}$$

where σ_x^2 denotes the variance of the signal, $x(n)$. Usually, the SNR is expressed in decibels (dB) as

$$
\begin{aligned}
\text{SNR} &= 10 \log_{10}\left(\frac{\sigma_x^2}{\sigma_e^2}\right) = 10 \log_{10}(3 \cdot 2^{2B}\sigma_x^2) \\
&= 10 \log_{10} 3 + 20B \log_{10} 2 + 10 \log_{10} \sigma_x^2 \\
&= 4.77 + 6.02B + 10 \log_{10} \sigma_x^2.
\end{aligned}
\tag{3.5.8}
$$

This equation indicates that for each additional bit used in the ADC, the converter provides about 6-dB signal-to-quantization-noise ratio gain. When using a 16-bit ADC ($B = 16$), the SNR is about 96 dB. Another important fact of (3.5.8) is that the SNR is proportional to σ_x^2. Therefore we want to keep the power of signal as large as possible. This is an important consideration when we discuss scaling issues in Section 3.6.

In digital audio applications, quantization errors arising from low-level signals are referred to as granulation noise. It can be eliminated using dither (low-level noise) added to the signal before quantization. However, dithering reduces the SNR. In many applications, the inherent analog audio components (microphones, amplifiers, or mixers) noise may already provide enough dithering, so adding additional dithers may not be necessary.

If the digital filter is a linear system, the effect of the input quantization noise alone on the output may be computed. For example, for the FIR filter defined in (3.1.16), the variance of the output noise due to the input quantization noise may be expressed as

$$\sigma_{y,e}^2 = \sigma_e^2 \sum_{l=0}^{L-1} b_l^2. \tag{3.5.9}$$

This noise is relatively small when compared with other numerical errors and is determined by the wordlength of ADC.

Example 3.5: Input quantization effects may be subjectively evaluated by observing and listening to the quantized speech. A speech file called `timit1.asc` (included in the software package) was digitized using $f_s = 8\,kHz$ and $B = 16$. This speech file can be viewed and played using the MATLAB script:

```
load(timit1.asc);
plot(timit1);
soundsc(timit1, 8000, 16);
```

where the MATLAB function `soundsc` autoscales and plays the vector as sound. We can simulate the quantization of data with 8-bit wordlength by

```
qx = round(timit1/256);
```

where the function, `round`, rounds the real number to the nearest integer. We then evaluate the quantization effects by

```
plot(qx);
soundsc(qx, 8000, 16);
```

By comparing the graph and sound of `timit1` and `qx`, the quantization effects may be understood.

3.5.2 Coefficient Quantization Noise

When implementing a digital filter, the filter coefficients are quantized to the wordlength of the DSP hardware so that they can be stored in the memory. The filter coefficients, b_l and a_m, of the digital filter defined by (3.2.18) are determined by a filter design package such as MATLAB for given specifications. These coefficients are usually represented using the floating-point format and have to be encoded using a finite number of bits for a given fixed-point processor. Let b_l' and a_m' denote the quantized values corresponding to b_l and a_m, respectively. The difference equation that can actually be implemented becomes

$$y(n) = \sum_{l=0}^{L-1} b_l' x(n - l) - \sum_{m=1}^{M} a_m' y(n - m). \qquad (3.5.10)$$

This means that the performance of the digital filter implemented on the DSP hardware will be slightly different from its design specification. Design and implementation of digital filters for real-time applications will be discussed in Chapter 5 for FIR filters and Chapter 6 for IIR filters.

If the wordlength B is not large enough, there will be undesirable effects. The coefficient quantization effects become more significant when tighter specifications are used. This generally affects IIR filters more than it affects FIR filters. In many applications, it is desirable for a pole (or poles) of IIR filters to lie close to the unit circle.

Coefficient quantization can cause serious problems if the poles of desired filters are too close to the unit circle because those poles may be shifted on or outside the unit circle due to coefficient quantization, resulting in an unstable implementation. Such undesirable effects due to coefficient quantization are far more pronounced when high-order systems (where L and M are large) are directly implemented since a change in the value of a particular coefficient can affect all the poles. If the poles are tightly clustered for a lowpass or bandpass filter with narrow bandwidth, the poles of the direct-form realization are sensitive to coefficient quantization errors. The greater the number of clustered poles, the greater the sensitivity.

The coefficient quantization noise is also affected by the different structures for the implementation of digital filters. For example, the direct-form implementation of IIR filters is more sensitive to coefficient quantization errors than the cascade structure consisting of sections of first- or second-order IIR filters. This problem will be further discussed in Chapter 6.

3.5.3 Roundoff Noise

As shown in Figure 3.3 and (3.1.11), we may need to compute the product $y(n) = \alpha x(n)$ in a DSP system. Assuming the wordlength associated with α and $x(n)$ is B bits, the multiplication yields $2B$ bits product $y(n)$. For example, a 16-bit number times another 16-bit number will produce a 32-bit product. In most applications, this product may have to be stored in memory or output as a B-bit word. The $2B$-bit product can be either truncated or rounded to B bits. Since truncation causes an undesired bias effect, we should restrict our attention to the rounding case.

In C programming, rounding a real number to an integer number can be implemented by adding 0.5 to the real number and then truncating the fractional part. For example, the following C statement

```
y = (int)(x+0.5);
```

rounds the real number x to the nearest integer y. As shown in Example 3.5, MATLAB provides the function round for rounding a real number.

In TMS320C55x implementation, the CPU rounds the operands enclosed by the rnd( ) expression qualifier. For example,

```
mov rnd(HI(AC0)), *AR1
```

This instruction will round the content of the high portion of AC0(31:16) and the rounded 16-bit value is stored in the memory location pointed at by AR1. Another key word, R (or r), when used with the operation code, also performs rounding operation on the operands. The following is an example that rounds the product of AC0 and AC1 and stores the rounded result in the upper portion of the accumulator AC1(31:16) and the lower portion of the accumulator AC1(15:0) is cleared:

```
mpyr AC0, AC1
```

The process of rounding a $2B$-bit product to B bits is very similar to that of quantizing discrete-time samples using a B-bit quantizer. Similar to (3.5.4), the nonlinear operation of product roundoff can be modeled as the linear process shown in Figure 3.13. That is,

$$y(n) = Q[\alpha x(n)] = \alpha x(n) + e(n), \qquad (3.5.11)$$

where $\alpha x(n)$ is the $2B$-bit product and $e(n)$ is the roundoff noise due to rounding $2B$-bit product to B-bit. The roundoff noise is a uniformly distributed random process in the interval defined in (3.5.3). Thus it has a zero-mean and its power is defined in (3.5.6).

It is important to note that most commercially available fixed-point DSP devices such as the TMS320C55x have double-precision ($2B$-bit) accumulator(s). As long as the program is carefully written, it is quite possible to ensure that rounding occurs only at the final stage of calculation. For example, consider the computation of FIR filter output given in (3.1.16). We can keep all the temporary products, $b_l x(n - l)$ for $l = 0, 1, \ldots, L - 1$, in the double-precision accumulator. Rounding is only performed when computation is completed and the sum of products is saved to memory with B-bit wordlength.

3.6 Overflow and Solutions

Assuming that the input signals and filter coefficients have been properly normalized (in the range of -1 and 1) for fixed-point arithmetic, the addition of these two B-bit numbers will always produce a B-bit sum. Therefore no roundoff error is introduced by addition. Unfortunately, when these two B-bit numbers are added, the sum may fall outside the range of -1 and 1. The term overflow is a condition in which the result of an arithmetic operation exceeds the capacity of the register used to hold that result. For example, assuming a 3-bit fixed-point hardware with fractional 2's complement data format (see Table 3.4) is used. If $x_1 = 0.75$ (011 in binary form) and $x_2 = 0.25$ (001), the binary sum of $x_1 + x_2$ is 100. The decimal value of the binary number 100 is -1, not the correct answer $+1$. That is, when the result exceeds the full-scale range of the register, overflow occurs and unacceptable error is produced. Similarly, subtraction may result in underflow.

When using a fixed-point processor, the range of numbers must be carefully examined and adjusted in order to avoid overflow. For the FIR filter defined in (3.1.16), this overflow results in the severe distortion of the output $y(n)$. For the IIR filter defined in (3.2.18), the effect is much more serious because the errors are fed back and render the filter useless. The problem of overflow may be eliminated using saturation arithmetic and proper scaling (or constraining) signals at each node within the filter to maintain the magnitude of the signal.

3.6.1 Saturation Arithmetic

Most commercially available DSP devices (such as the TMS320C55x) have mechanisms that protect against overflow and indicate if it occurs. Saturation arithmetic prevents a

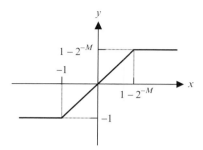

Figure 3.14   Transfer characteristic of saturation arithmetic

result from overflowing by keeping the result at a maximum (or minimum for an underflow) value. Saturation logic is illustrated in Figure 3.14 and can be expressed as

$$y = \begin{cases} 1 - 2^{-M}, & x \geq 1 - 2^{-M} \\ x, & -1 \leq x < 1 \\ -1, & x < -1, \end{cases} \tag{3.6.1}$$

where x is the original addition result and y is the saturated adder output. If the adder is under saturation mode, the undesired overflow can be avoided since the 32-bit accumulator fills to its maximum (or minimum) value, but does not roll over. Similar to the previous example, when 3-bit hardware with saturation arithmetic is used, the addition result of $x_1 + x_2$ is 011, or 0.75 in decimal value. Compared with the correct answer 1, there is an error of 0.25. However, the result is much better than the hardware without saturation arithmetic.

     Saturation arithmetic has a similar effect to 'clipping' the desired waveform. This is a nonlinear operation that will add undesired nonlinear components into the signal. Therefore saturation arithmetic can only be used to guarantee that overflow will not occur. It is not the best, nor the only solution, for solving overflow problems.

3.6.2 Overflow Handling

As mentioned earlier, the TMS320C55x supports the data saturation logic in the data computation unit (DU) to prevent data computation from overflowing. The logic is enabled when the overflow mode bit (SATD) in status register ST1 is set (SATD = 1). When the mode is set, the accumulators are loaded with either the largest positive 32-bit value (0x00 7FFF FFFF) or the smallest negative 32-bit value (0xFF 8000 0000) if the result overflows. The overflow mode bit can be set with the instruction

```
bset SATD
```

and reset (disabled) with the instruction

```
bclr SATD
```

The TMS320C55x provides overflow flags that indicate whether or not an arithmetic operation has exceeded the capability of the corresponding register. The overflow flag ACOVx, ($x = 0$, 1, 2, or 3) is set to 1 when an overflow occurs in the corresponding accumulator ACx. The corresponding overflow flag will remain set until a reset is performed or when a status bit clear instruction is implemented. If a conditional instruction that tests overflow status (such as a branch, a return, a call, or a conditional execution) is executed, the overflow flag is cleared. The overflow flags can be tested and cleared using instructions.

3.6.3 Scaling of Signals

The most effective technique in preventing overflow is by scaling down the magnitudes of signals at certain nodes in the DSP system and then scaling the result back up to the original level. For example, consider the simple FIR filter illustrated in Figure 3.15(a). Let $x(n) = 0.8$ and $x(n-1) = 0.6$, the filter output $y(n) = 1.2$. When this filter is implemented on a fixed-point DSP hardware without saturation arithmetic, undesired overflow occurs and we get a negative number as a result.

As illustrated in Figure 3.15(b), the scaling factor, $\beta < 1$, can be used to scale down the input signal and prevent overflowing. For example, when $\beta = 0.5$ is used, we have $x(n) = 0.4$ and $x(n-1) = 0.3$, and the result $y(n) = 0.6$. This effectively prevents the hardware overflow. Note that $\beta = 0.5$ can be implemented by right shifting 1 bit.

If the signal $x(n)$ is scaled by β, the corresponding signal variance changes to $\beta^2 \sigma_x^2$. Thus the signal-to-quantization-noise ratio given in (3.5.8) changes to

$$\text{SNR} = 10 \log_{10} \left(\frac{\beta^2 \sigma_x^2}{\sigma_e^2} \right) = 4.77 + 6.02B + 10 \log_{10} \sigma_x^2 + 20 \log_{10} \beta. \qquad (3.6.2)$$

Since we perform fractional arithmetic, $\beta < 1$ is used to scale down the input signal. The term $20 \log_{10} \beta$ has negative value. Thus scaling down the amplitude of the signal reduces the SNR. For example, when $\beta = 0.5$, $20 \log_{10} \beta = -6.02$ dB, thus reducing the SNR of the input signal by about 6 dB. This is equivalent to losing 1-bit in representing the signal.

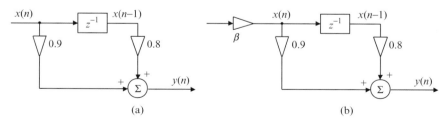

(a)   (b)

Figure 3.15 Block diagram of simple FIR filters: (a) without scaling, and (b) with scaling factor β

Therefore we have to keep signals as large as possible without overflow. In the FIR filter given in Figure 3.6, a scaling factor, β, can be applied to the input signal, $x(n)$, to prevent overflow during the computation of $y(n)$ defined in (3.1.16). The value of signal $y(n)$ can be bounded as

$$|y(n)| = \beta \left| \sum_{l=0}^{L-1} b_l x(n-l) \right| \leq \beta M_x \sum_{l=0}^{L-1} |b_l|, \tag{3.6.3}$$

where M_x is the peak value of $x(n)$ defined in (3.2.3). Therefore we can ensure that $|y(n)| < 1$ by choosing

$$\beta \leq \frac{1}{M_x \sum_{l=0}^{L-1} |b_l|}. \tag{3.6.4}$$

Note that the input signal is bounded and $|x(n)| \leq 1$, thus $M_x \leq 1$. The sum $\sum_{l=0}^{L-1} |b_l|$ can be calculated using the MATLAB statement

```
bsum = sum(abs(b));
```

where b is the coefficient vector.

Scaling the input by the scaling factor given in (3.6.4) guarantees that overflow never occurs in the FIR filter. However, the constraint on β is overly conservative for most signals of practical interest. We can use a more relaxed condition

$$\beta \leq \frac{1}{M_x \sqrt{\sum_{l=0}^{L-1} b_l^2}}. \tag{3.6.5}$$

Other scaling factors that may be used are based on the frequency response of the filter (will be discussed in Chapter 4). Assuming that the reference signal is narrowband, overflow can be avoided for all sinusoidal signals if the input is scaled by the maximum magnitude response of the filter. This scaling factor is perhaps the easiest to use, especially for IIR filters. It involves calculating the magnitude response and then selecting its maximum value.

An IIR filter designed by a filter design package such as MATLAB may have some of its filter coefficients greater than 1.0. To implement a filter with coefficients larger than 1.0, we can also scale the filter coefficients instead of changing the incoming signal. One common solution is to use a different Q format instead of the Q15 format to represent the filter coefficients. After the filtering operation is completed, the filter output needs to be scaled back to the original signal level. This issue will be discussed in the C55x experiment given in Section 3.8.5.

The TMS320C55x provides four 40-bit accumulators as introduced in Chapter 2. Each accumulator is split into three parts as illustrated in Figure 3.16. The guard bits are used as a head-margin for computations. These guard bits prevent overflow in iterative computations such as the FIR filtering of $L \leq 256$ defined in (3.1.16).

b39–b32	b31–b16	b15–b0
G	H	L
Guard bits	High-order bits	Low-order bits

Figure 3.16   TMS320C55x accumulator configuration

Because of the potential overflow in a fixed-point processor, engineers need to be concerned with the dynamic range of numbers. This requirement usually demands a greater coding effort and testing using real data for a given application. In general, the optimum solution for the overflow problem is by combining scaling factors, guard bits, and saturation arithmetic. The scaling factors are set as large as possible (close to but smaller than 1) and occasional overflow can be avoided by using guard bits and saturation arithmetic.

3.7   Implementation Procedure for Real-Time Applications

The digital filters and algorithms can be implemented on a DSP chip such as the TMS320C55x following a four-stage procedure to minimize the amount of time spent on finite wordlength analysis and real-time debugging. Figure 3.17 shows a flowchart of this procedure.

In the first stage, algorithm design and study is performed on a general-purpose computer in a non-real-time environment using a high-level MATLAB or C program with floating-point coefficients and arithmetic. This stage produces an 'ideal' system.

In the second stage, we develop the C (or MATLAB) program in a way that emulates the same sequence of operations that will be implemented on the DSP chip, using the same parameters and state variables. For example, we can define the data samples and filter coefficients as 16-bit integers to mimic the wordlength of 16-bit DSP chips. It is carefully redesigned and restructured, tailoring it to the architecture, the I/O timing structure, and the memory constraints of the DSP device. This program can also serve as a detailed outline for the DSP assembly language program or may be compiled using the DSP chip's C compiler. This stage produces a 'practical' system.

The quantization errors due to fixed-point representation and arithmetic can be evaluated using the simulation technique illustrated in Figure 3.18. The testing data $x(n)$ is applied to both the ideal system designed in stage 1 and the practical system developed in stage 2. The output difference, $e(n)$, between these two systems is due to finite-precision effects. We can re-optimize the structure and algorithm of the practical system in order to minimize finite-precision errors.

The third stage develops the DSP assembly programs (or mixes C programs with assembly routines) and tests the programs on a general-purpose computer using a DSP software simulator (CCS with simulator or EVM) with test data from a disk file. This test data is either a shortened version of the data used in stage 2, which can be generated internally by the program or read in as digitized data emulating a real application. Output from the simulator is saved as another disk file and is compared to the corresponding output of the C program in the second stage. Once a one-to-one agreement is obtained between these two outputs, we are assured that the DSP assembly program is essentially correct.

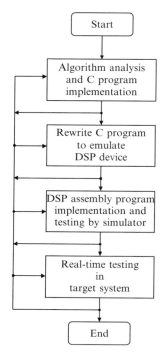

Figure 3.17   Implementation procedure for real-time DSP applications

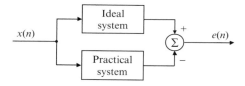

Figure 3.18   An efficient technique to study finite wordlength effects

The final stage downloads the compiled (or assembled) and linked program into the target hardware (such as EVM) and brings it to a real-time operation. Thus the real-time debugging process is primarily constrained to debugging the I/O timing structure and testing the long-term stability of the algorithm. Once the algorithm is running, we can again 'tune' the parameters of the systems in a real-time environment.

3.8   Experiments of Fixed-Point Implementations

The purposes of experiments in this section are to learn input quantization effects and to determine the proper fixed-point representation for a DSP system.

3.8.1   Experiment 3A – Quantization of Sinusoidal Signals

To experiment with input quantization effects, we shift off (right) bits of input signal and then evaluate the shifted samples. Altering the number of bits for shifting right, we can obtain an output stream that corresponds to a wordlength of 14 bits, 12 bits, and so on. The example given in Table 3.5 simulates an A/D converter of different wordlength. Instead of shifting the samples, we mask out the least significant 4 (or 8, or 10) bits of each sample, resulting in the 12 (8 or 6) bits data having comparable amplitude to the 16-bit data.

1. Copy the C function `exp3a.c` and the linker command file `exp3.cmd` from the software package to A:\Experiment3 directory, create project `exp3a` to simulate 16, 12, 8, and 6 bits A/D converters. Use the run-time support library `rts55.lib` and build the project.

2. Use the CCS graphic display function to plot all four output buffers: `out16`, `out12`, `out8`, and `out6`. Examples of the plots and graphic settings are shown in Figure 3.19 and Figure 3.20, respectively.

3. Compare the graphic results of each output stream, and describe the differences between waveforms represented by different wordlength.

Table 3.5   Program listing of quantizing a sinusoid, `exp3a.c`

```
#define BUF_SIZE 40
const int sineTable[BUF_SIZE] =
{ 0x0000, 0x01E0, 0x03C0, 0x05A0, 0x0740, 0x08C0, 0x0A00, 0x0B20,
0x0BE0, 0x0C40, 0x0C60, 0x0C40, 0x0BE0, 0x0B20, 0x0A00, 0x08C0,
0x0740, 0x05A0, 0x03C0, 0x01E0, 0x0000, 0xFE20, 0xFC40, 0xFA60,
0xF8C0, 0xF740, 0xF600, 0xF4E0, 0xF420, 0xF3C0, 0xF3A0, 0xF3C0,
0xF420, 0xF4E0, 0xF600, 0xF740, 0xF8C0, 0xFA60, 0xFC40, 0x0000};
int out16[BUF_SIZE];          /* 16 bits output sample buffer */
int out12[BUF_SIZE];          /* 12 bits output sample buffer */
int out8[BUF_SIZE];           /* 8 bits output sample buffer  */
int out6[BUF_SIZE];           /* 6 bits output sample buffer  */
void main()
{
  int i;
    for (i = 0; i < BUF_SIZE-1; i++)
    {
        out16[i] = sineTable[i];        /* 16-bit data    */
        out12[i] = sineTable[i]&0xfff0; /* Mask off 4-bit */
        out8[i] = sineTable[i]&0xff00;  /* Mask off 8-bit */
        out6[i] = sineTable[i]&0xfc00;  /* Mask off 10-bit */
    }

}
```

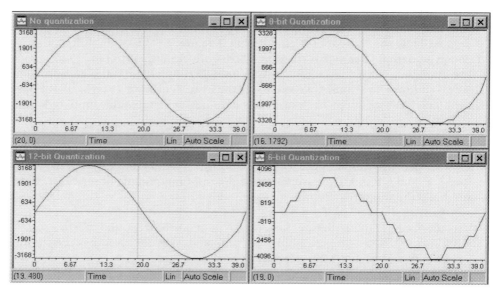

Figure 3.19 Quantizing 16-bit data (top-left) into 12-bit (bottom-left), 8-bit (top-right), and 6-bit (bottom-right)

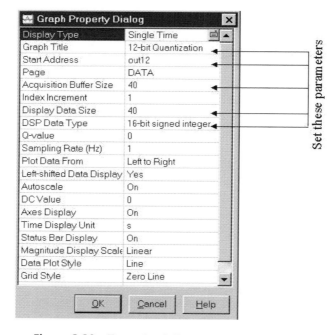

Figure 3.20 Example of displaying graphic settings

4. Find the mean and variance of quantization noise for the 12-, 8-, and 6-bit A/D converters.

3.8.2 Experiment 3B – Quantization of Speech Signals

There are many practical applications from cellular phones to MP3 players that process speech (audio) signals using DSP. To understand the quantization effects of speech signals, we use a digitized speech file, `timit1.asc`, as the input for this experiment. An experiment code for this experiment, `exp3b.c`, is listed in Table 3.6.

1. Refer to the program listed in Table 3.6. Write a C function called `exp3b.c` to simulate 16, 12, 8, and 4 bits A/D converters (or copy the file `exp3b.c` from the software package). Use the digitized speech file `timit1.asc` (included in the software package) as the input signal for the experiment. Create the project `exp3b`, add `exp3b.c` and `exp3.cmd` into the project.

2. Use CCS probe points to connect disk files as described in Chapter 1. In this experiment, we use a probe point to connect the input speech to the variable named `indata`. We also connect four output variables `out16`, `out12`, `out8`, and `out4` to generate quantized output files. As mentioned in Chapter 1, we need to add a header line to the input data file, `timit1.asc`. The header information is formatted in the following syntax using hexadecimal numbers:

Magic Number	Format	Starting Address	Page Number	Length
1651	2	C4	1	1

where

- the magic number is always set to 1651

Table 3.6 Program listing of quantizing a speech signal

```
#define LENGTH 27956      /* Length of input file timit1.asc */
int indata, out16, out12, out8, out4;
void main ( )
{
  int  i;
  for(i = 0; i < LENGTH; i++)
    out16 = indata;           /* Simulate 16-bit A/D */
    out12 = indata&0xfff0;  /* Simulate 12-bit A/D */
    out8 = indata&0xff00;   /* Simulate 8-bit A/D  */
    out4 = indata&0xf000;   /* Simulate 4-bit A/D  */
}
```

- the format is 2 for ASCII data file format

- the starting address is the address of the data variable where we want to connect the input file to, such as C4 in the above example

- the page number is 1 for data

- the length defines the number of data to pass each time

3. Invoke the CCS and set probe points for input and outputs in exp3b.c. Use probe points to output the speech signals with wordlength of 16, 12, 8, and 4 bits to data files. Because the output file generated by CCS probe points have a header line, we need to remove this header. If we want to use MATLAB to listen to the output files, we have to set the CCS output file in integer format. We can load the output file out12.dat and listen to it using the following MATLAB commands:

```
load out12.dat;          % Read data file
soundsc(out12, 8000, 16);  % Play at 8 kHz
```

Listen to the quantization effects between the files with different wordlength.

3.8.3   Experiment 3C – Overflow and Saturation Arithmetic

As discussed in Section 3.6, overflow may occur when DSP algorithms perform accumulations such as FIR filtering. When the number exceeds the maximum capacity of an accumulator, overflow occurs. Sometimes an overflow occurs when data is transferred to memory even though the accumulator does not overflow. This is because the C55x accumulators (AC0–AC3) have 40 bits, while the memory space is usually defined as a 16-bit word. There are several ways to handle the overflow. As introduced in Section 3.6, the C55x has a built-in overflow-protection unit that will saturate the data value if overflow occurs.

In this experiment, we will use an assembly routine, ovf_sat.asm (included in the software package), to evaluate the results with and without overflow protection. Table 3.7 lists a portion of ovf_sat.asm.

In the program, the following code repeatedly adds the constant 0x140 to AC0:

```
     rptblocal add_loop_end−1
     add #0x140≪#16, AC0
     mov hi(AC0), *AR5+
add_loop_end
```

The updated value is stored at the buffer pointed at by AR5. The content of AC0 will grow larger and larger and eventually the accumulator will overflow. When the overflow occurs, a positive number in AC0 suddenly becomes negative. However, when the C55x saturation mode is enabled, the overflowed positive number will be limited to 0x7FFFFFFF.

Table 3.7 List of assembly program to observe overflow and saturation

```
.def _ovftest
.bss buff, (0x100)
.bss buff1, (0x100)
;
; Code start
;
_ovftest
    bclr    SATD                    ; Clear saturation bit if set
    xcc     start, T0 != #0         ; If T0 != 0, set saturation bit
    bset    SATD
start
    pshboth XAR5                    ; Save XAR5
    ...  ...                        ; Some instructions omitted here
    mov     #0, AC0
    mov     #0x80-1, BRC0           ; Initialize loop counts for addition
    amov    #buff+0x80, XAR5        ; Initialize buffer pointer
    rptblocal add_loop_end-1
    add     #0x140≪#16, AC0         ; Use AC0 as a ramp up counter
    mov     hi(AC0), *AR5+          ; Save the counter to buffer
add_loop_end
    ...  ...                        ; Some instructions omitted here
    mov     #0x100-1, BRC0          ; Init loop counts for sinewave
    amov    #buff1, XAR5            ; Initialize buffer pointer
    mov     mmap(@AR0), BSA01       ; Initialize base register
    mov     #40, BK03              ; Set buffer size to 40
    mov     #20, AR0               ; Start with an offset of 20
    bset    AR0LC                   ; Activate circular buffer
    rptblocal sine_loop_end-1
    mov     *ar0+ ≪#16, AC0         ; Get sine value into high AC0
    sfts    AC0, #9                 ; Scale the sine value
    mov     hi(AC0), *AR5+          ; Save scaled value
sine_loop_end
    mov     #0, T0                  ; Return 0 if no overflow
    xcc     set_ovf_flag, overflow(AC0)
    mov     #1, T0                  ; Return 1 if overflow detected
set_ovf_flag
    bclr    AR0LC                   ; Reset circular buffer bit
    bclr    SATD                    ; Reset saturation bit
    popboth XAR5                    ; Restore AR5
    ret
```

The second portion of the code stores the left-shifted sinewave values to data memory locations. Without saturation protection, this shift will cause some of the shifted values to overflow.

As we mentioned in Chapter 2, circular addressing is a very useful addressing mode. The following segment of code in the example shows how to set up and use the circular addressing mode:

```
mov  #sineTable, BSA01  ; Initialize base register
mov  #40, BK03          ; Set buffer size to 40
mov  #20, AR0           ; Start with an offset of 20
bset AR0LC              ; Activate circular buffer
```

The first instruction sets up the circular buffer base register (BSAxx). The second instruction initializes the size of the circular buffer. The third instruction initializes the offset from the base as the starting point. In this case, the offset is set to 20 words from the base `sineTable[]`. The last instruction enables AR0 as the circular pointer. The program `exp3c.c` included in the software package is used to call the assembly routine `ovf_sat.asm`, and is shown in Figure 3.21.

1. Create the project `exp3c` that uses `exp3c.c` and `ovf_sat.asm` (included in the software package) for this experiment.

2. Use the graphic function to display the sinewave (top) and the ramp counter (bottom) as shown in Figure 3.22.

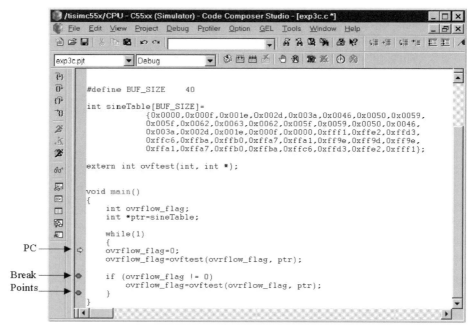

Figure 3.21   Experiment of C program `exp3c.c`

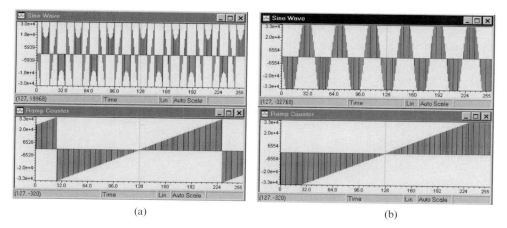

Figure 3.22 C55x data saturation example: (a) without saturation protection, and (b) with saturation protection enabled

3.8.4 Experiment 3D – Quantization of Coefficients

Filters are widely used for DSP applications. The C55x implementation of FIR (or IIR) filters often use 16-bit numbers to represent filter coefficients. Due to the quantization of coefficients, the filter implemented using fixed-point hardware will not have the exact same response as the filter that is obtained by a filter design package, such as MATLAB, which uses the floating-point numbers to represent coefficients. Since filter design will be discussed in Chapters 5 and 6, we only briefly describe the fourth-order IIR filter used in this experiment.

Table 3.8 shows an assembly language program that implements a fourth-order IIR lowpass filter. This filter is designed for 8 kHz sampling frequency with cut-off frequency 1800 Hz. The routine, _init_iir4, initializes the memory locations of x and y buffers to 0. The IIR filter routine, _iir4, filters the input signal. The coefficient data pointer (CDP) is used to point to the filter coefficients. The auxiliary registers, AR5 and AR6, are pointing to the x and y data buffers, respectively. After each sample is processed, both the x and y buffers are updated by shifting the data in the tapped-delay-line.

1. Write a C function, exp3d.c, to call the routine _ii4() to perform lowpass filter operation. The initialization needs to be done only once, while the routine _ii4() will be called for filtering every sample, these files are also included in the software package.

2. The filter coefficient quantization effects can be observed by modifying the MASK value defined in assembly routine _ii4(). Adjusting the MASK to generate 12, 10, and 8 bits quantized coefficients. Interface the C function, exp3d.c with a signal source and use either the simulator (or EVM) to observe the quantization effects due to the limited wordlength representing filter coefficients.

Table 3.8   List of C55x assembly program for a fourth-order IIR filter

```
;  exp3d_IIR.asm — fourth-order IIR filter
;
MASK .set 0xFFFF

  .def _iir4
  .def _init_iir4
;
; Original coefficients of fourth-order IIR LPF
; with sampling frequency of 8000 Hz
;
; int b[5] = {0.0072, 0.00287, 0.0431, 0.0287, 0.0072} ;
; int a[5] = {1.0000, −2.16860, 2.0097, −0.8766, 0.1505} ;
;
  .data;     ;Q13 formatted coefficients
coeff   ; b0, b1, b2, b3, b4
        .word 0x003B&MASK, 0x00EB&MASK
        .word 0x0161&MASK, 0x00EB&MASK, 0x003B&MASK
        ; −a1, −a2, −a3, −a4
        .word 0x4564&MASK,  −0x404F&MASK
        .word 0xC0D&MASK,   −0x04D1&MASK
        .bss x, 5          ; x delay line
        .bss y, 4          ; y delay line
  .text
_init_iir4
  pshboth XAR5
  amov    #x, XAR5
  rpt     #4
  mov     #0, *AR5+
  amov    #y, XAR5
  rpt     #3
  mov     #0, *AR5+
  popboth XAR5
  ret
;
;    Fourth-order IIR filter
;    Entry T0 = sample
;    Exit T0 = filtered sample
;
_iir4
  pshboth XAR5
  pshboth XAR6
  bset    SATD
  bset    SXM
  amov    #x, XAR5
  amov    #y, XAR6
```

Table 3.8 (*continued*)

```
   amov     #coeff, XCDP
   bset     FRCT
|| mov      T0, *AR5            ; x[0] = indata
;
; Perform IIR filtering
;
   mpym *AR5+, *CDP+, AC0   ; AC0 = x[0] * bn[0]
|| rpt   #3                 ; i = 1, 2, 3, 4
   macm *AR5+, *CDP+, AC0   ;     AC0 += x[i] * bn[i]
   rpt   #3                 ; i = 0, 1, 2, 3
   macm *AR6+, *CDP+, AC0   ;     AC0 += y[i] * an[i]
   amov #y+2, XAR5
   amov #y+3, XAR6
   sfts AC0, #2             ; Scale to Q15 format
|| rpt   #2
   mov  *AR5-, *AR6-        ; Update y[]
   mov  hi(AC0), *AR6
|| mov   hi(AC0), T0        ; Return y[0] in T0
   amov #x+3, XAR5
   amov #x+4, XAR6
   bclr FRCT
|| rpt   #3
   mov  *AR5-, *AR6-        ; Update x[]
   popboth XAR6
   popboth XAR5
   bclr SXM
   bclr SATD
|| ret
   .end
```

3.8.5 Experiment 3E – Synthesizing Sine Function

For many DSP applications, signals and system parameters, such as filter coefficients, are usually normalized in the range of -1 and 1 using fractional numbers. In Section 3.4, we introduced the fixed-point representation of fractional numbers and in Section 3.6, we discussed the overflow problems and present some solutions. The experiment in this section used polynomial approximation of the sinusoid function as an example to understand the fixed-point arithmetic operation and overflow control.

The cosine and sine functions can be expressed as the infinite power (Taylor) series expansion

$$\cos(\theta) = 1 - \frac{1}{2!}\theta^2 + \frac{1}{4!}\theta^4 - \frac{1}{6!}\theta^6 + \cdots, \qquad (3.8.1a)$$

$$\sin(\theta) = \theta - \frac{1}{3!}\theta^3 + \frac{1}{5!}\theta^5 - \frac{1}{7!}\theta^7 + \cdots, \tag{3.8.1b}$$

where θ is in radians and '!' represents the factorial operation. The accuracy of the approximation depends on how many terms are used in the series. Usually more terms are needed to provide reasonable accuracy for larger values of θ. However, in real-time DSP application, only a limited number of terms can be used. Using a function approximation approach such as the Chebyshev approximation, $\cos(\theta)$ and $\sin(\theta)$ can be computed as

$$\cos(\theta) = 1 - 0.0019220\theta - 4.9001474\theta^2 - 0.2648920\theta^3$$
$$+ 5.0454100\theta^4 + 1.8002930\theta^5, \tag{3.8.2a}$$

$$\sin(\theta) = 3.1406250\theta + 0.0202636700\theta^2 - 5.3251960\theta^3$$
$$+ 0.5446788\theta^4 + 1.8002930\theta^5, \tag{3.8.2b}$$

where the value θ is defined in the first quadrant. That is, $0 \le \theta < \pi/2$. For θ in the other quadrants, the following properties can be used to transfer it to the first quadrant:

$$\sin(180° - \theta) = \sin(\theta), \quad \cos(180° - \theta) = -\cos(\theta), \tag{3.8.3}$$

$$\sin(-180° + \theta) = -\sin(\theta), \quad \cos(-180° + \theta) = -\cos(\theta) \tag{3.8.4}$$

and

$$\sin(-\theta) = -\sin(\theta), \quad \cos(-\theta) = \cos(\theta). \tag{3.8.5}$$

The C55x assembly routine given in Table 3.9 synthesizes the sine and cosine functions, which can be used to calculate the angle θ from $-180°$ to $180°$.

As shown in Figure 3.11, data in the Q15 format is within the range defined in (3.4.4). Since the absolute value of the largest coefficient given in this experiment is 5.325196, we cannot use the Q15 format to represent this number. To properly represent the coefficients, we have to scale the coefficient, or use a different Q format that represents both the fractional numbers and the integers. We can achieve this by assuming the binary point to be three bits further to the right. This is called the Q12 format, which has one sign-bit, three integer bits, and 12 fraction bits, as illustrated in Figure 3.23(a). The Q12 format covers the range -8 to 8. In the given example, we use the Q12 format for all the coefficients, and map the angle $-\pi \le \theta \le \pi$ to a signed 16-bit number (0x8000 $\le x \le$ 0x7FFF), as shown in Figure 3.23(b).

When the `sine_cos` subroutine is called, a 16-bit mapped angle (function argument) is passed to the assembly routine in register T0 following the C calling conversion described in Chapter 2. The quadrant information is tested and stored in TC1 and TC2. If TC1 (bit 14) is set, the angle is located in either quadrant II or IV. We use the 2's complement to convert the angle to the first or third quadrant. We mask out the sign-bit to calculate the third quadrant angle in the first quadrant, and the negation changes the fourth quadrant angle to the first quadrant. Therefore the angle to be

Table 3.9   Sine function approximation routine for the C55x

```
;   sine_cos: 16-bit sine(x) and cos(x) approximation function
;
;   Entry:   T0 = x, in the range of [-pi = 0x8000 pi = 0x7fff]
;            AR0 -> Pointer
;   Return:  None
;     Update:
;     AR0 -> [0] = cos(x) = d0 + d1*x + d2*x^2 + d3*x^3 + d4*x^4 + d5*x^5
;     AR0 -> [1] = sin(x) = c1*x + c2*x^2 + c3*x^3 + c4*x^4 + c5*x^5
;

      .def _sine_cos
;
;   Approximation coefficients in Q12 (4096) format
;
      .data
coeff ; Sine approximation coefficients
      .word 0x3240          ; c1 =   3.140625
      .word 0x0053          ; c2 =   0.02026367
      .word 0xaacc          ; c3 = -5.325196
      .word 0x08b7          ; c4 =   0.54467780
      .word 0x1cce          ; c5 =   1.80029300
      ; Cosine approximation coefficients
      .word 0x1000          ; d0 =   1.0000
      .word 0xfff8          ; d1 = -0.001922133
      .word 0xb199          ; d2 = -4.90014738
      .word 0xfbc3          ; d3 = -0.2648921
      .word 0x50ba          ; d4 =   5.0454103
      .word 0xe332          ; d5 = -1.800293
;
;   Function starts
;
      .text
_sine_cos
   pshboth XAR5            ; Save AR5
   amov    #14, AR5
   btstp   AR5, T0         ; Test bit 15 and 14
;
;   Start cos(x)
;
   amov    #coeff+10, XAR5 ; Pointer to the end of coefficients
   xcc     _neg_x, TC1
   neg     T0              ; Negate if bit 14 is set
_neg_x
   and     #0x7fff, T0     ; Mask out sign bit
   mov     *AR5-<<#16, AC0 ; AC0 = d5
```

continues overleaf

Table 3.9 (*continued*)

```
|| bset    SATD                 ; Set Saturate bit
   mov     *AR5−≪#16, AC1       ; AC1 = d4
|| bset    FRCT                 ; Set up fractional bit
   mac     AC0, T0, AC1         ; AC1 = (d5*x + c4)
|| mov     *AR5−≪#16,AC0        ; AC0 = d3
   mac     AC1,T0, AC0          ; AC0 = (d5*x^2 + d4*x + d3)
|| mov     *AR5−≪#16, AC1       ; AC1 = d2
   mac     AC0, T0, AC1         ; AC1 = (d5*x^3 + d4*x^2 + d3*x + d2)
|| mov     *AR5−≪#16, AC0       ; AC0 = d1
   mac     AC1, T0, AC0         ; AC0 = (d5*x^4 + d4*x^3 + d3*x^2
                                ;          + d2*x + d1)
|| mov     *AR5−≪#16, AC1       ; AC1 = d0
   macr    AC0, T0, AC1         ; AC1 = (d5*x^4 + d4*x^3 + d3*x^2 + d2*x
                                ;          + d1)*x + d0
|| xcc     _neg_result1, TC2
   neg     AC1
_neg_result1
   mov     *AR5−≪#16, AC0       ; AC0 = c5
|| xcc     _neg_result2, TC1
   neg     AC1
_neg_result2
   mov     hi(saturate(AC1≪#3)), *AR0+ ; Return cos(x) in Q15
;
;  Start sin(x) computation
;
   mov     *AR5−≪#16, AC1       ; AC1 = c4
   mac     AC0, T0, AC1         ; AC1 = (c5*x + c4)
|| mov     *AR5−≪#16, AC0       ; AC0 = c3
   mac     AC1, T0, AC0         ; AC0 = (c5*x^2 + c4*x + c3)
|| mov     *AR5−≪#16, AC1       ; AC1 = c2
   mac     AC0, T0, AC1         ; AC1 = (c5*x^3 + c4*x^2 + c3*x + c2)
|| mov     *AR5−≪#16, AC0       ; AC0 = c1
   mac     AC1, T0, AC0         ; AC0 = (c5*x^4 + c4*x^3 + c3*x^2
                                ;          + c2*x + c1)
|| popboth XAR5                 ; Restore AR5
   mpyr    T0, AC0, AC1         ; AC1 = (c5*x^4 + c4*x^3 + c3*x^2
                                ;          + c2*x + c1)*x
|| xcc     _neg_result3, TC2
   neg     AC1
_neg_result3
   mov     hi(saturate(AC1≪#3)), *AR0−   ; Return sin(x) in Q15
|| bclr    FRCT                          ; Reset fractional bit
   bclr    SATD                          ; Reset saturate bit
   ret
   .end
```

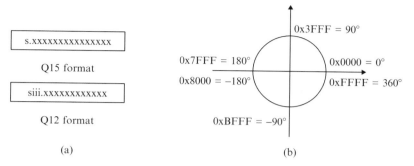

Figure 3.23 Scaled fixed-point number representation: (a) Q formats, and (b) Map angle value to 16-bit signed integer

calculated is always located in the first quadrant. Because we use the Q12 format coefficients, the computed result needs to be left shifted 3 bits to become the Q15 format.

1. Write a C function `exp3e.c` to call the sinusoid approximation function, `sine_cos`, written in assembly listed in Table 3.9 (These programs can be found in the software package). Calculate the angles in the following table.

θ	30°	45°	60°	90°	120°	135°	150°	180°
$\cos(\theta)$								
$\sin(\theta)$								
θ	−150°	−135°	−120°	−90°	−60°	−45°	−30°	360°
$\cos(\theta)$								
$\sin(\theta)$								

2. In the above implementation of sine approximation, what will the following C55x assembly instructions do? What may happen to the approximation result if we do not set these control bits?

(a) `.bset FRCT`
(b) `.bset SATD`
(c) `.bclr FRCT`
(d) `.bclr SATD`

References

[1] N. Ahmed and T. Natarajan, *Discrete-Time Signals and Systems*, Englewood Cliffs, NJ: Prentice-Hall, 1983.
[2] *MATLAB User's Guide*, Math Works, 1992.

[3] *MATLAB Reference Guide*, Math Works, 1992.

[4] A. V. Oppenheim and R. W. Schafer, *Discrete-Time Signal Processing*, Englewood Cliffs, NJ: Prentice-Hall, 1989.

[5] S. J. Orfanidis, *Introduction to Signal Processing*, Englewood Cliffs, NJ: Prentice-Hall, 1996.

[6] J. G. Proakis and D. G. Manolakis, *Digital Signal Processing – Principles, Algorithms, and Applications*, 3rd Ed., Englewood Cliffs, NJ: Prentice-Hall, 1996.

[7] P. Peebles, *Probability, Random Variables, and Random Signal Principles*, New York, NY: McGraw-Hill, 1980.

[8] A Bateman and W. Yates, *Digital Signal Processing Design*, New York: Computer Science Press, 1989.

[9] S. M. Kuo and D. R. Morgan, *Active Noise Control Systems – Algorithms and DSP Implementations*, New York: Wiley, 1996.

[10] C. Marven and G. Ewers, *A Simple Approach to Digital Signal Processing*, New York: Wiley, 1996.

[11] J. H. McClellan, R. W. Schafer, and M. A. Yoder, *DSP First: A Multimedia Approach*, 2nd Ed., Englewood Cliffs, NJ: Prentice-Hall, 1998.

[12] D. Grover and J. R. Deller, *Digital Signal Processing and the Microcontroller*, Upper Saddle River, NJ: Prentice-Hall, 1999.

Exercises

Part A

1. Compute the impulse response $h(n)$ for $n \geq 0$ of the digital systems defined by the following I/O difference equations:

    (a) $y(n) = x(n) - 0.75y(n-1)$

    (b) $y(n) - 0.5y(n-1) = 2x(n) - x(n-1)$

    (c) $y(n) = 2x(n) - 0.75x(n-1) + 1.5x(n-2)$

2. Construct detailed flow diagrams for the three digital systems defined in Problem 1.

3. Similar to the signal flow diagram for the FIR filter shown Figure 3.6, construct a detailed signal flow diagram for the IIR filter defined in (3.2.18).

4. A discrete-time system is time invariant (or shift invariant) if its input–output characteristics do not change with time. Otherwise this system is time varying. A digital system with input signal $x(n)$ is time invariant if and only if the output signal

    $$y(n-k) = F[x(n-k)]$$

    for any time shift k, i.e., when an input is delayed (shifted) by k, the output is delayed by the same amount. Show that the system defined in (3.2.18) is time-invariant system if the coefficients a_m and b_l are constant.

5. A linear system is a system that satisfies the superposition principle, which states that the response of the system to a weighted sum of signals is equal to the corresponding weighted sum of the responses of the system to each of the individual input signals. That is, a system is linear if and only if

$$F[a_1x_1(n) + a_2x_2(n)] = a_1F[x_1(n)] + a_2F[x_2(n)]$$

for any arbitrary input signals $x_1(n)$ and $x_2(n)$, and for any arbitrary constants a_1 and a_2. If the input is the sum (superposition) of two or more scaled sequences, we can find the output due to each sequence acting alone and then add (superimpose) the separate scaled outputs. Check whether the following systems are linear or nonlinear:

(a) $y(n) = 0.5x(n) + 0.75y(n-1)$

(b) $y(n) = x(n)x(n-1) + 0.5y(n-1)$

(c) $y(n) = 0.75x(n) + x(n)y(n-1)$

(d) $y(n) = 0.5x(n) + 0.25x^2(n)$

6. Show that a real-valued sinewave of amplitude A defined in (3.1.6) has the power $P_x = 0.5A^2$.

7. Equation (3.3.12) shows that the power is equal to the variance for a zero-mean random variable. Show that if the mean of the random variable x is m_x, the power of x is given by $P_x = m_x^2 + \sigma_x^2$.

8. An exponential random variable is defined by the probability density function

$$f(x) = \frac{\lambda}{2}e^{-\lambda|x|}.$$

Show that the mean value is 0 and the variance is $2/\lambda^2$.

9. Find the fixed-point 2's complement representation with $B = 8$ for the decimal numbers 0.152 and -0.738. Round the binary numbers to 6 bits and compute the corresponding roundoff errors.

10. If the quantization process uses truncation instead of rounding, show that the truncation error, $e(n) = x(n) - x(nT)$, will be in the interval $-\Delta < e(n) < 0$. Assuming that the truncation error is uniformly distributed in the interval $(-\Delta, 0)$, compute the mean and the variance of $e(n)$.

11. Identify the various types of finite wordlength effects that can occur in a digital filter defined by the I/O equation (3.2.18).

12. Consider the causal system with I/O equation

$$y(n) = x(n) - 0.5y(n-1)$$

and the input signal given as

$$x(n) = \begin{cases} 0.5, & n = 0 \\ 0, & n > 0. \end{cases}$$

(a) Compute $y(0)$, $y(1)$, $y(2)$, $y(3)$, and $y(\infty)$.

(b) Assume the DSP hardware has 4-bit worldlength $(B = 4)$, compute $y(n)$ for $n = 0, 1, 2, 3, \ldots, \infty$. In this case, show that $y(n)$ oscillates between ± 0.125 for $n \geq 2$.

(c) Repeat part (b) but use wordlength $B = 5$. Show that the output $y(n)$ oscillates between ± 0.0625 for $n \geq 3$.

Part B

13. Generate and plot (20 samples) the following sinusoidal signals using MATLAB:

    (a) $A = 1$, $f = 100$ Hz, and $f_s = 1$ kHz

    (b) $A = 1$, $f = 400$ Hz, and $f_s = 1$ kHz

    (c) Discuss the difference of results between (a) and (b)

    (d) $A = 1$, $f = 600$ Hz, and $f_s = 1$ kHz

    (e) Compare and explain the results (b) and (d).

14. Generate 1024 samples of pseudo-random numbers with mean 0 and variance 1 using the MATLAB function rand. Then use MATLAB functions mean, std, and hist to verify the results.

15. Generate 1024 samples of sinusoidal signal at frequency 1000 Hz, amplitude equal to unity, and the sampling rate is 8000 Hz. Mix the generated sinewave with the zero-mean pseudo-random number of variance 0.2.

16. Write a C program to implement the moving-average filter defined in (3.2.2). Test the filter using the corrupted sinewave generated in Problem 15 as input for different L. Plot both the input and output waveforms. Discuss the results related to the filter order L.

17. Given the difference equations in Problem 1, calculate and plot the impulse response $h(n)$, $n = 0, 1, \ldots, 127$ using MATLAB.

18. Assuming that $\hat{P}_x(0) = 1$, use MATLAB to estimate the power of $x(n)$ generated in Problem 15 by using the recursive power estimator given in (3.2.11). Plot $\hat{P}_x(n)$ for $n = 0, 1, \ldots, 1023$.

Part C

19. Using EVM (or other DSP boards) to conduct the quantization experiment in real-time:

    (a) Generate an analog input signal, such as a sinewave, using a signal generator. Both the input and output channels of the DSP are displayed on an oscilloscope. Assuming the ADC has 16-bit resolution and adjusting the amplitude of input signal to the full scale of ADC without clipping the waveform. Vary the number of bits (by shifting out or masking) to 14, 12, 10, etc. to represent the signal and output the signal to DAC. Observe the output waveform using the oscilloscope.

    (b) Replace the input source with a microphone, radio line output, or CD player, and send DSP output to a loudspeaker for audio play back. Vary the number of bits (16, 12, 8, 4, etc.) for the output signal, and listen to the output sound. Depending upon the type of loudspeaker being used, we may need to use an amplifier to drive the loudspeaker.

20. Implement the following square-root approximation equation in C55x assembly language:

$$\sqrt{x} = 0.2075806 + 1.454895x - 1.34491x^2 + 1.106812x^3 - 0.536499x^4 + 0.1121216x^5$$

This equation approximates an input variable within the range of $0.5 \le x \le 1$. Based on the values in the following table, calculate $\sqrt{x}$.

x	0.5	0.6	0.7	0.8	0.9
$\sqrt{x}$					

21. Write a C55x assembly function to implement the inverse square-root approximation equation as following:

$$1/\sqrt{x} = 1.84293985 - 2.57658958x + 2.11866164x^2 - 0.67824984x^3.$$

This equation approximates an input variable in the range of $0.5 \le x \le 1$. Use this approximation equation to compute $1/\sqrt{x}$ in the following table:

x	0.5	0.6	0.7	0.8	0.9
$1/\sqrt{x}$					

Note that $1/\sqrt{x}$ will result in a number greater than 1.0. Try to use Q14 data format. That is, use 0x3FFF for 1 and 0x2000 for 0.5, and scale back to Q15 after calculation.

4

Frequency Analysis

Frequency analysis of any given signal involves the transformation of a time-domain signal into its frequency components. The need for describing a signal in the frequency domain exists because signal processing is generally accomplished using systems that are described in terms of frequency response. Converting the time-domain signals and systems into the frequency domain is extremely helpful in understanding the characteristics of both signals and systems.

In Section 4.1, the Fourier series and Fourier transform will be introduced. The Fourier series is an effective technique for handling periodic functions. It provides a method for expressing a periodic function as the linear combination of sinusoidal functions. The Fourier transform is needed to develop the concept of frequency-domain signal processing. Section 4.2 introduces the z-transform, its important properties, and its inverse transform. Section 4.3 shows the analysis and implementation of digital systems using the z-transform. Basic concepts of discrete Fourier transforms will be introduced in Section 4.4, but detailed treatments will be presented in Chapter 7. The application of frequency analysis techniques using MATLAB to design notch filters and analyze room acoustics will be presented in Section 4.5. Finally, real-time experiments using the TMS320C55x will be presented in Section 4.6.

4.1 Fourier Series and Transform

In this section, we will introduce the representation of analog periodic signals using Fourier series. We will then expand the analysis to the Fourier transform representation of broad classes of finite energy signals.

4.1.1 Fourier Series

Any periodic signal, $x(t)$, can be represented as the sum of an infinite number of harmonically related sinusoids and complex exponentials. The basic mathematical representation of periodic signal $x(t)$ with period T_0 (in seconds) is the Fourier series defined as

$$x(t) = \sum_{k=-\infty}^{\infty} c_k e^{jk\Omega_0 t}, \tag{4.1.1}$$

where c_k is the Fourier series coefficient, and $\Omega_0 = 2\pi/T_0$ is the fundamental frequency (in radians per second). The Fourier series describes a periodic signal in terms of infinite sinusoids. The sinusoidal component of frequency $k\Omega_0$ is known as the kth harmonic.

The kth Fourier coefficient, c_k, is expressed as

$$c_k = \frac{1}{T_0} \int_{T_0} x(t) e^{-jk\Omega_0 t} dt. \tag{4.1.2}$$

This integral can be evaluated over any interval of length T_0. For an odd function, it is easier to integrate from 0 to T_0. For an even function, integration from $-T_0/2$ to $T_0/2$ is commonly used. The term with $k = 0$ is referred to as the DC component because $c_0 = \frac{1}{T_0} \int_{T_0} x(t) dt$ equals the average value of $x(t)$ over one period.

Example 4.1: The waveform of a rectangular pulse train shown in Figure 4.1 is a periodic signal with period T_0, and can be expressed as

$$x(t) = \begin{cases} A, & kT_0 - \tau/2 \le t \le kT_0 + \tau/2 \\ 0, & \text{otherwise,} \end{cases} \tag{4.1.3}$$

where $k = 0, \pm 1, \pm 2, \ldots$, and $\tau < T_0$. Since $x(t)$ is an even signal, it is convenient to select the integration from $-T_0/2$ to $T_0/2$. From (4.1.2), we have

$$c_k = \frac{1}{T_0} \int_{-\frac{T_0}{2}}^{\frac{T_0}{2}} A e^{-jk\Omega_0 t} dt = \frac{A}{T_0} \left[\frac{e^{-jk\Omega_0 t}}{-jk\Omega_0} \Big|_{-\frac{\tau}{2}}^{\frac{\tau}{2}} \right] = \frac{A\tau}{T_0} \frac{\sin\left(\frac{k\Omega_0 \tau}{2}\right)}{\frac{k\Omega_0 \tau}{2}}. \tag{4.1.4}$$

This equation shows that c_k has a maximum value $A\tau/T_0$ at $\Omega_0 = 0$, decays to 0 as $\Omega_0 \to \pm\infty$, and equals 0 at frequencies that are multiples of π. Because the periodic signal $x(t)$ is an even function, the Fourier coefficients c_k are real values.

For the rectangular pulse train with a fixed period T_0, the effect of decreasing τ is to spread the signal power over the frequency range. On the other hand, when τ is fixed but the period T_0 increases, the spacing between adjacent spectral lines decreases.

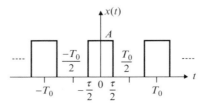

Figure 4.1   Rectangular pulse train

A periodic signal has infinite energy and finite power, which is defined by Parseval's theorem as

$$P_x = \frac{1}{T_0} \int_{T_0} |x(t)|^2 dt = \sum_{k=-\infty}^{\infty} |c_k|^2. \tag{4.1.5}$$

Since $|c_k|^2$ represents the power of the kth harmonic component of the signal, the total power of the periodic signal is simply the sum of the powers of all harmonics.

The complex-valued Fourier coefficients, c_k, can be expressed as

$$c_k = |c_k| e^{j\phi_k}. \tag{4.1.6}$$

A plot of $|c_k|$ versus the frequency index k is called the amplitude (magnitude) spectrum, and a plot of ϕ_k versus k is called the phase spectrum. If the periodic signal $x(t)$ is real valued, it is easy to show that c_0 is real valued and that c_k and c_{-k} are complex conjugates. That is,

$$c_k = c_{-k}^*, \quad |c_{-k}| = |c_k| \quad \text{and} \quad \phi_{-k} = -\phi_k. \tag{4.1.7}$$

Therefore the amplitude spectrum is an even function of frequency Ω, and the phase spectrum is an odd function of Ω for a real-valued periodic signal.

If we plot $|c_k|^2$ as a function of the discrete frequencies $k\Omega_0$, we can show that the power of the periodic signal is distributed among the various frequency components. This plot is called the power density spectrum of the periodic signal $x(t)$. Since the power in a periodic signal exists only at discrete values of frequencies $k\Omega_0$, the signal has a line spectrum. The spacing between two consecutive spectral lines is equal to the fundamental frequency Ω_0.

Example 4.2: Consider the output of an ideal oscillator as the perfect sinewave expressed as

$$x(t) = \sin(2\pi f_0 t), \quad f_0 = \frac{\Omega_0}{2\pi}.$$

We can then calculate the Fourier series coefficients using Euler's formula (Appendix A.3) as

$$\sin(2\pi f_0 t) = \frac{1}{2j} \left(e^{j2\pi f_0 t} - e^{-j2\pi f_0 t} \right) = \sum_{k=-\infty}^{\infty} c_k e^{jk2\pi f_0 t}.$$

We have

$$c_k = \begin{cases} 1/2j, & k = 1 \\ -1/2j, & k = -1 \\ 0, & \text{otherwise.} \end{cases} \tag{4.1.8}$$

This equation indicates that there is no power in any of the harmonic $k \neq \pm 1$. Therefore Fourier series analysis is a useful tool for determining the quality (purity) of a sinusoidal signal.

4.1.2 Fourier Transform

We have shown that a periodic signal has a line spectrum and that the spacing between two consecutive spectral lines is equal to the fundamental frequency $\Omega_0 = 2\pi/T_0$. The number of frequency components increases as T_0 is increased, whereas the envelope of the magnitude of the spectral components remains the same. If we increase the period without limit (i.e., $T_0 \to \infty$), the line spacing tends toward 0. The discrete frequency components converge into a continuum of frequency components whose magnitudes have the same shape as the envelope of the discrete spectra. In other words, when the period T_0 approaches infinity, the pulse train shown in Figure 4.1 reduces to a single pulse, which is no longer periodic. Thus the signal becomes non-periodic and its spectrum becomes continuous.

In real applications, most signals such as speech signals are not periodic. Consider the signal that is not periodic ($\Omega_0 \to 0$ or $T_0 \to \infty$), the number of exponential components in (4.1.1) tends toward infinity and the summation becomes integration over the entire continuous range $(-\infty, \infty)$. Thus (4.1.1) can be rewritten as

$$x(t) = \frac{1}{2\pi} \int_{-\infty}^{\infty} X(\Omega)e^{j\Omega t} d\Omega. \tag{4.1.9}$$

This integral is called the inverse Fourier transform. Similarly, (4.1.2) can be rewritten as

$$X(\Omega) = \int_{-\infty}^{\infty} x(t)e^{-j\Omega t} dt, \tag{4.1.10}$$

which is called the Fourier transform (FT) of $x(t)$. Note that the time functions are represented using lowercase letters, and the corresponding frequency functions are denoted by using capital letters. A sufficient condition for a function $x(t)$ that possesses a Fourier transform is

$$\int_{-\infty}^{\infty} |x(t)| dt < \infty. \tag{4.1.11}$$

That is, $x(t)$ is absolutely integrable.

Example 4.3: Calculate the Fourier transform of the function $x(t) = e^{-at}u(t)$, where $a > 0$ and $u(t)$ is the unit step function. From (4.1.10), we have

$$X(\Omega) = \int_{-\infty}^{\infty} e^{-at} u(t) e^{-j\Omega t} dt$$

$$= \int_{0}^{\infty} e^{-(a+j\Omega)t} dt$$

$$= \frac{1}{a + j\Omega}.$$

The Fourier transform $X(\Omega)$ is also called the spectrum of the analog signal $x(t)$. The spectrum $X(\Omega)$ is a complex-valued function of frequency Ω, and can be expressed as

$$X(\Omega) = |X(\Omega)| e^{j\phi(\Omega)}, \qquad (4.1.12)$$

where $|X(\Omega)|$ is the magnitude spectrum of $x(t)$, and $\phi(\Omega)$ is the phase spectrum of $x(t)$. In the frequency domain, $|X(\Omega)|^2$ reveals the distribution of energy with respect to the frequency and is called the energy density spectrum of the signal. When $x(t)$ is any finite energy signal, its energy is

$$E_x = \int_{-\infty}^{\infty} |x(t)|^2 dt = \frac{1}{2\pi} \int_{-\infty}^{\infty} |X(\Omega)|^2 d\Omega. \qquad (4.1.13)$$

This is called Parseval's theorem for finite energy signals, which expresses the principle of conservation of energy in time and frequency domains.

For a function $x(t)$ defined over a finite interval T_0, i.e., $x(t) = 0$ for $|t| > T_0/2$, the Fourier series coefficients c_k can be expressed in terms of $X(\Omega)$ using (4.1.2) and (4.1.10) as

$$c_k = \frac{1}{T_0} X(k\Omega_0). \qquad (4.1.14)$$

For a given finite interval function, its Fourier transform at a set of equally spaced points on the Ω-axis is specified exactly by the Fourier series coefficients. The distance between adjacent points on the Ω-axis is $2\pi/T_0$ radians.

If $x(t)$ is a real-valued signal, we can show from (4.1.9) and (4.1.10) that

$$FT[x(-t)] = X^*(\Omega) \quad \text{and} \quad X(-\Omega) = X^*(\Omega). \qquad (4.1.15)$$

It follows that

$$|X(-\Omega)| = |X(\Omega)| \quad \text{and} \quad \phi(-\Omega) = -\phi(\Omega). \qquad (4.1.16)$$

Therefore the amplitude spectrum $|X(\Omega)|$ is an even function of Ω, and the phase spectrum is an odd function.

If the time signal $x(t)$ is a delta function $\delta(t)$, its Fourier transform can be calculated as

$$X(\Omega) = \int_{-\infty}^{\infty} \delta(t) e^{-j\Omega t} dt = 1. \qquad (4.1.17)$$

This indicates that the delta function has frequency components at all frequencies. In fact, the narrower the time waveform, the greater the range of frequencies where the signal has significant frequency components.

Some useful functions and their Fourier transforms are summarized in Table 4.1. We may find the Fourier transforms of other functions using the Fourier transform properties listed in Table 4.2.

Table 4.1 Common Fourier transform pairs

Time function $x(t)$	Fourier transform $X(\Omega)$
$\delta(t)$	1
$\delta(t - \tau)$	$e^{-j\Omega\tau}$
1	$2\pi\delta(\Omega)$
$e^{-at}u(t)$	$\dfrac{1}{a + j\Omega}$
$e^{j\Omega_0 t}$	$2\pi\delta(\Omega - \Omega_0)$
$\sin(\Omega_0 t)$	$j\pi[\delta(\Omega + \Omega_0) - \delta(\Omega - \Omega_0)]$
$\cos(\Omega_0 t)$	$\pi[\delta(\Omega + \Omega_0) + \delta(\Omega - \Omega_0)]$
$\mathrm{sgn}(t) = \begin{cases} 1, & t \geq 0 \\ -1, & t < 0 \end{cases}$	$\dfrac{2}{j\Omega}$

Table 4.2 Useful properties of the Fourier transform

Time function $x(t)$	Property	Fourier transform $X(\Omega)$
$a_1 x_1(t) + a_2 x_2(t)$	Linearity	$a_1 X_1(\Omega) + a_2 X_2(\Omega)$
$\dfrac{dx(t)}{dt}$	Differentiation in time domain	$j\Omega X(\Omega)$
$tx(t)$	Differentiation in frequency domain	$j\dfrac{dX(\Omega)}{d\Omega}$
$x(-t)$	Time reversal	$X(-\Omega)$
$x(t - a)$	Time shifting	$e^{-j\Omega a} X(\Omega)$
$x(at)$	Time scaling	$\dfrac{1}{|a|} X\left(\dfrac{\Omega}{a}\right)$
$x(t) \sin(\Omega_0 t)$	Modulation	$\dfrac{1}{2j}[X(\Omega - \Omega_0) - X(\Omega + \Omega_0)]$
$x(t) \cos(\Omega_0 t)$	Modulation	$\dfrac{1}{2}[X(\Omega + \Omega_0) + X(\Omega - \Omega_0)]$
$e^{-at} x(t)$	Frequency shifting	$X(\Omega + a)$

Example 4.4: Find the Fourier transform of the time function

$$y(t) = e^{-a|t|}, \quad a > 0.$$

This equation can be written as

$$y(t) = x(-t) + x(t),$$

where

$$x(t) = e^{-at}u(t), \quad a > 0.$$

From Table 4.1, we have $X(\Omega) = 1/(a + j\Omega)$. From Table 4.2, we have $Y(\Omega) = X(-\Omega) + X(\Omega)$. This results in

$$Y(\Omega) = \frac{1}{a - j\Omega} + \frac{1}{a + j\Omega} = \frac{2a}{a^2 + \Omega^2}.$$

4.2 The z-Transform

Continuous-time signals and systems are commonly analyzed using the Fourier transform and the Laplace transform (will be introduced in Chapter 6). For discrete-time systems, the transform corresponding to the Laplace transform is the z-transform. The z-transform yields a frequency-domain description of discrete-time signals and systems, and provides a powerful tool in the design and implementation of digital filters. In this section, we will introduce the z-transform, discuss some important properties, and show its importance in the analysis of linear time-invariant (LTI) systems.

4.2.1 Definitions and Basic Properties

The z-transform (ZT) of a digital signal, $x(n)$, $-\infty < n < \infty$, is defined as the power series

$$X(z) = \sum_{n=-\infty}^{\infty} x(n)z^{-n}, \tag{4.2.1}$$

where $X(z)$ represents the z-transform of $x(n)$. The variable z is a complex variable, and can be expressed in polar form as

$$z = re^{j\theta}, \tag{4.2.2}$$

where r is the magnitude (radius) of z, and θ is the angle of z. When $r = 1$, $|z| = 1$ is called the unit circle on the z-plane. Since the z-transform involves an infinite power series, it exists only for those values of z where the power series defined in (4.2.1) converges. The region on the complex z-plane in which the power series converges is called the region of convergence (ROC).

As discussed in Section 3.1, the signal $x(n)$ encountered in most practical applications is causal. For this type of signal, the two-sided z-transform defined in (4.2.1) becomes a one-sided z-transform expressed as

$$X(z) = \sum_{n=0}^{\infty} x(n)z^{-n}. \tag{4.2.3}$$

Clearly if $x(n)$ is causal, the one-sided and two-sided z-transforms are equivalent.

Example 4.5: Consider the exponential function

$$x(n) = a^n u(n).$$

The z-transform can be computed as

$$X(z) = \sum_{n=-\infty}^{\infty} a^n z^{-n} u(n) = \sum_{n=0}^{\infty} (az^{-1})^n.$$

Using the infinite geometric series given in Appendix A.2, we have

$$X(z) = \frac{1}{1 - az^{-1}} \quad \text{if } |az^{-1}| < 1.$$

The equivalent condition for convergence (or ROC) is

$$|z| > |a|.$$

Thus we obtain $X(z)$ as

$$X(z) = \frac{z}{z - a}, \quad |z| > |a|.$$

There is a zero at the origin $z = 0$ and a pole at $z = a$. The ROC and the pole–zero plot are illustrated in Figure 4.2 for $0 < a < 1$, where '×' marks the position of the pole and 'o' denotes the position of the zero. The ROC is the region outside the circle with radius a. Therefore the ROC is always bounded by a circle since the convergence condition is on the magnitude $|z|$. A causal signal is characterized by an ROC that is outside the maximum pole circle and does not contain any pole.

The properties of the z-transform are extremely useful for the analysis of discrete-time LTI systems. These properties are summarized as follows:

1. *Linearity* (superposition). The z-transform is a linear transformation. Therefore the z-transform of the sum of two sequences is the sum of the z-transforms of the individual sequences. That is,

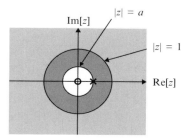

Figure 4.2 Pole, zero, and ROC (shaded area) on the z-plane

$$ZT[a_1 x_1(n) + a_2 x_2(n)] = a_1 ZT[x_1(n)] + a_2 ZT[x_2(n)]$$
$$= a_1 X_1(z) + a_2 X_2(z), \qquad (4.2.4)$$

where a_1 and a_2 are constants, and $X_1(z)$ and $X_2(z)$ are the z-transforms of the signals $x_1(n)$ and $x_2(n)$, respectively. This linearity property can be generalized for an arbitrary number of signals.

2. *Time shifting*. The z-transform of the shifted (delayed) signal $y(n) = x(n - k)$ is

$$Y(z) = ZT[x(n - k)] = z^{-k} X(z), \qquad (4.2.5)$$

where the minus sign corresponds to a delay of k samples. This delay property states that the effect of delaying a signal by k samples is equivalent to multiplying its z-transform by a factor of z^{-k}. For example, $ZT[x(n - 1)] = z^{-1} X(z)$. Thus the unit delay z^{-1} in the z-domain corresponds to a time shift of one sampling period in the time domain.

3. *Convolution*. Consider the signal

$$x(n) = x_1(n) * x_2(n), \qquad (4.2.6)$$

where $*$ denotes the linear convolution introduced in Chapter 3, we have

$$X(z) = X_1(z) X_2(z). \qquad (4.2.7)$$

Therefore the z-transform converts the convolution of two time-domain signals to the multiplication of their corresponding z-transforms.

Some of the commonly used signals and their z-transforms are summarized in Table 4.3.

Table 4.3 Some common z-transform pairs

$x(n)$, $n \geq 0$, c is constant	$X(z)$
c	$\dfrac{cz}{z-1}$
cn	$\dfrac{cz}{(z-1)^2}$
c^n	$\dfrac{z}{z-c}$
nc^n	$\dfrac{cz}{(z-c)^2}$
ce^{-an}	$\dfrac{cz}{z-e^{-a}}$
$\sin(\omega_0 n)$	$\dfrac{z\sin(\omega_0)}{z^2-2z\cos(\omega_0)+1}$
$\cos(\omega_0 n)$	$\dfrac{z[z-\cos(\omega_0)]}{z^2-2z\cos(\omega_0)+1}$

4.2.2 Inverse z-transform

The inverse z-transform can be expressed as

$$x(n) = \text{ZT}^{-1}[X(z)] = \frac{1}{2\pi j}\oint_C X(z)z^{n-1}dz, \tag{4.2.8}$$

where C denotes the closed contour in the ROC of $X(z)$ taken in a counterclockwise direction. Several methods are available for finding the inverse z-transform. We will discuss the three most commonly used methods – long division, partial-fraction expansion, and residue method.

Given the z-transform $X(z)$ of a causal sequence, it can be expanded into an infinite series in z^{-1} or z by long division. To use the long-division method, we express $X(z)$ as the ratio of two polynomials such as

$$X(z) = \frac{B(z)}{A(z)} = \frac{\displaystyle\sum_{l=0}^{L-1} b_l z^{-l}}{\displaystyle\sum_{m=0}^{M} a_m z^{-m}}, \tag{4.2.9}$$

where $A(z)$ and $B(z)$ are expressed in either descending powers of z, or ascending powers of z^{-1}. Dividing $B(z)$ by $A(z)$ obtains a series of negative powers of z if a positive-time sequence is indicated by the ROC. If a negative-time function is indicated, we express $X(z)$ as a series of positive powers of z. The method will not work for a sequence defined in both positive and negative time. In addition, it is difficult to obtain a closed-form solution of the time-domain signal $x(n)$ via the long-division method.

The long-division method can be performed recursively. That is,

$$x(n) = \left[b_n - \sum_{m=1}^{n} x(n-m)a_m \right] \bigg/ a_0, \quad n = 1, 2, \ldots \tag{4.2.10}$$

where

$$x(0) = b_0/a_0. \tag{4.2.11}$$

This recursive equation can be implemented on a computer to obtain $x(n)$.

Example 4.6: Given

$$X(z) = \frac{1 + 2z^{-1} + z^{-2}}{1 - z^{-1} + 0.3561z^{-2}}$$

using the recursive equation given in (4.2.10), we have

$$x(0) = b_0/a_0 = 1,$$
$$x(1) = [b_1 - x(0)a_1]/a_0 = 3,$$
$$x(2) = [b_2 - x(1)a_1 - x(0)a_2]/a_0 = 3.6439,$$
$$\cdots$$

This yields the time domain signal $x(n) = \{1, 3, 3.6439, \ldots\}$ obtained from long division.

The partial-fraction-expansion method factors the denominator of $X(z)$ if it is not already in factored form, then expands $X(z)$ into a sum of simple partial fractions. The inverse z-transform of each partial fraction is obtained from the z-transform tables such as Table 4.3, and then added to give the overall inverse z-transform. In many practical cases, the z-transform is given as a ratio of polynomials in z or z^{-1} as shown in (4.2.9). If the poles of $X(z)$ are of first order and $M = L - 1$, then $X(z)$ can be expanded as

$$X(z) = c_0 + \sum_{l=1}^{L-1} \frac{c_l}{1 - p_l z^{-1}} = c_0 + \sum_{l=1}^{L-1} \frac{c_l z}{z - p_l}, \tag{4.2.12}$$

where p_l are the distinct poles of $X(z)$ and c_l are the partial-fraction coefficients. The coefficient c_l associated with the pole p_l may be obtained with

$$c_l = \frac{X(z)}{z}(z - p_l)\bigg|_{z=p_l}. \tag{4.2.13}$$

If the order of the numerator $B(z)$ is less than that of the denominator $A(z)$ in (4.2.9), that is $L - 1 < M$, then $c_0 = 0$. If $L - 1 > M$, then $X(z)$ must be reduced first in order to make $L - 1 \leq M$ by long division with the numerator and denominator polynomials written in descending power of z^{-1}.

Example 4.7: For the z-transform

$$X(z) = \frac{z^{-1}}{1 - 0.25z^{-1} - 0.375z^{-2}}$$

we can first express $X(z)$ in positive powers of z, expressed as

$$X(z) = \frac{z}{z^2 - 0.25z - 0.375} = \frac{z}{(z - 0.75)(z + 0.5)} = \frac{c_1 z}{z - 0.75} + \frac{c_2 z}{z + 0.5}.$$

The two coefficients are obtained by (4.2.13) as follows:

$$c_1 = \frac{X(z)}{z}(z - 0.75)\bigg|_{z=0.75} = \frac{1}{z + 0.5}\bigg|_{z=0.75} = 0.8$$

and

$$c_2 = \frac{1}{z - 0.75}\bigg|_{z=-0.5} = -0.8.$$

Thus we have

$$X(z) = \frac{0.8z}{z - 0.75} - \frac{0.8z}{z + 0.5}.$$

The overall inverse z-transform $x(n)$ is the sum of the two inverse z-transforms. From entry 3 of Table 4.3, we obtain

$$x(n) = 0.8[(0.75)^n - (-0.5)^n], \quad n \geq 0.$$

The MATLAB function residuez finds the residues, poles and direct terms of the partial-fraction expansion of $B(z)/A(z)$ given in (4.2.9). Assuming that the numerator and denominator polynomials are in ascending powers of z^{-1}, the function

```
[c, p, g] = residuez (b, a);
```

finds the partial-fraction expansion coefficients, c_l, and the poles, p_l, in the returned vectors c and p, respectively. The vector g contains the direct (or polynomial) terms of the rational function in z^{-1} if $L - 1 \geq M$. The vectors b and a represent the coefficients of polynomials $B(z)$ and $A(z)$, respectively.

If $X(z)$ contains one or more multiple-order poles, the partial-fraction expansion must include extra terms of the form $\sum_{j=1}^{m} \frac{g_j}{(z-p_l)^j}$ for an mth order pole at $z = p_l$. The coefficients g_j may be obtained with

$$g_j = \frac{1}{(m-j)!} \frac{d^{m-j}}{dz^{m-j}} \left[\frac{(z-p_l)^m X(z)}{z} \right]_{z=p_l}. \tag{4.2.14}$$

Example 4.8: Consider the function

$$X(z) = \frac{z^2 + z}{(z-1)^2}.$$

We first express $X(z)$ as

$$X(z) = \frac{g_1}{(z-1)} + \frac{g_2}{(z-1)^2}.$$

From (4.2.14), we have

$$g_1 = \frac{d}{dz} \left[\frac{(z-1)^2 X(z)}{z} \right]_{z=1} = \frac{d}{dz} (z+1) \Big|_{z=1} = 1,$$

$$g_2 = \frac{(z-1)^2 X(z)}{z} \Big|_{z=1} = (z+1)|_{z=1} = 2.$$

Thus

$$X(z) = \frac{z}{(z-1)} + \frac{2z}{(z-1)^2}.$$

From Table 4.3, we obtain

$$x(n) = \text{ZT}^{-1} \left[\frac{z}{z-1} \right] + \text{ZT}^{-1} \left[\frac{2z}{(z-1)^2} \right] = 1 + 2n, \quad n \geq 0.$$

The residue method is based on Cauchy's integral theorem expressed as

$$\frac{1}{2\pi j} \oint_c z^{k-m-1} dz = \begin{cases} 1 & \text{if } k = m \\ 0 & \text{if } k \neq m. \end{cases} \tag{4.2.15}$$

Thus the inversion integral in (4.2.8) can be easily evaluated using Cauchy's residue theorem expressed as

$$x(n) = \frac{1}{2\pi j} \oint_c X(z) z^{n-1} dz$$

$$= \sum \text{residues of } X(z) z^{n-1} \text{ at poles of } X(z) z^{n-1} \text{ within } C. \tag{4.2.16}$$

The residue of $X(z)z^{n-1}$ at a given pole at $z = p_l$ can be calculated using the formula

$$R_{z=p_l} = \frac{d^{m-1}}{dz^{m-1}} \left[\frac{(z - p_l)^m}{(m-1)!} X(z)z^{n-1} \right]_{z=p_l}, \quad m \geq 1, \quad (4.2.17)$$

where m is the order of the pole at $z = p_l$. For a simple pole, Equation (4.2.17) reduces to

$$R_{z=p_l} = (z - p_l)X(z)z^{n-1}\big|_{z=p_l}. \quad (4.2.18)$$

Example 4.9: Given the following z-transform function:

$$X(z) = \frac{1}{(z-1)(z-0.5)},$$

we have

$$X(z)z^{n-1} = \frac{z^{n-1}}{(z-1)(z-0.5)}.$$

This function has a simple pole at $z = 0$ when $n = 0$, and no pole at $z = 0$ for $n \geq 1$. For the case $n = 0$,

$$X(z)z^{n-1} = \frac{1}{z(z-1)(z-0.5)}.$$

The residue theorem gives

$$\begin{aligned}
x(n) &= R_{z=0} + R_{z=1} + R_{z=0.5} \\
&= zX(z)z^{n-1}\big|_{z=0} + (z-1)X(z)z^{n-1}\big|_{z=1} + (z-0.5)X(z)z^{n-1}\big|_{z=0.5} \\
&= 2 + 2 + (-4) = 0.
\end{aligned}$$

For the case that $n \geq 1$, the residue theorem is applied to obtain

$$\begin{aligned}
x(n) &= R_{z=1} + R_{z=0.5} \\
&= (z-1)X(z)z^{n-1}\big|_{z=1} + (z-0.5)X(z)z^{n-1}\big|_{z=0.5} \\
&= 2 - 2(0.5)^{n-1} = 2\left[1 - (0.5)^{n-1}\right], \quad n \geq 1.
\end{aligned}$$

We have discussed three methods for obtaining the inverse z-transform. A limitation of the long-division method is that it does not lead to a closed-form solution. However, it is simple and lends itself to software implementation. Because of its recursive nature, care should be taken to minimize possible accumulation of numerical errors when the number of data points in the inverse z-transform is large. Both the partial-fraction-expansion and the residue methods lead to closed-form solutions. The main disadvantage with both methods is the need to factor the denominator polynomial, which is done by finding the poles of $X(z)$. If the order of $X(z)$ is high, finding the poles of $X(z)$ may be a difficult task. Both methods may also involve high-order differentiation if $X(z)$ contains multiple-order poles. The partial-fraction-expansion method is useful in generating the coefficients of parallel structures for digital filters. Another application of z-transforms and inverse z-transforms is to solve linear difference equations with constant coefficients.

4.3 Systems Concepts

As mentioned earlier, the z-transform is a powerful tool in analyzing digital systems. In this section, we introduce several techniques for describing and characterizing digital systems.

4.3.1 Transfer Functions

Consider the discrete-time LTI system illustrated in Figure 3.8. The system output is computed by the convolution sum defined as $y(n) = x(n) * h(n)$. Using the convolution property and letting $ZT[x(n)] = X(z)$ and $ZT[y(n)] = Y(z)$, we have

$$Y(z) = X(z)H(z), \tag{4.3.1}$$

where $H(z) = ZT[h(n)]$ is the z-transform of the impulse response of the system. The frequency-domain representation of LTI system is illustrated in Figure 4.3.

The transfer (system) function $H(z)$ of an LTI system may be expressed in terms of the system's input and output. From (4.3.1), we have

$$H(z) = \frac{Y(z)}{X(z)} = ZT[h(n)] = \sum_{n=-\infty}^{\infty} h(n)z^{-n}. \tag{4.3.2}$$

Therefore the transfer function of the LTI system is the rational function of two polynomials $Y(z)$ and $X(z)$. If the input $x(n)$ is the unit impulse $\delta(n)$, the z-transform of such an input is unity (i.e., $X(z) = 1$), and the corresponding output $Y(z) = H(z)$.

One of the main applications of the z-transform in filter design is that the z-transform can be used in creating alternative filters that have exactly the same input–output behavior. An important example is the cascade or parallel connection of two or more

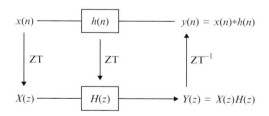

Figure 4.3   A block diagram of LTI system in both time-domain and z-domain

systems, as illustrated in Figure 4.4. In the cascade (series) interconnection, the output of the first system, $y_1(n)$, is the input of the second system, and the output of the second system, $y(n)$, is the overall system output. From Figure 4.4(a), we have

$$Y_1(z) = X(z)H_1(z) \quad \text{and} \quad Y(z) = Y_1(z)H_2(z).$$

Thus

$$Y(z) = X(z)H_1(z)H_2(z).$$

Therefore the overall transfer function of the cascade of the two systems is

$$H(z) = \frac{Y(z)}{X(z)} = H_1(z)H_2(z). \tag{4.3.3}$$

Since multiplication is commutative, $H_1(z)H_2(z) = H_2(z)H_1(z)$, the two systems can be cascaded in either order to obtain the same overall system response. The overall impulse response of the system is

$$h(n) = h_1(n) * h_2(n) = h_2(n) * h_1(n). \tag{4.3.4}$$

Similarly, the overall impulse response and the transfer function of the parallel connection of two LTI systems shown in Figure 4.4(b) are given by

$$h(n) = h_1(n) + h_2(n) \tag{4.3.5}$$

and

$$H(z) = H_1(z) + H_2(z). \tag{4.3.6}$$

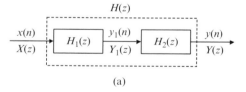

(a)

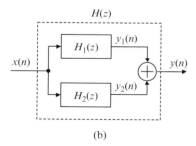

(b)

Figure 4.4   Interconnect of digital systems: (a) cascade form, and (b) parallel form

If we can multiply several z-transforms to get a higher-order system, we can also factor z-transform polynomials to break down a large system into smaller sections. Since a cascading system is equivalent to multiplying each individual system transfer function, the factors of a higher-order polynomial, $H(z)$, would represent component systems that make up $H(z)$ in a cascade connection. The concept of parallel and cascade implementation will be further discussed in the realization of IIR filters in Chapter 6.

Example 4.10: The following LTI system has the transfer function:

$$H(z) = 1 - 2z^{-1} + z^{-3}.$$

This transfer function can be factored as

$$H(z) = \left(1 - z^{-1}\right)\left(1 - z^{-1} - z^{-2}\right) = H_1(z)H_2(z).$$

Thus the overall system $H(z)$ can be realized as the cascade of the first-order system $H_1(z) = 1 - z^{-1}$ and the second-order system $H_2(z) = 1 - z^{-1} - z^{-2}$.

4.3.2 Digital Filters

The general I/O difference equation of an FIR filter is given in (3.1.16). Taking the z-transform of both sides, we have

$$Y(z) = b_0 X(z) + b_1 z^{-1} X(z) + \cdots + b_{L-1} z^{-(L-1)} X(z)$$
$$= \left[b_0 + b_1 z^{-1} + \cdots + b_{L-1} z^{-(L-1)}\right] X(z). \qquad (4.3.7)$$

Therefore the transfer function of the FIR filter is expressed as

$$H(z) = \frac{Y(z)}{X(z)} = b_0 + b_1 z^{-1} + \cdots + b_{L-1} z^{-(L-1)} = \sum_{l=0}^{L-1} b_l z^{-1}. \qquad (4.3.8)$$

The signal-flow diagram of the FIR filter is shown in Figure 3.6. FIR filters can be implemented using the I/O difference equation given in (3.1.16), the transfer function defined in (4.3.8), or the signal-flow diagram illustrated in Figure 3.6.

Similarly, taking the z-transform of both sides of the IIR filter defined in (3.2.18) yields

$$Y(z) = b_0 X(z) + b_1 z^{-1} X(z) + \cdots + b_{L-1} z^{-L+1} X(z) - a_1 z^{-1} Y(z) - \cdots - a_M z^{-M} Y(z)$$
$$= \left(\sum_{l=0}^{L-1} b_l z^{-l}\right) X(z) - \left(\sum_{m=1}^{M} a_m z^{-m}\right) Y(z). \qquad (4.3.9)$$

By rearranging the terms, we can derive the transfer function of an IIR filter as

$$H(z) = \frac{Y(z)}{X(z)} = \frac{\displaystyle\sum_{l=0}^{L-1} b_l z^{-l}}{1 + \displaystyle\sum_{m=1}^{M} a_m z^{-m}} = \frac{B(z)}{1 + A(z)}, \tag{4.3.10}$$

where $B(z) = \sum_{l=0}^{L-1} b_l z^{-l}$ and $A(z) = \sum_{m=1}^{M} a_m z^{-m}$. Note that if all $a_m = 0$, the IIR filter given in (4.3.10) is equivalent to the FIR filter described in (4.3.8).

The block diagram of the IIR filter defined in (4.3.10) can be illustrated in Figure 4.5, where $A(z)$ and $B(z)$ are the FIR filters as shown in Figure 3.6. The numerator coefficients b_l and the denominator coefficients a_m are referred to as the feedforward and feedback coefficients of the IIR filter defined in (4.3.10). A more detailed signal-flow diagram of an IIR filter is illustrated in Figure 4.6 assuming that $M = L - 1$. IIR filters can be implemented using the I/O difference equation expressed in (3.2.18), the transfer function given in (4.3.10), or the signal-flow diagram shown in Figure 4.6.

4.3.3 Poles and Zeros

Factoring the numerator and denominator polynomials of $H(z)$, Equation (4.3.10) can be further expressed as the rational function

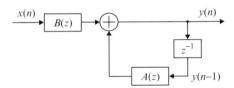

Figure 4.5 IIR filter $H(z)$ consists of two FIR filters $A(z)$ and $B(z)$

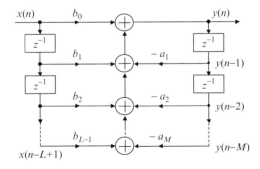

Figure 4.6 Detailed signal-flow diagram of IIR filter

$$H(z) = \frac{b_0}{a_0} z^{M-L+1} \frac{\prod_{l=1}^{L-1}(z - z_l)}{\prod_{m=1}^{M}(z - p_m)}, \tag{4.3.11}$$

where $a_0 = 1$. Without loss of generality, we let $M = L - 1$ in (4.3.11) in order to obtain

$$H(z) = b_0 \frac{\prod_{l=1}^{M}(z - z_l)}{\prod_{m=1}^{M}(z - p_m)} = \frac{b_0(z - z_1)(z - z_2) \cdots (z - z_M)}{(z - p_1)(z - p_2) \cdots (z - p_M)}. \tag{4.3.12}$$

The roots of the numerator polynomial are called the zeros of the transfer function $H(z)$. In other words, the zeros of $H(z)$ are the values of z for which $H(z) = 0$, i.e., $B(z) = 0$. Thus $H(z)$ given in (4.3.12) has M zeros at $z = z_1, z_2, \ldots, z_M$. The roots of the denominator polynomial are called the poles, and there are M poles at $z = p_1, p_2, \ldots, p_M$. The poles of $H(z)$ are the values of z such that $H(z) = \infty$. The LTI system described in (4.3.12) is a pole–zero system, while the system described in (4.3.8) is an all-zero system. The poles and zeros of $H(z)$ may be real or complex, and some poles and zeros may be identical. When they are complex, they occur in complex-conjugate pairs to ensure that the coefficients a_m and b_l are real.

Example 4.11: Consider the simple moving-average filter given in (3.2.1). Taking the z-transform of both sides, we have

$$Y(z) = \frac{1}{L} \sum_{l=0}^{L-1} z^{-l} X(z).$$

Using the geometric series defined in Appendix A.2, the transfer function of the filter can be expressed as

$$H(z) = \frac{Y(z)}{X(z)} = \frac{1}{L} \sum_{l=0}^{L-1} z^{-l} = \frac{1}{L} \left[\frac{1 - z^{-L}}{1 - z^{-1}} \right]. \tag{4.3.13}$$

This equation can be rearranged as

$$Y(z) = z^{-1} Y(z) + \frac{1}{L} \left[X(z) - z^{-L} X(z) \right].$$

Taking the inverse z-transform of both sides and rearranging terms, we obtain

$$y(n) = y(n-1) + \frac{1}{L} [x(n) - x(n-L)].$$

This is an effective way of deriving (3.2.2) from (3.2.1).

The roots of the numerator polynomial $z^L - 1 = 0$ determine the zeros of $H(z)$ defined in (4.3.13). Using the complex arithmetic given in Appendix A.3, we have

$$z_k = e^{j(2\pi/L)k}, \quad k = 0, 1, \ldots, L - 1. \tag{4.3.14}$$

Therefore there are L zeros on the unit circle $|z| = 1$. Similarly, the poles of $H(z)$ are determined by the roots of the denominator $z^{L-1}(z - 1)$. Thus there are $L - 1$ poles at the origin $z = 0$ and one pole at $z = 1$. A pole–zero diagram of $H(z)$ given in (4.3.13) for $L = 8$ on the complex plane is illustrated in Figure 4.7. The pole–zero diagram provides an insight into the properties of a given LTI system.

Describing the z-transform $H(z)$ in terms of its poles and zeros will require finding the roots of the denominator and numerator polynomials. For higher-order polynomials, finding the roots is a difficult task. To find poles and zeros of a rational function $H(z)$, we can use the MATLAB function `roots` on both the numerator and denominator polynomials. Another useful MATLAB function for analyzing transfer function is `zplane(b, a)`, which displays the pole–zero diagram of $H(z)$.

Example 4.12: Consider the IIR filter with the transfer function

$$H(z) = \frac{1}{1 - z^{-1} + 0.9z^{-2}}.$$

We can plot the pole–zero diagram using the following MATLAB script:

```
b =[1]; a =[1, -1, 0.9];
zplane (b, a);
```

Similarly, we can plot Figure 4.7 using the following MATLAB script:

```
b =[1, 0, 0, 0, 0, 0, 0, 0, 1]; a =[1, -1];
zplane (b, a);
```

As shown in Figure 4.7, the system has a single pole at $z = 1$, which is at the same location as one of the eight zeros. This pole is canceled by the zero at $z = 1$. In this case, the pole–zero cancellation occurs in the system transfer function itself. Since the system

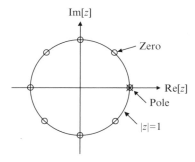

Figure 4.7   Pole–zero diagram of the moving-averaging filter, $L = 8$

output $Y(z) = X(z)H(z)$, the pole–zero cancelation may occur in the product of system transfer function $H(z)$ with the z-transform of the input signal $X(z)$. By proper selection of the zeros of the system transfer function, it is possible to suppress one or more poles of the input signal from the output of the system, or vice versa. When the zero is located very close to the pole but not exactly at the same location to cancel the pole, the system response has a very small amplitude.

The portion of the output $y(n)$ that is due to the poles of $X(z)$ is called the forced response of the system. The portion of the output that is due to the poles of $H(z)$ is called the natural response. If a system has all its poles within the unit circle, then its natural response dies down as $n \to \infty$, and this is referred to as the transient response. If the input to such a system is a periodic signal, then the corresponding forced response is called the steady-state response.

Consider the recursive power estimator given in (3.2.11) as an LTI system $H(z)$ with input $w(n) = x^2(n)$ and output $y(n) = \hat{P}_x(n)$. As illustrated in Figure 4.8, Equation (3.2.11) can be rewritten as

$$y(n) = (1 - \alpha)y(n - 1) + \alpha w(n).$$

Taking the z-transform of both sides, we obtain the transfer function that describes this efficient power estimator as

$$H(z) = \frac{Y(z)}{W(z)} = \frac{\alpha}{1 - (1 - \alpha)z^{-1}}. \tag{4.3.15}$$

This is a simple first-order IIR filter with a zero at the origin and a pole at $z = 1 - \alpha$. A pole–zero plot of $H(z)$ given in (4.3.15) is illustrated in Figure 4.9. Note that $\alpha = 1/L$ results in $1 - \alpha = (L - 1)/L$, which is slightly less than 1. When L is large, i.e., a longer window, the pole is closer to the unit circle.

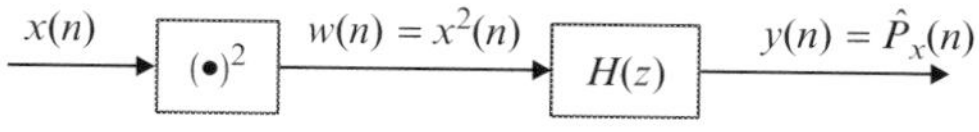

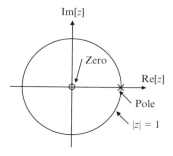

Figure 4.8 Block diagram of recursive power estimator

Figure 4.9 Pole–zero diagram of the recursive power estimator

An LTI system $H(z)$ is stable if and only if all the poles are inside the unit circle. That is,

$$|p_m| < 1 \quad \text{for all } m. \tag{4.3.16}$$

In this case, $\lim_{n \to \infty} \{h(n)\} = 0$. In other words, an LTI system is stable if and only if the unit circle is inside the ROC of $H(z)$.

Example 4.13: Given an LTI system with transfer function

$$H(z) = \frac{z}{z - a}.$$

There is a pole at $z = a$. From Table 4.3, we show that

$$h(n) = a^n, \quad n \geq 0.$$

When $|a| > 1$, i.e., the pole at $z = a$ is outside the unit circle, we have

$$\lim_{n \to \infty} h(n) \to \infty.$$

that is an unstable system. However, when $|a| < 1$, the pole is inside the unit circle, we have

$$\lim_{n \to \infty} h(n) \to 0,$$

which is a stable system.

The power estimator described in (4.3.15) is stable since the pole at $1 - \alpha = (L - 1)/L < 1$ is inside the unit circle. A system is unstable if $H(z)$ has pole(s) outside the unit circle or multiple-order pole(s) on the unit circle. For example, if $H(z) = z/(z - 1)^2$, then $h(n) = n$, which is unstable. A system is marginally stable, or oscillatory bounded, if $H(z)$ has first-order pole(s) that lie on the unit circle. For example, if $H(z) = z/(z + 1)$, then $h(n) = (-1)^n, n \geq 0$.

4.3.4 Frequency Responses

The frequency response of a digital system can be readily obtained from its transfer function. If we set $z = e^{j\omega}$ in $H(z)$, we have

$$H(z)\big|_{z = e^{j\omega}} = \sum_{n=-\infty}^{\infty} h(n)z^{-n}\big|_{z = e^{j\omega}} = \sum_{n=-\infty}^{\infty} h(n)e^{-j\omega n} = H(\omega). \tag{4.3.17}$$

Thus the frequency response of the system is obtained by evaluating the transfer function on the unit circle $|z| = |e^{j\omega}| = 1$. As summarized in Table 3.1, the digital frequency $\omega = 2\pi f/f_s$ is in the range $-\pi \leq \omega \leq \pi$.

The characteristics of the system can be described using the frequency response of the frequency ω. In general, $H(\omega)$ is a complex-valued function. It can be expressed in polar form as

$$H(\omega) = |H(\omega)|e^{j\phi(\omega)}, \tag{4.3.18}$$

where $|H(\omega)|$ is the magnitude (or amplitude) response and $\phi(\omega)$ is the phase shift (phase response) of the system at frequency ω. The magnitude response $|H(\omega)|$ is an even function of ω, and the phase response $\phi(\omega)$ is an odd function of ω. We only need to know that these two functions are in the frequency region $0 \leq \omega \leq \pi$. The quantity $|H(\omega)|^2$ is referred to as the squared-magnitude response. The value of $|H(\omega_0)|$ for a given $H(\omega)$ is called the system gain at frequency ω_0.

Example 4.14: The simple moving-average filter expressed as

$$y(n) = \frac{1}{2}[x(n) + x(n-1)], \quad n \geq 0$$

is a first-order FIR filter. Taking the z-transform of both sides and re-arranging the terms, we obtain

$$H(z) = \frac{1}{2}(1 + z^{-1}).$$

From (4.3.17), we have

$$H(\omega) = \frac{1}{2}(1 + e^{-j\omega}) = \frac{1}{2}(1 + \cos\omega - j\sin\omega),$$

$$|H(\omega)|^2 = \{\mathrm{Re}[H(\omega)]\}^2 + \{\mathrm{Im}[H(\omega)]\}^2 = \frac{1}{2}(1 + \cos\omega),$$

$$\phi(\omega) = \tan^{-1}\left\{\frac{\mathrm{Im}[H(\omega)]}{\mathrm{Re}[H(\omega)]}\right\} = \tan^{-1}\left(\frac{-\sin\omega}{1 + \cos\omega}\right).$$

From Appendix A.1, we have

$$\sin\omega = 2\sin\left(\frac{\omega}{2}\right)\cos\left(\frac{\omega}{2}\right) \quad \text{and} \quad \cos\omega = 2\cos^2\left(\frac{\omega}{2}\right) - 1.$$

Therefore the phase response is

$$\phi(\omega) = \tan^{-1}\left[-\tan\left(\frac{\omega}{2}\right)\right] = -\frac{\omega}{2}.$$

As discussed earlier, MATLAB is an excellent tool for analyzing signals in the frequency domain. For a given transfer function, $H(z)$, expressed in a general form in (4.3.10), the frequency response can be analyzed with the MATLAB function

```
[H, w] = freqz (b, a, N);
```

which returns the N-point frequency vector w and the N-point complex frequency response vector H, given its numerator and denominator coefficients in vectors b and a, respectively.

Example 4.15: Consider the difference equation of IIR filter defined as

$$y(n) = x(n) + y(n-1) - 0.9y(n-2). \qquad (4.3.19a)$$

This is equivalent to the IIR filter with the transfer function

$$H(z) = \frac{1}{1 - z^{-1} + 0.9z^{-2}}. \qquad (4.3.19b)$$

The MATLAB script to analyze the magnitude and phase responses of this IIR filter is listed (exam 4_15.m in the software package) as follows:

```
b = [1]; a = [1, -1, 0.9];
[H, w] = freqz(b, a, 128);
magH = abs(H); angH = angle(H);
subplot(2, 1, 1), plot(magH), subplot(2, 1, 2), plot(angH);
```

The MATLAB function abs(H) returns the absolute value of the elements of H and angle(H) returns the phase angles in radians.

A simple, but useful, method of obtaining the brief frequency response of an LTI system is based on the geometric evaluation of its pole–zero diagram. For example, consider a second-order IIR filter expressed as

$$H(z) = \frac{b_0 + b_1 z^{-1} + b_2 z^{-2}}{1 + a_1 z^{-1} + a_2 z^{-2}} = \frac{b_0 z^2 + b_1 z + b_2}{z^2 + a_1 z + a_2}. \qquad (4.3.20)$$

The roots of the characteristic equation

$$z^2 + a_1 z + a_2 = 0 \qquad (4.3.21)$$

are the poles of the filter, which may be either real or complex. For complex poles,

$$p_1 = re^{j\theta} \quad \text{and} \quad p_2 = re^{-j\theta}, \qquad (4.3.22)$$

where r is radius of the pole and θ is the angle of the pole. Therefore Equation (4.3.20) becomes

$$\left(z - re^{j\theta}\right)\left(z - re^{-j\theta}\right) = z^2 - 2r\cos\theta + r^2 = 0. \qquad (4.3.23)$$

Comparing this equation with (4.3.21), we have

$$r = \sqrt{a_2} \quad \text{and} \quad \theta = \cos^{-1}(-a_1/2r). \qquad (4.3.24)$$

The filter behaves as a digital resonator for r close to unity. The system with a pair of complex-conjugated poles as given in (4.3.22) is illustrated in Figure 4.10.

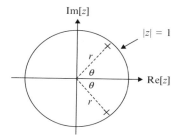

Figure 4.10 A second-order IIR filter with complex-conjugated poles

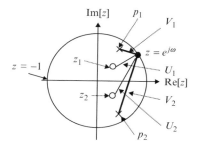

Figure 4.11 Geometric evaluation of the magnitude response from the pole–zero diagram

Similarly, we can obtain two zeros, z_1 and z_2, by evaluating $b_0 z^2 + b_1 z + b_2 = 0$. Thus the transfer function defined in (4.3.20) can be expressed as

$$H(z) = \frac{b_0(z - z_1)(z - z_2)}{(z - p_1)(z - p_2)}.$$ (4.3.25)

In this case, the frequency response is given by

$$H(\omega) = \frac{b_0(e^{j\omega} - z_1)(e^{j\omega} - z_2)}{(e^{j\omega} - p_1)(e^{j\omega} - p_2)}.$$ (4.3.26)

Assuming that $b_0 = 1$, the magnitude response of the system can be shown as

$$|H(\omega)| = \frac{U_1 U_2}{V_1 V_2},$$ (4.3.27)

where U_1 and U_2 represent the distances from the zeros z_1 and z_2 to the point $z = e^{j\omega}$, and V_1 and V_2 are the distances of the poles p_1 and p_2, to the same point as illustrated in Figure 4.11. The complete magnitude response can be obtained by evaluating $|H(\omega)|$ as the point z moves from $z = 0$ to $z = -1$ on the unit circle. As the point z moves closer to the pole p_1, the length of the vector V_1 decreases, and the magnitude response increases.

When the pole p_1 is close to the unit circle, V_1 becomes very small when z is on the same radial line with pole p_1 ($\omega = \theta$). The magnitude response has a peak at this resonant frequency. The closer r is to the unity, the sharper the peak. The digital resonator is an elementary bandpass filter with its passband centered at the resonant frequency θ. On the other hand, as the point z moves closer to the zero z_1, the zero vector U_1 decreases as does the magnitude response. The magnitude response exhibits a peak at the pole angle, whereas the magnitude response falls to the valley at the zero.

4.4   Discrete Fourier Transform

In Section 4.1, we developed the Fourier series representation for continuous-time periodic signals and the Fourier transform for finite-energy aperiodic signals. In this section, we will repeat similar developments for discrete-time signals. The discrete-time signals to be represented in practice are of finite duration. An alternative transformation called the discrete Fourier transform (DFT) for a finite-length signal, which is discrete in frequency, also will be introduced in this section.

4.4.1   Discrete-Time Fourier Series and Transform

As discussed in Section 4.1, the Fourier series representation of an analog periodic signal of period T_0 consists of an infinite number of frequency components, where the frequency spacing between two successive harmonics is $1/T_0$. However, as discussed in Chapter 3, the frequency range for discrete-time signals is defined over the interval $(-\pi, \pi)$. A periodic digital signal of fundamental period N samples consists of frequency components separated by $2\pi/N$ radians, or $1/N$ cycles. Therefore the Fourier series representation of the discrete-time signal will contain up to a maximum of N frequency components.

Similar to (4.1.1), given a periodic signal $x(n)$ with period N such that $x(n) = x(n - N)$, the Fourier series representation of $x(n)$ is expressed as

$$x(n) = \sum_{k=0}^{N-1} c_k e^{jk(2\pi/N)n}, \qquad (4.4.1)$$

which consists of N harmonically related exponentials functions $e^{jk(2\pi/N)n}$ for $k = 0, 1, \ldots, N - 1$. The Fourier coefficients, c_k, are defined as

$$c_k = \frac{1}{N} \sum_{n=0}^{N-1} x(n) e^{-jk(2\pi/N)n}. \qquad (4.4.2)$$

These Fourier coefficients form a periodic sequence of fundamental period N such that

$$c_{k+iN} = c_k, \qquad (4.4.3)$$

where i is an integer. Thus the spectrum of a periodic signal with period N is a periodic sequence with the same period N. The single period with frequency index $k = 0, 1, \ldots, N - 1$ corresponds to the frequency range $0 \le f \le f_s$ or $0 \le F \le 1$.

Similar to the case of analog aperiodic signals, the frequency analysis of discrete-time aperiodic signals involves the Fourier transform of the time-domain signal. In previous sections, we have used the z-transform to obtain the frequency characteristics of discrete signals and systems. As shown in (4.3.17), the z-transform becomes the evaluation of the Fourier transform on the unit circle $z = e^{j\omega}$. Similar to (4.1.10), the Fourier transform of a discrete-time signal $x(n)$ is defined as

$$X(\omega) = \sum_{n=-\infty}^{\infty} x(n)e^{-j\omega n}. \tag{4.4.4}$$

This is called the discrete-time Fourier transform (DTFT) of the discrete-time signal $x(n)$.

It is clear that $X(\omega)$ is a complex-valued continuous function of frequency ω, and $X(\omega)$ is periodic with period 2π. That is,

$$X(\omega + 2\pi i) = X(\omega). \tag{4.4.5}$$

Thus the frequency range for a discrete-time signal is unique over the range $(-\pi, \pi)$ or $(0, 2\pi)$. For real-valued $x(n)$, $X(\omega)$ is complex-conjugate symmetric. That is,

$$X(-\omega) = X^*(\omega). \tag{4.4.6}$$

Similar to (4.1.9), the inverse discrete-time Fourier transform of $X(\omega)$ is given by

$$x(n) = \frac{1}{2\pi} \int_{-\pi}^{\pi} X(\omega)e^{j\omega n}d\omega. \tag{4.4.7}$$

Consider an LTI system $H(z)$ with input $x(n)$ and output $y(n)$. From (4.3.1) and letting $z = e^{j\omega}$, we can express the output spectrum of system in terms of its frequency response and the input spectrum. That is,

$$Y(\omega) = H(\omega)X(\omega), \tag{4.4.8}$$

where $X(\omega)$ and $Y(\omega)$ are the DTFT of the input $x(n)$ and output $y(n)$, respectively. Similar to (4.3.18), we can express $X(\omega)$ and $Y(\omega)$ as

$$X(\omega) = |X(\omega)|e^{j\phi_x(\omega)}, \tag{4.4.9a}$$

$$Y(\omega) = |Y(\omega)|e^{j\phi_y(\omega)}. \tag{4.4.9b}$$

Substituting (4.3.18) and (4.4.9) into (4.4.8), we have

$$|Y(\omega)|e^{j\phi_y(\omega)} = |H(\omega)||X(\omega)|e^{j[\phi(\omega)+\phi_x(\omega)]}. \tag{4.4.10}$$

This equation shows that

$$|Y(\omega)| = |H(\omega)||X(\omega)| \qquad (4.4.11)$$

and

$$\phi_y(\omega) = \phi(\omega) + \phi_x(\omega). \qquad (4.4.12)$$

Therefore the output magnitude spectrum $|Y(\omega)|$ is the product of the magnitude response $|H(\omega)|$ and the input magnitude spectrum $|X(\omega)|$. The output phase spectrum $\phi_y(\omega)$ is the sum of the system phase response $\phi(\omega)$ and the input phase spectrum $\phi_x(\omega)$.

For example, if the input signal is the sinusoidal signal at frequency ω_0 expressed as

$$x(n) = A\cos(\omega_0 n), \qquad (4.4.13)$$

its steady-state response can be expressed as

$$y(n) = A|H(\omega_0)|\cos[\omega_0 + \phi(\omega_0)], \qquad (4.4.14)$$

where $|H(\omega_0)|$ is the system amplitude gain at frequency ω_0 and $\phi(\omega_0)$ is the phase shift of the system at frequency ω_0. Therefore it is clear that the sinusoidal steady-state response has the same frequency as the input, but its amplitude and phase angle are determined by the system's magnitude response $|H(\omega)|$ and phase response $\phi(\omega)$ at any given frequency ω_0.

4.4.2 Aliasing and Folding

As discussed in Section 4.1, let $x(t)$ be an analog signal, and let $X(f)$ be its Fourier transform, defined as

$$X(f) = \int_{-\infty}^{\infty} x(t)e^{-j2\pi ft}dt, \qquad (4.4.15)$$

where f is the frequency in Hz. The sampling of $x(t)$ with sampling period T yields the discrete-time signal $x(n)$. Similar to (4.4.4), the DTFT of $x(n)$ can be expressed as

$$X(F) = \sum_{n=-\infty}^{\infty} x(n)e^{-j2\pi Fn}. \qquad (4.4.16)$$

The periodic sampling imposes a relationship between the independent variables t and n in the signals $x(t)$ and $x(n)$ as

$$t = nT = \frac{n}{f_s}. \qquad (4.4.17)$$

This relationship in time domain implies a corresponding relationship between the frequency variable f and F in $X(f)$ and $X(F)$, respectively. Note that $F = f/f_s$ is the normalized digital frequency defined in (3.1.8). It can be shown that

$$X(F) = \frac{1}{T} \sum_{k=-\infty}^{\infty} X(f - kf_s). \qquad (4.4.18)$$

This equation states that $X(F)$ is the sum of all repeated values of $X(f)$, scaled by $1/T$, and then frequency shifted to kf_s. It also states that $X(F)$ is a periodic function with period $T = 1/f_s$. This periodicity is necessary because the spectrum $X(F)$ of the discrete-time signal $x(n)$ is periodic with period $F = 1$ or $f = f_s$. Assume that a continuous-time signal $x(t)$ is bandlimited to f_M, i.e.,

$$|X(f)| = 0 \quad \text{for} \quad |f| \geq f_M, \qquad (4.4.19)$$

where f_M is the bandwidth of signal $x(t)$. The spectrum is 0 for $|f| \geq f_M$ as shown in Figure 4.12(a).

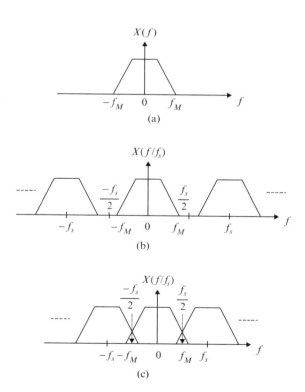

Figure 4.12 Spectrum replication caused by sampling: (a) spectrum of analog bandlimited signal $x(t)$, (b) sampling theorem is satisfied, and (c) overlap of spectral components

The effect of sampling is that it extends the spectrum of $X(f)$ repeatedly on both sides of the f-axis, as shown in Figure 4.12. When the sampling rate f_s is selected to be greater than $2f_M$, i.e., if $f_M \leq f_s/2$, the spectrum $X(f)$ is preserved in $X(F)$ as shown in Figure 4.12(b). Therefore when $f_s \geq 2f_M$, we have

$$X(F) = \frac{1}{T}X(f) \quad \text{for } |F| \leq \frac{1}{2} \quad \text{or} \quad |f| \leq f_N, \tag{4.4.20}$$

where $f_N = f_s/2$ is called the Nyquist frequency. In this case, there is no aliasing, and the spectrum of the discrete-time signal is identical (within the scale factor $1/T$) to the spectrum of the analog signal within the fundamental frequency range $|f| \leq f_N$ or $|F| \leq 1/2$. The analog signal $x(t)$ can be recovered from the discrete-time signal $x(n)$ by passing it through an ideal lowpass filter with bandwidth f_M and gain T. This fundamental result is the sampling theorem defined in (1.2.3). This sampling theorem states that a bandlimited analog signal $x(t)$ with its highest frequency (bandwidth) being f_M can be uniquely recovered from its digital samples, $x(n)$, provided that the sampling rate $f_s \geq 2f_M$.

However, if the sampling rate is selected such that $f_s < 2f_M$, the shifted replicas of $X(f)$ will overlap in $X(F)$, as shown in Figure 4.12(c). This phenomenon is called aliasing, since the frequency components in the overlapped region are corrupted when the signal is converted back to the analog form. As discussed in Section 1.1, we used an analog lowpass filter with cut-off frequency less than f_N before the A/D converter in order to prevent aliasing. The goal of filtering is to remove signal components that may corrupt desired signal components below f_N. Thus the lowpass filter is called the antialiasing filter.

Consider two sinewaves of frequencies $f_1 = 1\,\text{Hz}$ and $f_2 = 5\,\text{Hz}$ that are sampled at $f_s = 4\,\text{Hz}$, rather than at 10 Hz according to the sampling theorem. The analog waveforms are illustrated in Figure 4.13(a), while their digital samples and reconstructed waveforms are illustrated in Figure 4.13(b). As shown in the figures, we can reconstruct the original waveform from the digital samples for the sinewave of frequency $f_1 = 1\,\text{Hz}$. However, for the original sinewave of frequency $f_2 = 5\,\text{Hz}$, the reconstructed signal is identical to the sinewave of frequency 1 Hz. Therefore f_1 and f_2 are said to be aliased to one another, i.e., they cannot be distinguished by their discrete-time samples.

In general, the aliasing frequency f_2 related to f_1 for a given sampling frequency f_s can be expressed as

$$f_2 = if_s \pm f_1, \quad i \geq 1. \tag{4.4.21}$$

For example, if $f_1 = 1\,\text{Hz}$ and $f_s = 4\,\text{Hz}$, the set of aliased f_2 corresponding to f_1 is given by

$$f_2 = i \cdot 4 \pm 1, \quad i = 1, 2, 3, \ldots$$
$$= 3, 5, 7, 9, \ldots \tag{4.4.22}$$

The folding phenomenon can be illustrated as the aliasing diagram shown in Figure 4.14. From the aliasing diagram, it is apparent that when aliasing occurs, aliasing frequencies in $x(t)$ that are higher than f_N will fold over into the region $0 \leq f \leq f_N$.

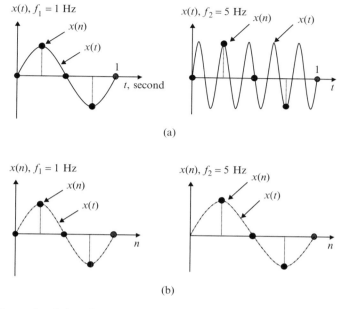

Figure 4.13 Example of the aliasing phenomenon: (a) original analog waveforms and digital samples for $f_1 = 1\,\text{Hz}$ and $f_2 = 5\,\text{Hz}$, and (b) digital samples of $f_1 = 1\,\text{Hz}$ and $f_2 = 5\,\text{Hz}$ and reconstructed waveforms

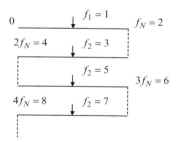

Figure 4.14 An example of aliasing diagram for $f_1 = 1\,\text{Hz}$ and $f_s = 4\,\text{Hz}$

4.4.3 Discrete Fourier Transform

To perform frequency analysis of a discrete-time signal $x(n)$, we convert the time-domain signal into frequency domain using the DTFT defined in (4.4.4). However, $X(\omega)$ is a continuous function of frequency, and it also requires an infinite number of time-domain samples $x(n)$ for calculation. Thus the DTFT is not computationally feasible using DSP hardware. In this section, we briefly introduce the discrete Fourier transform (DFT) of the finite-duration sequence $x(n)$ by sampling its spectrum $X(\omega)$. We will further discuss DFT in detail in Chapter 7.

The discrete Fourier transform for transforming a time-domain signal $\{x(0), x(1), x(2), \ldots, x(N-1)\}$ into frequency-domain samples $\{X(k)\}$ of length N is expressed as

$$X(k) = \sum_{n=0}^{N-1} x(n) e^{-j\left(\frac{2\pi}{N}\right)kn}, \quad k = 0, 1, \ldots, N-1, \qquad (4.4.23)$$

where n is the time index, k is the frequency index, and $X(k)$ is the kth DFT coefficient. The inverse discrete Fourier transform (IDFT) is defined as

$$x(n) = \frac{1}{N} \sum_{k=0}^{N-1} X(k) e^{j\left(\frac{2\pi}{N}\right)kn}, \quad n = 0, 1, \ldots, N-1. \qquad (4.4.24)$$

Equation (4.4.23) is called the analysis equation for calculating the spectrum from the signal, and (4.4.24) is called the synthesis equation used to reconstruct the signal from its spectrum. This pair of DFT and IDFT equations holds for any discrete-time signal that is periodic with period N.

When we define the twiddle factor as

$$W_N = e^{-j\left(\frac{2\pi}{N}\right)}, \qquad (4.4.25)$$

the DFT defined in (4.4.23) can be expressed as

$$X(k) = \sum_{n=0}^{N-1} x(n) W_N^{kn}, \quad k = 0, 1, \ldots, N-1. \qquad (4.4.26)$$

Similarly, the IDFT defined in (4.4.24) can be expressed as

$$x(n) = \frac{1}{N} \sum_{k=0}^{N-1} X(k) W_N^{-kn}, \quad n = 0, 1, \ldots, N-1. \qquad (4.4.27)$$

Note that W_N is the Nth root of unity since $e^{-j2\pi} = 1$. Because the W_N^{kn} are N-periodic, the DFT coefficients are N-periodic. The scalar $1/N$ that appears in the IDFT in (4.4.24) does not appear in the DFT. However, if we had chosen to define the DFT with the scalar $1/N$, it would not have appeared in the IDFT. Both forms of these definitions are equivalent.

Example 4.16: In this example, we develop a user-written MATALB function to implement DFT computations. MATLAB supports complex numbers indicated by the special functions i and j.

Consider the following M-file (dft.m in the software package):

```
function [Xk] = dft(xn, N)
% Discrete Fourier transform function
%    [Xk] = dft(xn, N)
% where xn is the time-domain signal x(n)
```

```
%      N is the length of sequence
%      Xk is the frequency-domain X(k)
n = [0:1:N-1];
k = [0:1:N-1];
WN = exp(-j*2*pi/N);    % Twiddle factor
nk = n'*k;              % N by N matrix
WNnk = WN.^nk;          % Twiddle factor matrix
Xk = xn*WNnk;           % DFT
```

In this M-file, the special character ' (prime or apostrophe) denotes the transpose of a matrix. The exam4_16.m (included in the software package) with the following statements:

```
n = [0:127]; N = 128;
xn = 1.5*sin(0.2*pi*n+0.25*pi);
Xk = dft(xn, N);
semilogy(abs(Xk));
axis([0 63 0 120]);
```

will display the magnitude spectrum of sinewave $x(n)$ in logarithmic scale, and the x-axis shows only the range from 0 to π.

4.4.4 Fast Fourier Transform

The DFT and IDFT play an important role in many DSP applications including linear filtering, correlation analysis, and spectrum analysis. To compute one of the $X(k)$ coefficients in (4.4.23), we need N complex multiplications and $N-1$ complex additions. To generate N coefficients, we need N^2 multiplications and $N^2 - N$ additions. The DFT can be manipulated to obtain a very efficient algorithm to compute it. Efficient algorithms for computing the DFT are called the fast Fourier transform (FFT) algorithms, which require a number of operations proportional to $N \log_2 N$ rather than N^2. The development, implementation, and application of FFT will be further discussed in Chapter 7.

MATLAB provides the built-in function fft(x), or fft(x, N) to compute the DFT of the signal vector x. If the argument N is omitted, then the length of the DFT is the length of x. When the sequence length is a power of 2, a high-speed radix-2 FFT algorithm is employed. The MATLAB function fft(x, N) performs N-point FFT. If the length of x is less than N, then x is padded with zeros at the end. If the length of x is greater than N, fft truncates the sequence x and performs DFT of the first N samples only. MATLAB also provides ifft(x) to compute the IDFT of the vector x, and ifft(x, N) to calculate the N-point IDFT.

The function fft(x, N) generates N DFT coefficients $X(k)$ for $k = 0, 1, \ldots N - 1$. The Nyquist frequency ($f_N = f_s/2$) corresponds to the frequency index $k = N/2$. The frequency resolution of the N-point DFT is

$$\Delta = \frac{f_s}{N}. \qquad (4.4.28)$$

The frequency f_k (in Hz) corresponding to the index k can be computed by

$$f_k = k\Delta = \frac{kf_s}{N}, \quad k = 0, 1, \ldots, N - 1. \tag{4.4.29}$$

Since the magnitude spectrum $|X(k)|$ is an even function of k, we only need to display the spectrum for $0 \leq k \leq N/2$ (or $0 \leq \omega_k \leq \pi$).

Example 4.17: By considering the sinewave given in Example 3.1, we can generate the time-domain signal and show the magnitude spectrum of signal by using the following MATLAB script (`exam4_17.m` in the software package):

```
N = 256;
n = [0:N−1];
omega = 0.25*pi;
xn = 2*sin(omega*n);
Xk = fft(xn, N);          % Perform FFT
absXk = abs(Xk);          % Compute magnitude spectrum
plot(absXk(1:(N/2)));     % Plot from 0 to π
```

The phase response can be obtained using the MATLAB function `phase = angle(Xk)`, which returns the phase angles in radians of the elements of complex vector `Xk`.

4.5 Applications

In this section, we will introduce two examples of using frequency analysis techniques for designing simple notch filters and analyzing room acoustics.

4.5.1 Design of Simple Notch Filters

A notch filter contains one or more deep notches (nulls) in its magnitude response. To create a null in the frequency response at frequency ω_0, we simply introduce a pair of complex-conjugate zeros on the unit circle at angle ω_0. That is, at

$$z = e^{\pm j\omega_0}. \tag{4.5.1}$$

Thus the transfer function for an FIR notch filter is

$$\begin{aligned} H(z) &= (1 - e^{j\omega_0}z^{-1})(1 - e^{-j\omega_0}z^{-1}) \\ &= 1 - 2\cos(\omega_0)z^{-1} + z^{-2}. \end{aligned} \tag{4.5.2}$$

This is the FIR filter of order 2.

The magnitude response of a notch filter described in (4.5.2) having a null at $\omega_0 = 0.2\pi$ can be obtained using the MATLAB script `notch.m` given in the software package. The magnitude response of the designed notch filter is illustrated in Figure 4.15.

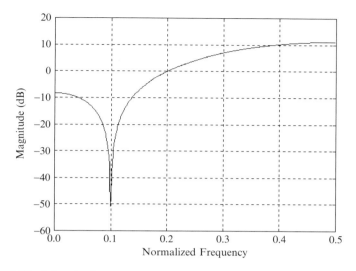

Figure 4.15   Magnitude response of a notch filter with zeros only for $\omega_0 = 0.2\pi$

Obviously, the second-order FIR notch filter has a relatively wide bandwidth, which means that other frequency components around the null are severely attenuated. To reduce the bandwidth of the null, we may introduce poles into the system. Suppose that we place a pair of complex-conjugate poles at

$$z_p = re^{\pm j\theta_0}, \tag{4.5.3}$$

where r and θ_0 are radius and angle of poles, respectively. The transfer function for the resulting filter is

$$
\begin{aligned}
H(z) &= \frac{(1 - e^{j\omega_0}z^{-1})(1 - e^{-j\omega_0}z^{-1})}{(1 - re^{j\theta_0}z^{-1})(1 - re^{-j\theta_0}z^{-1})} \\
&= \frac{1 - 2\cos(\omega_0)z^{-1} + z^{-2}}{1 - 2r\cos(\theta_0)z^{-1} + r^2z^{-2}}
\end{aligned}
\tag{4.5.4}
$$

The notch filter expressed in (4.5.4) is the second-order IIR filter.

The magnitude response of the filter defined by (4.5.4) is plotted in Figure 4.16 for $\omega_0 = \theta_0 = 0.2\pi$, and $r = 0.85$. When compared with the magnitude response of the FIR filter shown in Figure 4.15, we note that the effect of the pole is to reduce the bandwidth of the notch. The MATLAB script notch1.m in the software package is used to generate Figure 4.16.

Let θ_0 be fixed ($\theta_0 = \omega_0 = 0.2\pi$) and the value of r changed. The magnitude responses of the filter are shown in Figure 4.17 for $r = 0.75, 0.85$, and 0.95. Obviously, the closer the r value to 1 (poles are closer to the unit circle), the narrower the bandwidth. The MATLAB script notch2.m used to generate Figure 4.17 is available in the software package.

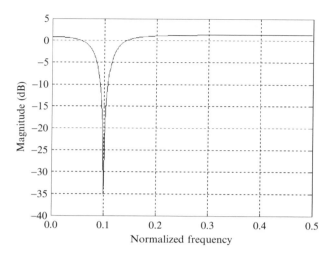

Figure 4.16 Magnitude response of a notch filter with zeros and poles, $\omega_0 = \theta_0 = 0.2\pi$ and $r = 0.85$

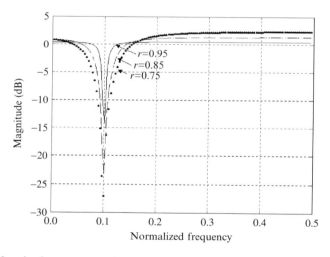

Figure 4.17 Magnitude response of notch filter with both zeros and poles, $\omega_0 = \theta_0 = 0.2\pi$ and different values of r

4.5.2 Analysis of Room Acoustics

A room transfer function (RTF) expresses the transmission characterstics of a sound between a source (loudspeaker) and a receiver (microphone) in a room, as illustrated in Figure 4.18, where $H(z)$ denotes the RTF. The output signal of the receiver can be expressed in the z-domain as

$$Y(z) = H(z)X(z).$$

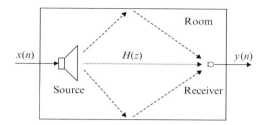

Figure 4.18   RTF between a source and a receiver in a room

The RTF includes the characterstics of the direct sound and all reflected sounds (reverberations) in the room.

An efficient model is required to represent the RTF with a few parameters for reducing memory and computation requirements. The first method for modeling an RTF is an all-zero model as defined in (4.3.8), with the coefficients corresponding to the impulse response of the RTF in the time domain. The all-zero model can be realized with an FIR filter. When the reverberation time is 500 ms, the FIR filter needs 4000 coefficients (at 8 kHz sampling rate) to represent the RTF. Furthermore, the RTF varies due to changes in the source and receiver positions.

The pole–zero model defined in (4.3.10) can also be used to model RTF. From a physical point of view, poles represent resonances, and zeros represent time delays and anti-resonances. Because the poles can represent a long impulse response caused by resonances with fewer parameters than the zeros, the pole–zero model seems to match a physical RTF better than the all-zero model. Because the acoustic poles corresponding to the resonance properties are invariant, the pole–zero model that has constant poles and variable zeros is cost effective.

It is also possible to use an all-pole modeling of room responses to reduce the equalizer length. The all-pole model of RTF can be expressed as

$$H(z) = \frac{1}{1 + A(z)} = \frac{1}{1 + \sum_{m=1}^{M} a_m z^{-m}}. \tag{4.5.5}$$

Acoustic poles correspond to the resonances of a room and do not change even if the source and receiver positions change or people move. This technique can be applied to dereverberation of recorded signals, acoustic echo cancellation, etc. In this section, we show how the MATLAB functions are used to model and analyze the room acoustics.

To evaluate the room transfer function, impulse responses of a rectangular room ($246 \times 143 \times 111$ cubic inches) were measured using the maximum-length sequence technique. The original data is sampled at 48 kHz, which is then bandlimited to 100–400 Hz and decimated to 1 kHz. An example of room impulse response is shown in Figure 4.19, which is generated using the following MATLAB script:

```
load imp.dat;
plot(imp(1:1000)), title('Room impulse response');
xlabel('Time' ), ylabel('Amplitude');
```

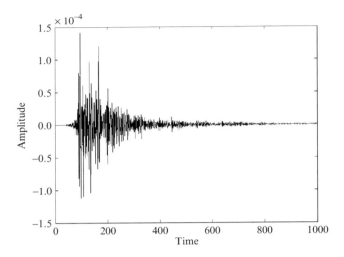

Figure 4.19   Example of a measure room impulse response

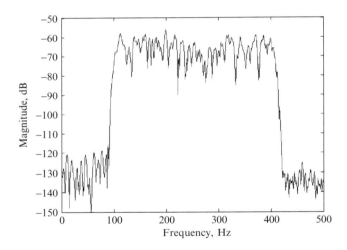

Figure 4.20   Magnitude response of measured RTF

where the room impulse response samples are stored in the ASCII file `imp.dat`. Both the MATLAB script `imprtf.m` and the data file `imp.dat` are included in the software package.

We can easily evaluate the magnitude response of the room transfer function using the MATLAB script `magrtf.m` available in the software package. The magnitude response is shown in Figure 4.20.

MATLAB provides a powerful function $a = \text{lpc}(x, N)$ to estimate the coefficients a_m of an Mth-order all-pole IIR filter. A user-written MATLAB function `all_pole.m` that shows the magnitude responses of the measured and modeled RTF is given in the software package. This MATLAB function can be invoked by using the following commands:

```
load imp.dat;
all_pole(imp, 120);
```

The impulse response of the RTF is modeled by the all-pole model defined in (4.5.5) by using the MATLAB function a = lpc(imp_x, pole_number), where the pole number M is selected as 120. In order to evaluate the accuracy of the model, the MATLAB function freqz(1, a, leng, 1000) is used to compute the frequency response of the estimated model. The magnitude response of the RTF model is then compared with the measured magnitude response of RTF from the measured room impulse response. It is shown that the all-pole model matches the peaks better than the valleys. Note that the higher the model order M, the better the model match can be obtained.

A pole–zero model for the RTF can be estimated by using Prony's method as follows:

```
[b, a] = prony(imp_x, nb, na);
```

where b and a are vectors containing the estimated numerator and denominator coefficients, and nb and na are orders of numerator b and denominator a.

4.6 Experiments Using the TMS320C55X

In Section 4.4.3, we introduced the DFT and implemented it in the MATLAB function dft.m. The C program that implements an N-point DFT defined by Equation (4.4.26) is listed in Table 4.4. The computation of the DFT involves nested loops, multiplication of complex data samples, and generation of complex twiddle factors. In the subsequent experiments, we will write assembly routines to implement the DFT function.

Assuming we have a complex data sample $x(n) = x_r(n) + jx_i(n)$ and a complex twiddle factor $W_N^{kn} = \cos(2\pi kn/N) - j\sin(2\pi kn/N) = W_r - jW_i$ defined in (4.4.25), the product of $x(n)$ and W_N^{kn} can be expressed as

$$x(n) W_N^{kn} = x_r(n) W_r + x_i(n) W_i + j[x_i(n) W_r - x_r(n) W_i], \qquad (4.6.1)$$

where the subscripts r and i denote the real and imaginary parts of complex variable. Equation (4.6.1) can be rewritten as

$$X[n] = X_r[n] + jX_i[n] \qquad (4.6.2)$$

for $n = 0, 1, \ldots, N - 1$, where

$$X_r[n] = x_r(n) W_r + x_i(n) W_i, \qquad (4.6.3a)$$

$$X_i[n] = x_i(n) W_r - x_r(n) W_i. \qquad (4.6.3b)$$

The C program listed in Table 4.4 uses two arrays, Xin[2*N] and Xout[2*N], to represent the complex (real and imaginary) input and output data samples. The input samples for the experiment are generated using the MATLAB script listed in Table 4.5.

Table 4.4 List of dft() in C

```c
#define PI 3.1415926536
void dft(float Xin[], float Xout[])
{
    int n, k, j;
    float angle;
    float Xr[N], Xi[N];
    float Wn[2];
    for(k = 0; k < N; k++)
    {
      Xr[k] = 0;
      Xi[k] = 0;
      for(j = 0, n = 0; n < N; n++)
      {
        angle = (2.0*PI*k*n)/N;
        W[0] = cos(angle);
        W[1] = sin(angle);
        Xr[k] = Xr[k] + Xin[j]*W[0] + Xin[j+1]*W[1];
        Xi[k] = Xi[k] + Xin[j + 1]*W[0] - Xin[j]*W[1];
        j += 2;
      }
      Xout[n++] = Xr[k];
      Xout[n++] = Xi[k];
    }
}
```

Table 4.5 Generating signal using MATLAB

```matlab
fs = 8000;               % Sampling frequency in Hz
f1 = 500;                % 1st sinewave frequency in Hz
f2 = 1000;               % 2nd sinewave frequency in Hz
f3 = 2000;               % 3rd sinewave frequency in Hz
n = [0:127];             % n = 0, 1, ..., 127
w1 = 2*pi*f1/fs;
w2 = 2*pi*f2/fs;
w3 = 2*pi*f3/fs;
x = 0.3*sin(w1*n) + 0.3*sin(w2*n) + 0.3*sin(w3*n);
```

Table 4.5 (*continued*)

```
fid = fopen('input.dat', 'w' );   % Open file input.dat for write
fprintf(fid, 'int input[128] = {\n');
fprintf(fid, '%5d, \n', round(x(1:127)*32767));
fprintf(fid, '%5d}; \n', round(x(128)*32767));
fclose(fid);                       % Close file input.dat

fid = fopen('input.inc' , 'w' );   % Open file input.inc for write
fprintf(fid, '   .word %5d\n', round(x*32767));
fclose(fid);                       % Close file input.inc
```

This program generates 128 data samples. The data is then represented using the Q15 format for the experiments. They are stored in the data files input.dat and input.inc, and are included in the software package. The data file input.dat is used by the C program exp4a.c for Experiment 4A, and the data file input.inc is used by the assembly routine exp4b.asm for Experiment 4B.

4.6.1 Experiment 4A – Twiddle Factor Generation

The sine–cosine generator we implemented for the experiments in Chapter 3 can be used to generate the twiddle factors for comparing the DFT. Recall the assembly function sine_cos.asm developed in Section 3.8.5. This assembly routine is written as a C-callable function that follows the C55x C-calling convention. There are two arguments passed by the function as sine_cos(angle, Wn). The first argument is passed through the C55x temporary register T0 containing the input angle in radians. The second argument is passed by the auxiliary register AR0 as a pointer Wn to the memory locations, where the results of sin(angle) and cos(angle) will be stored upon return. The following C example shows how to use the assembly sine–cosine generator inside nested loops to generate the twiddle factors:

```
#define N 128
#define TWOPIN 0x7FFF≫6   /* 2πkn/N,  N = 128                    */
int n, k, angle;
int Wn[2];                 /* Wn[0] = cos(angle), Wn[1] = sin(angle) */
for(k = 0; k < N; k++)
{
   for(n = 0; n < N; n++)
   {
       angle = TWOPIN* k* n;
       sine_cos(angle, Wn);
   }
}
```

The assembly code that calls the subroutine `sine_cos` is listed as follows:

```
mov   *(angle), T0    ; Pass the first argument in T0
amov  #Wn,XAR0        ; Pass the second argument in AR0
call  sine_cos        ; Call sine_cos subroutine
```

In Chapter 2, we introduced how to write nested loops using block-repeat and single repeat instructions. Since the inner loop of the C code `dft()` contains multiple instructions, we will use the block-repeat instruction (`rptb`) to implement both of the inner and outer loops. The C55x has two block-repeat counters, registers BRC0 and BRC1. When implementing nested loops, the repeat counter BRC1 must be used as the inner-loop counter, while the BRC0 should be used as the outer-loop counter. Such an arrangement allows the C55x to automatically reload the inner-loop repeat counter BRC1 every time the outer-loop counter being updated. The following is an example of using BRC0 and BRC1 for nested loops N times:

```
mov  #N-1, BRC0
mov  #N-1, BRC1
rptb outer_loop-1
(more outer loop instructions...)
rptb inner_loop-1
(more inner loop instructions...)
inner_loop
(more outer loop instructions...)
outer_loop
```

The calculation of the angle depends on two variables, k and n, as

$$\text{angle} = (2\pi/N)kn. \tag{4.6.4}$$

As defined in Figure 3.23 of Section 3.8.5, the fixed-point representation of value π for sine–cosine generator is 0x7FFF. The angle used to generate the twiddle factors for DFT of $N = 128$ can be expressed as

$$\text{angle} = (2^*0x7FFF/128)^*k^*n = (0x1FF)^*k^*n, \tag{4.6.5}$$

where the inner-loop index $n = 0, 1, \ldots, N - 1$ and the outer-loop index $k = 0, 1, \ldots, N - 1$. The following is the assembly routine that calculates the angles for $N = 128$:

```
N            .set 128
TWOPIN    .set 0x7FFF ≫ 6    ; 2*PI/N, N = 128
        .bss Wn, 2             ; Wn[0] = Wr, Wn[1] = Wi
        .bss angle, 1          ; Angle for sine-cosine function
        mov  #N-1, BRC0        ; Repeat counter for outer-loop
        mov  #N-1, BRC1        ; Repeat counter for inner-loop
        mov  #0, T2            ; k = T2 = 0
```

```
        rptb outer_loop-1           ; for(k = 0; k < N;k++) {
        mov  #TWOPIN ≪#16, AC0      ; hi(AC0) = 2* PI/N
        mpy  T2, AC0
        mov  #0, * (angle)
        mov  AC0, T3                ; angle = 2* PI* k/N
        rptb inner_loop-1           ; for(n = 0 ; n < N;n++) {
        mov  * (angle), T0          ; T0 = 2* PI* k* n/N
        mov  * (angle), AC0
        add  T3, AC0
        mov  AC0, * (angle)         ; Update angle
        amov #Wn, XAR0              ; AR0 is the pointer to Wn
        call sine_cos               ; sine_cos(angle, Wn)
inner_loop
        add  #1, T2
outer_loop
```

4.6.2   Experiment 4B – Complex Data Operation

For the experiment, the complex data and twiddle factor vectors are arranged in order of the real and the imaginary pairs. That is, the input array $\text{Xin}[2N] = \{X_r, X_i, X_r, X_i, \ldots\}$ and the twiddle factor $[2] = \{W_r, W_i\}$. The computation of (4.6.3) is implemented in C as follows:

```
Xr[n] = 0;
Xi[n] = 0;
for(n = 0; n < N; n++)
{
    Xr[n] = Xr[n] + Xin[n] *Wr + Xin [n + 1] *Wi;
    Xi[n] = Xi[n] + Xin[n + 1] *Wr − Xin [n] *Wi;
}
```

The C55x assembly program implementation of (4.6.3) is listed as follows:

```
        mov     #0, AC2
        mov     #0, AC3
        rptb    inner_loop-1
        macm40  *AR5+, *AR0, AC2    ; Xin[n] *Wr
        macm40  *AR5−, *AR0+, AC3   ; Xin[n+1] *Wr
        masm40  *AR5+, *AR0, AC3    ; Xi[k] = Xin[n+1] *Wr − Xin [n] *Wi
        macm40  *AR5+, *AR0−, AC2   ; Xr[k] = Xin [n] *Wr + Xin[n+1] *Wi
inner_loop
```

Because the DFT function accumulates N intermediate results, the possible overflow during computation should be considered. The instruction `masm40` enables the use of accumulator guard bits that allow the intermediate multiply–accumulate result to be handled in a 40-bit accumulator. Finally, we can put all the pieces together to complete the routine, as listed in Table 4.6.

Table 4.6   List of DFT assembly routine `dft_128.asm`

```
; DFT_128 — 128-point DFT routine
;
; Entry T0:    AR0: pointer to complex input buffer
;              AR1: pointer to complex output buffer
; Return: None
;
         .def  dft_128
         .ref  sine_cos
N        .set  128
TWOPIN  .set  0x7fff ≫6          ; 2*PI/N, N = 128
         .bss  Wn, 2             ; Wn[0] = Wr, Wn[1] = Wi
         .bss  angle, 1          ; Angle for sine—cosine function
         .text
_dft_128
    pshboth XAR5                 ; Save AR5
    bset    SATD
    mov     #N−1, BRC0           ; Repeat counter for outer loop
    mov     #N−1, BRC1           ; Repeat counter for inner loop
    mov     XAR0, XAR5           ; AR5 pointer to sample buffer
    mov     XAR0, XAR3
    mov     #0, T2               ; k = T2 = 0
    rptb    outer_loop-1         ; for(k = 0; k < N; k++) {
    mov     XAR3, XAR5           ; Reset x[ ] pointer
    mov     #TWOPIN ≪#16, AC0    ; hi(AC0) = 2*PI/N
    mpy     T2, AC0
    mov     #0, AC2              ; Xr[k] = 0
    mov     #0, AC3              ; Xi[k] = 0
    mov     #0, *(angle)
    mov     AC0, T3              ; angle = 2*PI*k/N
    rptb    inner_loop-1         ; for(n = 0; n < N; n++) {
    mov     *(angle), T0         ; T0 = 2*PI*k*n/N
    mov     *(angle), AC0
    add     T3, AC0
    mov     AC0, *(angle)        ; Update angle
    amov    #Wn, XAR0            ; AR0 is the pointer to Wn
    call    _sine_cos            ; sine_cos(angle, Wn)
    bset    SATD                 ; sine_cos turn off FRCT & SATD
    macm40  *AR5+, *AR0, AC2     ; XR[k] + Xin[n]*Wr
    macm40  *AR5−, *AR0+, AC3    ; XI[k] + Xin[n + 1]*Wr
    masm40  *AR5+, *AR0, AC3     ; XI[k] + Xin[n + 1]*Wr − Xin[n]*Wi
    macm40  *AR5+, *AR0−, AC2    ; XR[k] + Xin[n]*Wr + Xin[n + 1]*Wi
inner_loop                      ; } end of inner loop
    mov     hi(AC2 ≪ #−5), *AR1+
    mov     hi(AC3 ≪ #−5), *AR1+
    add     #1, T2
outer_loop                      ; } end of outer loop
```

Table 4.6 (*continued*)

```
    popboth XAR5
    bclr    SATD
    ret
    .end
```

Table 4.7 List of C program `exp4a.c`

```c
/* Experiment 4A — exp4a.c */
#include "input.dat"
#define   N 128
extern void dft_128(int *, int *);
extern void mag_128(int *, int *);
int   Xin[2*N];
int   Xout[2*N];
int   Spectrum[N];
void main()
{
  int i, j;
  for(j = 0, i = 0; i < N; i++)
  {
    Xin[j++] = input[i];     /* Get real sample    */
    Xin[j++] = 0;            /* Imaginary sample = 0 */
  }
  dft_128(Xin, Xout);        /* DFT routine        */
  mag_128(Xout, Spectrum);   /* Compute spectrum   */
}
```

4.6.3 Experiment 4C – Implementation of DFT

We will complete the DFT routine of $N = 128$ and test it in this section. The C program listed in Table 4.7 calls for the assembly routine `dft_128()` to compute the 128-point DFT.

The data file, `input.dat`, is an ASCII file that contains 128 points of data sampled at 8 kHz, and is available in the software package. First, the program composes the complex input data array `Xin[2*N]` by zero-filling the imaginary parts. Then the DFT is carried out by the subroutine `dft_128()`. The 128 complex DFT samples are stored in the output data array `Xout[2*N]`. The subroutine `mag_128()` at the end of the program is used to compute the squared magnitude spectrum of the 128 complex DFT samples from the array `Xout[2*N]`. The magnitude is then stored in the array called `Spectrum[N]`, which will be used for graphic display later. The assembly routine, `mag_128.asm`, is listed in Table 4.8.

Table 4.8   The list of assembly program `mag_128.asm`

```
; Compute the magnitude response of the input
;
; Entry:   AR0: input buffer pointer
;          AR1: output buffer pointer
; Exit: None
;
        .def      _mag_128
N       .set      128
_mag_128
        bset      SATD
        pshboth   XAR5
        mov       #N-1, BRC0                    ; Set BRC0 for loop N times
        mov       XAR0, XAR5
        bset      FRCT
        rptblocal mag_loop-1
        mpym      *AR0+, *AR5+, AC0             ; Xr[i]*Xr[i]
        macm      *AR0+, *AR5+, AC0             ; Xr[i]*Xr[i] + Xi[i]*Xi[i]
        mov       hi(saturate(AC0)), *AR1+
        mag_loop
        popboth   XAR5
        bclr      SATD
        bclr      FRCT
        ret
        .end
```

Perform the following steps for Experiment 4C:

1. Write the C program `exp4a.c` based on the example (or copy from the software package) that will complete the following tasks:
   (a) Compose the complex input sample to `Xin[]`.
   (b) Call the subroutine `dft_128()` to perform DFT.
   (c) Call the subroutine `mag_128()` to compute the squared magnitude spectrum of the DFT.

2. Write the assembly routine `dft_128.asm` for the DFT, and write the assembly routine `mag_128.asm` for computing the magnitude spectrum (or copy these files from the software package).

3. Test and debug the programs. Plot the magnitude spectrum (`Spectrum[N]`) and the input samples as shown in Figure 4.21.

4. Profile the DFT routine and record the program and data memory usage. Also, record the clock cycles used for each subroutine.

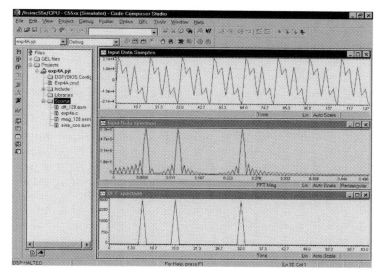

Figure 4.21 The plots of time-domain input signal (top), the input spectrum (middle) which shows three peaks are located at frequencies 0.5 kHz, 1 kHz, 2 kHz, and the DFT result (bottom)

4.6.4   Experiment 4D – Experiment Using Assembly Routines

In the previous experiments, we have written programs using either C programs or C combined with assembly programs. In some applications, it is desirable to develop the software using only assembly language. In this section, we will learn how to write and use assembly routines for an entire experiment.

For previous experiments, we linked the run-time support library rts55.lib to our programs. This library contains a routine called boot.asm for setting up the processor from the reset state. The settings include:

1. Initialize extended stack pointer XSP and extended system stack pointer XSSP.

2. Turn the sign extension mode on for arithmetic operations.

3. Turn the 40-bit accumulator mode off, and set the default as 32-bit mode.

4. Turn DU and AU saturate mode off.

5. Turn off the fractional mode.

6. Turn off the circular addressing mode.

When we replace the C function main() with an assembly program, we do not need to use the run-time support library rts55.lib. However, we have to create an assembly routine called vectors.asm that will set up the program starting address and begin the program execution from that point. The following is the list of assembly code:

```
;    vectors.asm
       .def    rsv
       .ref    start
       .sect   "vectors"
rsv    .ivec   start
```

where the assembly directive `.ivec` defines the starting address of the C55x interrupt vector table.

Interrupts are hardware- or software-driven signals that cause the C55x to suspend its current program and execute an interrupt service routine (ISR). Once the interrupt is acknowledged, the C55x executes the branch instruction at the corresponding interrupt vector table to perform an ISR. There are 32 interrupts and each interrupt uses 64 bits (4-word) in the C55x vector table. The first 32 bits contain the 24-bit program address of ISR. The second 32-bit can be ISR instructions. This 32-bit code will be executed before branching to the ISR. The label `start` is the entry point of our experiment program. At power up (or reset), the C55x program counter will be pointing to the first interrupt vector table, which is a branch instruction to the label `start` to begin executing the program. Since the vectors are fixed for the C55x, we need to map the address of interrupt-vector table `.ivec` to the program memory at address 0xFFFF00. The linker command file can be used to map the address of the vector table.

Because we do not use `boot.asm`, we are responsible for setting up the system before we can begin to perform our experiment. The stack pointer must be correctly set before any subroutine calls (or branch to ISR) can be made. Some of the C55x operation states/modes should also be set accordingly. The following example shows some of the settings at the beginning of our program:

```
stk_size   .set 0x100
stack      .usect ".stack",stk_size
sysstack   .usect ".stack",stk_size
      .def  start
      .sect .text
start
      bset   SATD
      bset   SATA
      bset   SXMD
      bclr   C54CM
      bclr   CPL
      amov   #(stack+stk_size), XSP
      mov    #(sysstack+stk_size), SSP
```

The label `start` in the code defines the entry point of the program. Since it will also be used by the `vectors.asm`, it needs to be defined as a global label. The first three bit-set instructions (`bset`) set up the saturation mode for both the DU and AU and the sign extension mode. The next two bit-clear instructions (`bclr`) turn off the C54x compatibility mode and the C compiler mode. The last two move instructions (`amov`/`mov`) initialize the stack pointers. In this example, the stack size is defined as 0x100 long and starts in the section named `.stack` in the data memory. When subroutine calls occur, the 24-bit program counter PC(23:0) will split into two portions.

The stack pointer SP is used for storing the lower 16-bit address of the program counter PC(15:0), and the system stack pointer SSP is used for the upper 8-bit of the PC(23:16).

In this experiment, we wrote an assembly program `exp4b.asm` listed in Table 4.9 to replace the `main()` function in the C program `exp4a.c`.

Table 4.9   List of assembly program `exp4b.asm`

```
;   exp4b.asm: Call dft_128 to compute DFT
;
N            .set 128
stk_size     .set 0x100
stack        .usect ".stack", stk_size
sysstack     .usect ".stack", stk_size
_Xin         .usect "in_data", 2*N
_Xout        .usect "out_data", 2*N
_Spectrum    .usect "out_data", N
        .sect .data
input .copy input.dat
        .def    start
        .def    _Xin, _Xout, _Spectrum
        .ref    _dft_128, _mag_128
        .sect .text
start
    bset   SATD                 ; Set up saturation for DU
    bset   SATA                 ; Set up saturation for AU
    bset   SXMD                 ; Set up sign extension mode
    bclr   C54CM                ; Disable C54x compatibility mode
    bclr   CPL                  ; Turn off C compiler mode
    amov   #(stack+stk_size), XSP      ; Setup DSP stack
    mov    #(sysstack+stk_size), SSP   ; Setup system stack
    mov    #N-1, BRC0           ; Init counter for loop N times
    amov   #input, XAR0         ; Input data array pointer
    amov   #Xin, XAR1           ; Xin array pointer
    rptblocal complex_data-1    ; Form complex data
    mov    *AR0+, *AR1+
    mov    #0, *AR1+
complex_data
    amov   #_Xin, XAR0          ; Xin array pointer
    amov   #_Xout, XAR1         ; Xout array pointer
    call   _dft_128             ; Perform 128-ponts DFT
    amov   #_Xout, XAR0         ; Xout pointer
    amov   #_Spectrum, XAR1     ; Spectrum array pointer
    call   _mag_128             ; Computer squared-mag response
here   b   here
    .end
```

Perform the following steps for Experiment 4D:

1. Write an assembly routine `exp4b.asm` to replace the C program `exp4a.c` (or copy it from the software package).

2. Use the `.usect "indata"` directive for array `Xin[ 256]`, `.usect "outdata"` for `Xout[256]` and `Spectrum[128]`, and use `sect.code` for the program section of the assembly routine `exp4b.asm`. Create a linker command file `exp4b.cmd` and add the above sections. The code section `.code` in the program memory starts at address 0x20400 with a length of 4096 bytes. The `indata` section starts in the data memory of word address 0x8000, and its length is 256 words. The `outdata` section starts in the data memory with staring address of 0x08800 and has the length of 512 words.

3. Test and debug the programs, verify the memory locations for sections `.code`, `indata`, and `outdata`, and compare the DFT results with experiment results obtained in Section 4.6.3.

References

[1] N. Ahmed and T. Natarajan, *Discrete-Time Signals and Systems*, Englewood Cliffs, NJ: Prentice-Hall, 1983.
[2] *MATLAB User's Guide*, Math Works, 1992.
[3] *MATLAB Reference Guide*, Math Works, 1992.
[4] A. V. Oppenheim and R. W. Schafer, *Discrete-Time Signal Processing*, Englewood Cliffs, NJ: Prentice-Hall, 1989.
[5] S. J. Orfanidis, *Introduction to Signal Processing*, Englewood Cliffs, NJ: Prentice-Hall, 1996.
[6] J. G. Proakis and D. G. Manolakis, *Digital Signal Processing – Principles, Algorithms, and Applications*, 3rd Ed., Englewood Cliffs, NJ: Prentice-Hall, 1996.
[7] A Bateman and W. Yates, *Digital Signal Processing Design*, New York: Computer Science Press, 1989.
[8] S. M. Kuo and D. R. Morgan, *Active Noise Control Systems – Algorithms and DSP Implementations*, New York: Wiley, 1996.
[9] H. P. Hsu, *Signals and Systems*, New York: McGraw-Hill, 1995.

Exercises

Part A

1. Similar to Example 4.1, assume the square wave is expressed as

$$x(t) = \begin{cases} A, & kT_0 \leq t \leq (k+0.5)T_0 \\ 0, & \text{otherwise,} \end{cases}$$

where k is an integer. Compute the Fourier series coefficients.

2. Similar to Example 4.2, compute the Fourier series coefficients for the signal

$$x(t) = \cos(\Omega_0 t).$$

3. Find and sketch the Fourier transform of the following signals:

   (a) The rectangular signal defined as

   $$x(t) = \begin{cases} A, & |t| \leq \tau \\ 0, & t > \tau. \end{cases}$$

   (b) The periodic impulse train defined as

   $$\delta(t) = \sum_{k=-\infty}^{\infty} \delta(t - kT_0).$$

   (c) $x(t) = \delta(t)$.

   (d) $x(t) = 1$.

4. Find the z-transform and ROC of the following sequences:

   (a) $x(n) = 1, \quad n \geq 0$.

   (b) $x(n) = e^{-an}, \quad n \geq 0$.

   (c) $x(n) = \begin{cases} -a^n, & n = -1, -2, \ldots, -\infty \\ 0, & n \geq 0. \end{cases}$

   (d) $x(n) = \sin(\omega n), \quad n \geq 0$.

5. The z-transform of an N-periodic sequence can be expressed in terms of the z-transform of its first period. That is, if

   $$x_1(n) = \begin{cases} x(n), & 0 \leq n \leq N - 1 \\ 0, & \text{elsewhere} \end{cases}$$

   denotes the sequence of the first period of the periodic sequence $x(n)$. Show that

   $$X(z) = \frac{X_1(z)}{1 - z^{-N}}, \quad |z^N| > 1,$$

   where $X_1(z) = \text{ZT}[x_1(n)]$.

6. Given the periodic signal

   $$x(n) = \{1, 1, 1, -1, -1, -1, 1, 1, 1, 1, -1, \ldots\}$$

   of period equal to 6. Find $X(z)$.

7. For a finite length (N) signal, (4.2.2) can be further simplified to

$$X(z) = \sum_{n=0}^{N-1} x(n)z^{-n}.$$

Consider the sequence

$$x(n) = \begin{cases} a^n, & 0 \le n \le N-1 \\ 0, & \text{otherwise} \end{cases}$$

where $a > 0$. Find $X(z)$ and plot the poles and zeros of $X(z)$.

8. Find the z-transform and ROC of

$$x(n) = \begin{cases} 1, & n < 0 \\ (0.5)^n, & n \ge 0. \end{cases}$$

9. Using partial-fraction-expansion method to find the inverse z-transform of

(a) $X(z) = \dfrac{z}{2z^2 - 3z + 1}$, $|z| < \dfrac{1}{2}$.

(b) $X(z) = \dfrac{z}{3z^2 - 4z + 1}$.

(c) $X(z) = \dfrac{2z^2}{(z+1)(z+2)^2}$.

10. Using residue method to find the inverse z-transform of the following functions:

(a) $X(z) = \dfrac{z^2 + z}{(z-1)^2}$.

(b) $X(z) = \dfrac{z^2 + 2z}{(z - 0.6)^3}$.

(c) $X(z) = \dfrac{1}{(z - 0.4)(z + 1)^2}$.

11. The first-order trapezoidal integration rule in numerical analysis is described by the I/O difference equation

$$y(n) = \frac{1}{2}[x(n) + x(n - 1)] + y(n - 1), \quad n \ge 0.$$

Treat this rule as an LTI system. Find

(a) The transfer function $H(z)$.

(b) The impulse response $h(n)$.

12. Consider the second-order IIR filter

$$y(n) = a_1 y(n-1) + a_2 y(n-2) + x(n), \quad n \geq 0.$$

(a) Compute $H(z)$.

(b) Discuss stability conditions related to coefficients a_1 and a_2.

13. Determine the stability of the following IIR filters:

(a) $H(z) = \dfrac{z(z-1)}{(z^2 - z + 1)(z + 0.8)}$.

(b) $3y(n) = 3.7y(n-1) - 0.7y(n-2) + x(n-1), \quad n \geq 0.$

14. In Figure 4.4(b), let $H_1(z)$ and $H_2(z)$ are the transfer functions of the two first-order IIR filters defined as

$$y_1(n) = x(n) - 0.5y_1(n-1), \quad n \geq 0$$

$$y_2(n) = x(n) + y_2(n-1), \quad n \geq 0.$$

(a) Find the overall transfer function $H(z) = H_1(z) + H_2(z)$.

(b) Find the output $y(n)$ if the input $x(n) = (-1)^n, \quad n \geq 0$.

15. Consider the first-order IIR system

$$y(n) = \frac{1-a}{2}[x(n) + x(n-1)] + ay(n-1), \quad n \geq 0,$$

Find the squared-magnitude response $|H(\omega)|^2$.

16. Consider a moving average filter defined in (3.2.1). Find the magnitude response $|H(\omega)|$ and the phase response $\phi(\omega)$.

17. Consider the FIR filter

$$y(n) = x(n) + 2x(n-1) + 4x(n-2) + 2x(n-3) + x(n-4), \quad n \geq 0.$$

Find the transfer function $H(z)$ and magnitude response $|H(\omega)|$.

18. Derive Equation (4.4.18).

Part B

19. Compute c_k given in (4.1.4) for $A = 1$, $T_0 = 0.1$, and $\tau = 0.05$, 0.01, 0.001, and 0.0001. Using MATLAB function stem to plot c_k for $k = 0, \pm 1, \ldots \pm 20$.

20. Repeat the Problem 19 for $A = 1$, $\tau = 0.001$ and $T_0 = 0.005$, 0.001, and 0.01.

Part C

21. The assembly program, dft_128.asm, can be further optimized. Use parallel instructions to improve the DFT performance. Profile the optimized code, and compare the cycle counts against the profile data obtained in experiment given in Section 4.6.3.

22. Find the clock rate of the TMS320C55x device, and use the profile data to calculate the total time the DFT routine spent to compute 128-point samples. Can this DFT routine be used for a real-time application? Why?

23. Why does the C55x DSP have two stack pointers, XSP and XSSP, and how are these pointers initialized? The C55x also uses RETA and CFCT during subroutine calls. What are these registers? Create and use a simple assembly program example to describe how the RETA, CFCT, XSP, and XSSP react to a nested subroutine call (hint: use references from the CCS help menu).

24. Modify experiments given in Section 4.6.4 to understand how the linker works:

    (a) Move all the programs in .text section (exp4b. asm, dft_128.asm and mag_128.asm) to a new section named .sect "dft_code", which starts at the program memory of address 0x020400. Adjust the section length if necessary.

    (b) Put all the data variables (exp4b. asm, dft_128.asm, and mag_128.asm) under the section named. usect "dft_vars", which starts at the data memory of address 0x08000. Again, adjust the section length if necessary.

    (c) Build and run the project. Examine memory segments in the map file. How does the linker handle each program and data section?

5

Design and Implementation of FIR Filters

A filter is a system that is designed to alter the spectral content of input signals in a specified manner. Common filtering objectives include improving signal quality, extracting information from signals, or separating signal components that have been previously combined. A digital filter is a mathematical algorithm implemented in hardware, firmware, and/or software that operates on a digital input signal to produce a digital output signal for achieving filtering objectives. A digital filter can be classified as being linear or nonlinear, time invariant or varying. This chapter is focused on the design and implementation of linear, time-invariant (LTI) finite impulse response (FIR) filters. The time-invariant infinite impulse response (IIR) filters will be discussed in Chapter 6, and the time-varying adaptive filters are introduced in Chapter 8.

The process of deriving the digital filter transfer function $H(z)$ that satisfies the given set of specifications is called digital filter design. Although many applications require only simple filters, the design of more complicated filters requires the use of sophisticated techniques. A number of computer-aided design tools (such as MATLAB) are available for designing digital filters. Even though such tools are widely available, we should understand the basic characteristics of digital filters and familiar with techniques used for implementing digital filters. Many DSP books devote substantial efforts to the theory of designing digital filters, especially approximation methods, reflecting the considerable work that has been done for calculating and optimizing filter coefficients.

5.1 Introduction to Digital Filters

As discussed in previous chapters, filters can be divided into two categories: analog filters and digital filters. Similar specifications are used for both analog and digital filters. In this chapter, we will discuss digital filters exclusively. The digital filters are assumed to have a single input $x(n)$, and a single output $y(n)$. Analog filters are used as design prototypes for digital IIR filters, and will be briefly introduced in Chapter 6.

5.1.1   Filter Characteristics

A digital filter is said to be linear if the output due to the application of input,

$$x(n) = a_1 x_1(n) + a_2 x_2(n),\qquad(5.1.1)$$

is equal to

$$y(n) = a_1 y_1(n) + a_2 y_2(n),\qquad(5.1.2)$$

where a_1 and a_2 are arbitrary constants, and $y_1(n)$ and $y_2(n)$ are the filter outputs due to the application of the inputs $x_1(n)$ and $x_2(n)$, respectively. The important property of linearity is that in the computation of $y(n)$ due to $x(n)$, we may decompose $x(n)$ into a summation of simpler components $x_i(n)$. We then compute the response $y_i(n)$ due to input $x_i(n)$. The summation of $y_i(n)$ will be equal to the output $y(n)$. This property is also called the superposition.

A time-invariant system is a system that remains unchanged over time. A digital filter is time-invariant if the output due to the application of delayed input $x(n - m)$ is equal to the delayed output $y(n - m)$, where m is a positive integer. It means that if the input signal is the same, the output signal will always be the same no matter what instant the input signal is applied. It also implies that the characteristics of a time-invariant filter will not change over time.

A digital filter is causal if the output of the filter at time n_0 does not depend on the input applied after n_0. It depends only on the input applied at and before n_0. On the contrary, the output of a non-causal filter depends not only on the past input, but also on the future input. This implies that a non-causal filter is able to predict the input that will be applied in the future. This is impossible for any real physical filter.

Linear, time-invariant filters are characterized by magnitude response, phase response, stability, rise time, settling time, and overshoot. Magnitude response specifies the gains (amplify, pass, or attenuate) of the filter at certain frequencies, while phase response indicates the amount of phase changed by the filter at different frequencies. Magnitude and phase responses determine the steady-state response of the filter. For an instantaneous change in input, the rise time specifies an output-changing rate. The settling time describes an amount of time for the output to settle down to a stable value, and the overshoot shows if the output exceeds the desired output value. The rise time, the settling time, and the overshoot specify the transient response of the filter in the time domain.

A digital filter is stable if, for every bounded input signal, the filter output is bounded. A signal $x(n)$ is bounded if its magnitude $|x(n)|$ does not go to infinity. A digital filter with the impulse response $h(n)$ is stable if and only if

$$\sum_{n=0}^{\infty} |h(n)| < \infty.\qquad(5.1.3)$$

Since an FIR filter has only a finite number of non-zero $h(n)$, the FIR filter is always stable. Stability is critical in DSP implementations because it guarantees that the filter output will never grow beyond bounds, thus avoiding numerical overflow in computing the convolution sums.

As mentioned earlier, filtering is a process that passes certain frequency components in a signal through the system and attenuates other frequency components. The range of frequencies that is allowed to pass through the filter is called the passband, and the range of frequencies that is attenuated by the filter is called the stopband. If a filter is defined in terms of its magnitude response, there are four different types of filters: lowpass, highpass, bandpass, and bandstop filters. Each ideal filter is characterized by a passband over which frequencies are passed unchanged (except with a delay) and a stopband over which frequencies are rejected completely. The two-level shape of the magnitude response gives these filters the name brickwall. Ideal filters help in analyzing and visualizing the processing of actual filters employed in signal processing. Achieving an ideal brickwall characteristic is not feasible, but ideal filters are useful for conceptualizing the impact of filters on signals.

As discussed in Chapter 3, there are two basic types of digital filters: FIR filters and IIR filters. An FIR filter of length L can be represented with its impulse response $h(n)$ that has only L non-zero samples. That is, $h(n) = 0$ for all $n \geq L$. An FIR filter is also called a transversal filter. Some advantages and disadvantages of FIR filters are summarized as follows:

1. Because there is no feedback of past outputs as defined in (3.1.16), the FIR filters are always stable. That is, a bounded input results in a bounded output. This inherent stability is also manifested in the absence of poles in the transfer function as defined in (4.3.8), except possibly at the origin.

2. The filter has finite memory because it 'forgets' all inputs before the $(L-1)$th previous one.

3. The design of linear phase filters can be guaranteed. In applications such as audio signal processing and data transmission, linear phase filters are preferred since they avoid phase distortion.

4. The finite-precision errors (discussed in Chapter 3) are less severe in FIR filters than in IIR filters.

5. FIR filters can be easily implemented on most DSP processors such as the TMS320C55x introduced in Chapter 2.

6. A relatively higher order FIR filter is required to obtain the same characteristics as compared with an IIR filter. Thus more computations are required, and/or longer time delay may be involved in the case of FIR filters.

5.1.2 Filter Types

An ideal frequency-selective filter passes certain frequency components without any change and completely stops the other frequencies. The range of frequencies that are passed without attenuation is the passband of the filter, and the range of frequencies that is attenuated is the stopband. Thus the magnitude response of an ideal filter is given by $|H(\omega)| = 1$ in the passband and $|H(\omega)| = 0$ in the stopband. Note that the frequency response $H(\omega)$ of a digital filter is a periodic function of ω, and the magnitude response $|H(\omega)|$ of a digital filter with real coefficients is an even function of ω. Therefore the digital filter specifications are given only for the range $0 \leq \omega \leq \pi$.

The magnitude response of an ideal lowpass filter is illustrated in Figure 5.1(a). The regions $0 \leq \omega \leq \omega_c$ and $\omega > \omega_c$ are referred to as the passband and stopband, respectively. The frequency that separates the passband and stopband is called the cut-off frequency ω_c. An ideal lowpass filter has magnitude response $|H(\omega)| = 1$ in the frequency range $0 \leq \omega \leq \omega_c$ and has $|H(\omega)| = 0$ for $\omega > \omega_c$. Thus a lowpass filter passes all low-frequency components below the cut-off frequency and attenuates all high-frequency components above ω_c. Lowpass filters are generally used when the signal components of interest are in the range of DC to the cut-off frequency, but other higher frequency components (or noise) are present.

The magnitude response of an ideal highpass filter is illustrated in Figure 5.1(b). The regions $\omega \geq \omega_c$ and $0 \leq \omega < \omega_c$ are referred to as the passband and stopband, respectively. A highpass filter passes all high-frequency components above the cut-off frequency ω_c and attenuates all low-frequency components below ω_c. As discussed in Chapter 1, highpass filters can be used to eliminate DC offset, 60 Hz hum, and other low frequency noises.

The magnitude response of an ideal bandpass filter is illustrated in Figure 5.1(c). The regions $\omega < \omega_a$ and $\omega > \omega_b$ are referred to as the stopband. The frequencies ω_a and ω_b are called the lower cut-off frequency and the upper cut-off frequency, respectively. The

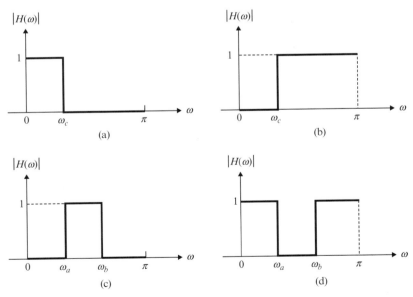

Figure 5.1   Magnitude response of ideal filters: (a) lowpass, (b) highpass, (c) bandpass, and (d) bandstop

region $w_a \leq w \leq w_b$ is called the passband. A bandpass filter passes all frequency components between the two cut-off frequencies w_a and w_b, and attenuates all frequency components below the frequency w_a and above the frequency w_b. If the passband is narrow, it is more common to specify the center frequency and the bandwidth of the passband. A narrow bandpass filter may be called a resonator (or peaking) filter.

The magnitude response of an ideal bandstop (or band-reject) filter is illustrated in Figure 5.1(d). The regions $w \leq w_a$ and $w \geq w_b$ are referred to as the passband. The region $w_a < w < w_b$ is called the stopband. A bandstop filter attenuates all frequency components between the two cutoff frequencies w_a and w_b, and passes all frequency components below the frequency w_a and above the frequency w_b. A narrow bandstop filter designed to attenuate a single frequency component is called a notch filter. For example, a common source of noise is a power line generating a 60 Hz sinusoidal signal. This noise can be removed by passing the corrupted signal through a notch filter with notch frequency at 60 Hz. The design of simple notch filters was introduced in Section 4.5.1.

In addition to these frequency-selective filters, an allpass filter provides frequency response $|H(w)| = 1$ for all frequency w, thus passing all frequencies with uniform gain. These filters do not remove frequency components, but alter the phase response. The principal use of allpass filters is to correct the phase distortion introduced by the physical system and/or other filters. For example, it is used as a delay equalizer. In this application, it is designed such that when cascaded with another digital system, the overall system has a constant group delay in the frequency range of interest. A very special case of the allpass filter is the ideal Hilbert transformer, which produces a 90° phase shift of input signals.

5.1.3 Filter Specifications

In practice, we cannot achieve the infinitely sharp cutoff implied by the ideal filters shown in Figure 5.1. This will be shown later by considering the impulse response of the ideal lowpass filter that is non-causal and hence not physically realizable. Instead we must compromise and accept a more gradual cutoff between passband and stopband, as well as specify a transition band between the passband and stopband. The design is based on magnitude response specifications only, so the phase response of the filter is not controlled. Whether this is important depends on the application. Realizable filters do not exhibit the flat passband or the perfect linear phase characteristic. The deviation of $|H(w)|$ from unity (0 dB) in the passband is called magnitude distortion, and the deviation from the linear phase of the phase response $H(w)$ is called phase distortion.

The characteristics of digital filters are often specified in the frequency domain. For frequency-selective filters, the magnitude response specifications of a digital filter are often given in the form of tolerance (or ripple) schemes. In addition, a transition band is specified between the passband and the stopband to permit the magnitude drop off smoothly. A typical magnitude response of lowpass filter is shown in Figure 5.2. The dotted horizontal lines in the figure indicate the tolerance limits. In the passband, the magnitude response has a peak deviation δ_p and in the stopband, it has a maximum deviation δ_s. The frequencies w_p and w_s are the passband edge (cut-off) frequency and the stopband edge frequency, respectively.

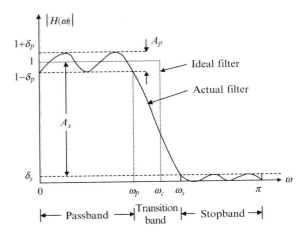

Figure 5.2 Magnitude response and performance measurement of a lowpass filter

As shown in Figure 5.2, the magnitude of passband defined by $0 \leq \omega \leq \omega_p$ approximates unity with an error of $\pm\delta_p$. That is,

$$1 - \delta_p \leq |H(\omega)| \leq 1 + \delta_p, \quad 0 \leq \omega \leq \omega_p. \tag{5.1.4}$$

The passband ripple, δ_p, is a measure of the allowed variation in magnitude response in the passband of the filter. Note that the gain of the magnitude response is normalized to 1 (0 dB). In practical applications, it is easy to scale the filter output by multiplying the output by a constant, which is equivalent to multiplying the whole magnitude response by the same constant gain.

In the stopband, the magnitude approximates 0 with an error δ_s. That is,

$$|H(\omega)| \leq \delta_s, \quad \omega_s \leq \omega \leq \pi. \tag{5.1.5}$$

The stopband ripple (or attenuation) describes the maximum gain (or minimum attenuation) for signal components above the ω_s.

Passband and stopband deviations may be expressed in decibels. The peak passband ripple, δ_p, and the minimum stopband attenuation, δ_s, in decibels are given as

$$A_p = 20 \log_{10} \left(\frac{1 + \delta_p}{1 - \delta_p} \right) \text{ dB} \tag{5.1.6}$$

and

$$A_s = -20 \log_{10} \delta_s \text{ dB}. \tag{5.1.7}$$

Thus we have

$$\delta_p = \frac{10^{A_p/20} - 1}{10^{A_p/20} + 1} \qquad\qquad (5.1.8)$$

and

$$\delta_s = 10^{-A_s/20}. \qquad\qquad (5.1.9)$$

Example 5.1: Consider a filter specified as having a magnitude response in the passband within ± 0.01. That is, $\delta_p = 0.01$. From (5.1.6), we have

$$A_p = 20\log_{10}\left(\frac{1.01}{0.99}\right) = 0.1737\,\text{dB}.$$

When the minimum stopband attenuation is given as $\delta_s = 0.01$, we have

$$A_s = -20\log_{10}(0.01) = 40\,\text{dB}.$$

The transition band is the area between the passband edge frequency ω_p and the stopband edge frequency ω_s. The magnitude response decreases monotonically from the passband to the stopband in this region. Generally, the magnitude in the transition band is left unspecified. The width of the transition band determines how sharp the filter is. It is possible to design filters that have minimum ripple over the passband, but a certain level of ripple in this region is commonly accepted in exchange for a faster roll-off of gain in the transition band. The stopband is chosen by the design specifications. Generally, the smaller δ_p and δ_s are, and the narrower the transition band, the more complicated (higher order) the designed filter becomes.

An example of a narrow bandpass filter is illustrated in Figure 5.3. The center frequency ω_m is the point of maximum gain (or maximum attenuation for a notch filter). If a logarithm scale is used for frequency such as in many audio applications, the center frequency at the geometric mean is expressed as

$$\omega_m = \sqrt{\omega_a \omega_b}, \qquad\qquad (5.1.10a)$$

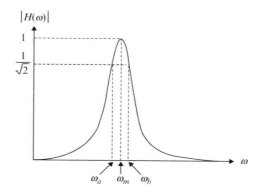

Figure 5.3   Magnitude response of bandpass filter with narrow bandwidth

where ω_a and ω_b are the lower and upper cut-off frequencies, respectively. The bandwidth is the difference between the two cut-off frequencies for a bandpass filter. That is,

$$BW = \omega_b - \omega_a. \qquad (5.1.10b)$$

The 3-dB bandwidth commonly used in practice is defined as

$$|H(\omega_a)| = |H(\omega_b)| = \frac{1}{\sqrt{2}} \cong 0.707. \qquad (5.1.11)$$

Another way of describing a resonator (or notch) filter is the quality factor defined as

$$Q = \frac{\omega_m}{2\pi BW}. \qquad (5.1.12)$$

There are many applications that require high Q filters.

When a signal passes through a filter, it is modified both in amplitude and phase. The phase response is an important filter characteristic because it affects time delay of the different frequency components passing through the filter. If we consider a signal that consists of several frequency components, the phase delay of the filter is the average time delay the composite signal suffers at each frequency. The group delay function is defined as

$$T_d(\omega) = \frac{-d\phi(\omega)}{d\omega}, \qquad (5.1.13)$$

where $\phi(\omega)$ is the phase response of the filter.

A filter is said to have a linear phase if its phase response satisfies

$$\phi(\omega) = -\alpha\omega, \quad -\pi \le \omega \le \pi \qquad (5.1.14)$$

or

$$\phi(\omega) = \beta - \alpha\omega, \quad -\pi \le \omega \le \pi \qquad (5.1.15)$$

These equations show that for a filter with a linear phase, the group delay $T_d(\omega)$ given in (5.1.13) is a constant α for all frequencies. This filter avoids phase distortion because all sinusoidal components in the input are delayed by the same amount. A filter with a nonlinear phase will cause a phase distortion in the signal that passes through it. This is because the frequency components in the signal will each be delayed by a different amount, thereby altering their harmonic relationships. Linear phase is important in data communications, audio, and other applications where the temporal relationships between different frequency components are critical.

The specifications on the magnitude and phase (or group delay) of $H(\omega)$ are based on the steady-state response of the filter. Therefore they are called the steady-state specifications. The speed of the response concerns the rate at which the filter reaches the steady-state response. The transient performance is defined for the response right after the application of an input signal. A well-designed filter should have a fast response, a small rise time, a small settling time, and a small overshoot.

In theory, both the steady-state and transient performance should be considered in the design of a digital filter. However, it is difficult to consider these two specifications simultaneously. In practice, we first design a filter to meet the magnitude specifications. Once this filter is obtained, we check its phase response and transient performance. If they are satisfactory, the design is completed. Otherwise, we must repeat the design process. Once the transfer function has been determined, we can obtain a realization of the filter. This will be discussed later.

5.2 FIR Filtering

The signal-flow diagram of the FIR filter is shown in Figure 3.6. As discussed in Chapter 3, the general I/O difference equation of FIR filter is expressed as

$$y(n) = b_0 x(n) + b_1 x(n-1) + \cdots + b_{L-1} x(n-L+1) = \sum_{l=0}^{L-1} b_l x(n-l), \qquad (5.2.1)$$

where b_l are the impulse response coefficients of the FIR filter. This equation describes the output of the FIR filter as a convolution sum of the input with the impulse response of the system. The transfer function of the FIR filter defined in (5.2.1) is given by

$$H(z) = b_0 + b_1 z^{-1} + \cdots + b_{L-1} z^{-(L-1)} = \sum_{l=0}^{L-1} b_l z^{-l}. \qquad (5.2.2)$$

5.2.1 Linear Convolution

As discussed in Section 3.2.2, the output of the linear system defined by the impulse response $h(n)$ for an input signal $x(n)$ can be expressed as

$$y(n) = x(n) * h(n) = \sum_{l=-\infty}^{\infty} h(l) x(n-l). \qquad (5.2.3)$$

Thus the output of the LTI system at any given time is the sum of the input samples convoluted by the impulse response coefficients of the system. The output at time n_0 is given as

$$y(n_0) = \sum_{l=-\infty}^{\infty} h(l) x(n_0 - l). \qquad (5.2.4)$$

Assuming that n_0 is positive, the process of computing the linear convolution involves the following four steps:

1. *Folding*. Fold $x(l)$ about $l = 0$ to obtain $x(-l)$.

2. *Shifting*. Shift $x(-l)$ by n_0 samples to the right to obtain $x(n_0 - l)$.

3. *Multiplication*. Multiply $h(l)$ by $x(n_0 - l)$ to obtain the products $h(l) \cdot x(n_0 - l)$ for all l.

4. *Summation*. Sum all the products to obtain the output $y(n_0)$ at time n_0.

Repeat steps 2–4 in computing the output of the system at other time instants n_0.

This general procedure of computing convolution sums can be applied to (5.2.1) for calculating the FIR filter output $y(n)$. As defined in (3.2.15), the impulse response of the FIR filter is

$$h(l) = \begin{cases} 0, & l < 0 \\ b_l, & 0 \le l < L \\ 0, & l \ge L. \end{cases} \tag{5.2.5}$$

If the input signal is causal, the general linear convolution equation defined in (5.2.3) can be simplified to (5.2.1). Note that the convolution of the length M input with the length L impulse response results in length $L + M - 1$ output.

Example 5.2: Consider an FIR filter that consists of four coefficients b_0, b_1, b_2, and b_3. From (5.2.1), we have

$$y(n) = \sum_{l=0}^{3} b_l x(n - l), \quad n \ge 0.$$

This yields

$$\begin{aligned} n = 0, \quad & y(0) = b_0 x(0), \\ n = 1, \quad & y(1) = b_0 x(1) + b_1 x(0), \\ n = 2, \quad & y(2) = b_0 x(2) + b_1 x(1) + b_2 x(0), \\ n = 3, \quad & y(3) = b_0 x(3) + b_1 x(2) + b_2 x(1) + b_3 x(0). \end{aligned}$$

In general, we have

$$y(n) = b_0 x(n) + b_1 x(n - 1) + b_2 x(n - 2) + b_3 x(n - 3), \quad n \ge 3.$$

The graphical interpretation is illustrated in Figure 5.4.

As shown in Figure 5.4, the input sequence is flipped around (folding) and then shifted to the right over the filter coefficients. At each time instant, the output value is the sum of products of overlapped coefficients with the corresponding input data aligned below it. This flip-and-slide form of linear convolution can be illustrated in Figure 5.5. Note that shifting $x(-l)$ to the right is equivalent to shift b_l to the left one unit at each sampling period.

As shown in Figure 5.5, the input sequence is extended by padding $L - 1$ zeros to its right. At time $n = 0$, the only non-zero product comes from b_0 and $x(0)$ which are time aligned. It takes the filter $L - 1$ iterations before it is completely overlapped with the input sequence. The first $L - 1$ outputs correspond to the transient behavior of the FIR filter. For $n \geq L - 1$, the filter aligns over the non-zero portion of the input sequence. That is, the signal buffer of FIR filter is full and the filter is in the steady state. If the input is a finite-length sequence of M samples, there are $L + M - 1$ output samples and the last $L - 1$ samples also correspond to transients.

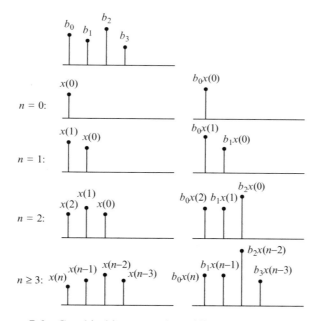

Figure 5.4 Graphical interpretation of linear convolution, $L = 4$

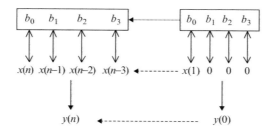

Figure 5.5 Flip-and-slide process of linear convolution

5.2.2 Some Simple FIR Filters

A multiband filter has more than one passband and stopband. A special case of the multiband filter is the comb filter. A comb filter has evenly spaced zeros, with the shape of the magnitude response resembling a comb in order to block frequencies that are integral multiples of a fundamental frequency. A difference equation of a comb filter is given as

$$y(n) = x(n) - x(n - L),\qquad\qquad(5.2.6)$$

where the number of delay L is an integer. The transfer function of this multiplier-free FIR filter is

$$H(z) = 1 - z^{-L} = \frac{z^L - 1}{z^L}.\qquad\qquad(5.2.7)$$

Thus the comb filter has L poles at the origin (trivial poles) and L zeros equally spaced on the unit circle at

$$z_l = e^{j(2\pi/L)l},\quad l = 0, 1, \ldots, L - 1.\qquad\qquad(5.2.8)$$

Example 5.3: The zeros and the frequency response of a comb filter can be computed and plotted using the following MATLAB script for $L = 8$:

```
b = [1 0 0 0 0 0 0 0 −1];
zplane(b, 1)
freqz(b, 1, 128);
```

The zeros on the z-plane are shown in Figure 5.6(a) and the characteristic of comb shape is shown in Figure 5.6(b). The center of the passband lies halfway between the zeros of the response, that is at frequencies $\dfrac{(2l + 1)\pi}{L}$, $l = 0, 1, \ldots, L - 1$.

Because there is not a large attenuation in the stopband, the comb filter can only be used as a crude bandstop filter to remove harmonics at frequencies

$$\omega_l = 2\pi l/L,\quad l = 0, 1, \ldots, L - 1.\qquad\qquad(5.2.9)$$

Comb filters are useful for passing or eliminating specific frequencies and their harmonics. Periodic signals have harmonics and using comb filters are more efficient than having individual filters for each harmonic. For example, the constant humming sound produced by large transformers located in electric utility substations are composed of even-numbered harmonics (120 Hz, 240 Hz, 360 Hz, etc.) of the 60 Hz power frequency. When a desired signal is corrupted by the transformer noise, the comb filter with notches at the multiples of 120 Hz can be used to eliminate undesired harmonics.

We can selectively cancel one or more zeros in a comb filter with corresponding poles. Canceling the zero provides a passband, while the remaining zeros provide attenuation for a stopband. For example, we can add a pole at $z = 1$. Thus the transfer function given in (5.2.7) is changed to

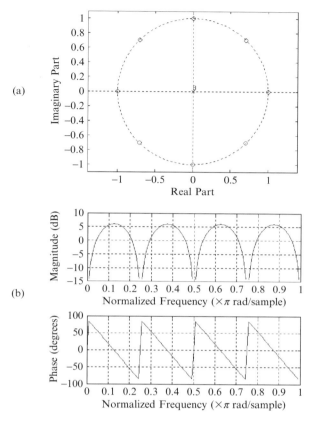

Figure 5.6 Zeros of a simple comb filter ($L = 8$) and its frequency response: (a) zeros, and (b) magnitude (top) and phase (bottom) responses

$$H(z) = \frac{1 - z^{-L}}{1 - z^{-1}}.$$ (5.2.10)

This is a lowpass filter with passband centered at $z = 1$, where the pole–zero cancellation occurs. Since the pole at $z = 1$ is canceled by the zero at $z = 1$, the system defined by (5.2.10) is still the FIR filter. Note that canceling the zero at $z = 1$ produces a lowpass filter, canceling the zeros at $z = \pm j$ produces a bandpass filter, and canceling the zero at $z = -1$ produces a highpass filter.

Applying the scaling factor $1/L$ to (5.2.10), the transfer function becomes

$$H(z) = \frac{1}{L}\left(\frac{1 - z^{-L}}{1 - z^{-1}}\right).$$ (5.2.11)

This is the moving-average filter introduced in Chapter 3 with the I/O difference equation expressed as

$$y(n) = \frac{1}{L}[x(n) - x(n-L) + y(n-1)]$$

$$= \frac{1}{L}\sum_{l=0}^{L-1} x(n-l). \tag{5.2.12}$$

The moving-average filter is a very simple lowpass filtering operation that passes the zero-frequency (or the mean) component. However, there are disadvantages of this type of filter such as the passband cut-off frequency is a function of L and the sampling rate f_s, and the stopband attenuation is fixed by L.

Example 5.4: Consider a simple moving-average filter

$$y(n) = \frac{1}{2}[x(n) + x(n-1)], \quad n \geq 0.$$

The transfer function of the filter can be expressed as

$$H(z) = \frac{1}{2}(1 + z^{-1}),$$

which has a single zero at $z = -1$ and the frequency response is given by

$$H(\omega) = \frac{1}{2}(1 + e^{-j\omega}) = \frac{1}{2}e^{-j\omega/2}\left[e^{j\omega/2} + e^{-j\omega/2}\right] = e^{-j\omega/2}\cos(\omega/2).$$

Therefore, we have

$$|H(\omega)|^2 = \cos\left(\frac{\omega}{2}\right)^2 = \frac{1}{2}[1 + \cos(\omega)].$$

This is lowpass-filter response, which falls off monotonically to 0 at $\omega = \pi$. We can show that

$$\phi(\omega) = \frac{-\omega}{2},$$

thus the filter has linear phase.

5.2.3 Linear Phase FIR Filters

In many practical applications, it is required that a digital filter has a linear phase. In particular, it is important for phase-sensitive signals such as speech, music, images, and data transmission where nonlinear phase would give unacceptable frequency distortion. FIR filters can be designed to obtain exact linear phase.

If L is an odd number, we define $M = (L-1)/2$. If we define $h_l = b_{l+M}$, then (5.2.1) can be written as

$$B(z) = \sum_{l=0}^{2M} b_l z^{-l} = \sum_{l=-M}^{M} b_{l+M} z^{-(l+M)} = z^{-M} \left[\sum_{l=-M}^{M} h_l z^{-l} \right] = z^{-M} H(z), \qquad (5.2.13)$$

where

$$H(z) = \sum_{l=-M}^{M} h_l z^{-l}. \qquad (5.2.14)$$

Let h_l have the symmetry property expressed as

$$h_l = h_{-l}, \quad l = 0, 1, \ldots, M. \qquad (5.2.15)$$

From (5.2.13), the frequency response $B(\omega)$ can be written as

$$
\begin{aligned}
B(\omega) &= B(z)|_{z = e^{j\omega}} = e^{-j\omega M} H(\omega) \\
&= e^{-j\omega M} \left[\sum_{l=-M}^{M} h_l e^{-j\omega l} \right] = e^{-j\omega M} \left[h_0 + \sum_{l=1}^{M} h_l \left(e^{j\omega l} + e^{-j\omega l} \right) \right] \\
&= e^{-j\omega M} \left[h_0 + 2 \sum_{l=1}^{M} h_l \cos(\omega l) \right].
\end{aligned}
\qquad (5.2.16)
$$

If h_l is real, then $H(\omega)$ is a real function of ω. If $H(\omega) \geq 0$, then the phase of $B(\omega)$ is equal to

$$\phi(\omega) = -\omega M, \qquad (5.2.17)$$

which is a linear function of ω. However, if $H(\omega) < 0$, then the phase of $B(\omega)$ is equal to $\pi - \omega M$. Thus, if there are sign changes in $H(\omega)$, there are corresponding 180° phase shifts in $B(\omega)$, and $B(\omega)$ is only piecewise linear. However, it is still simple to refer to the filter as having linear phase.

If h_l has the anti-symmetry property expressed as

$$h_l = -h_{-l}, \quad l = 0, 1, \ldots, M, \qquad (5.2.18)$$

this implies $h(0) = 0$. Following the derivation of (5.2.16), we obtain

$$
\begin{aligned}
B(\omega) &= e^{-j\omega M} H(\omega) = e^{-j\omega M} \left[\sum_{l=1}^{M} h_l \left(e^{j\omega l} + e^{-j\omega l} \right) \right] \\
&= e^{-j\omega M} \left[-2j \sum_{l=1}^{M} h_l \sin(\omega l) \right].
\end{aligned}
\qquad (5.2.19)
$$

If h_l is real, then $H(\omega)$ is pure imaginary and the phase of $B(z)$ is a linear function of ω.

The filter order L is assumed to be an odd integer in the above derivations. If L is an even integer and $M = L/2$, then the derivations of (5.2.16) and (5.2.19) have to be modified slightly. In conclusion, an FIR filter has linear phase if its coefficients satisfy the following (positive) symmetric condition:

$$b_l = b_{L-1-l}, \quad l = 0, 1, \ldots, L-1, \tag{5.2.20}$$

or, the anti-symmetric (negative symmetry) condition

$$b_l = -b_{L-1-l}, \quad l = 0, 1, \ldots, L-1. \tag{5.2.21}$$

There are four types of linear phase FIR filters, depending on whether L is even or odd and whether b_l has positive or negative symmetry as illustrated in Figure 5.7. The group delay of a symmetric (or anti-symmetric) FIR filter is $T_d(\omega) = L/2$, which corresponds to the midpoint of the FIR filter. The frequency response of the type I

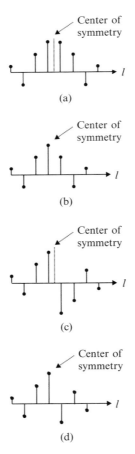

Figure 5.7 Coefficients of the four types of linear phase FIR filters: (a) type I: L even $(L = 8)$, positive symmetry, (b) type II: L odd $(L = 7)$, positive symmetry, (c) type III: L even $(L = 8)$, negative symmetry, and (d) type IV: L odd $(L = 7)$, negative symmetry

(*L* even, positive symmetry) filter is always 0 at the Nyquist frequency. This type of filter is unsuitable for a highpass filter. Type III (*L* even, negative symmetry) and IV (*L* odd, negative symmetry) filters introduce a 90° phase shift, thus they are often used to design Hilbert transformers. The frequency response is always 0 at DC frequency, making them unsuitable for lowpass filters. In addition, type III response is always 0 at the Nyquist frequency, also making it unsuitable for a highpass filter.

The symmetry (or anti-symmetry) property of a linear-phase FIR filter can be exploited to reduce the total number of multiplications into almost half. Consider the realization of FIR filter with an even length *L* and positive symmetric impulse response as given in (5.2.20), Equation (5.2.2) can be combined as

$$H(z) = b_0\left(1 + z^{-L+1}\right) + b_1\left(z^{-1} + z^{-L+2}\right) + \cdots + b_{L/2-1}\left(z^{-L/2+1} + z^{-L/2}\right). \quad (5.2.22)$$

The I/O difference equation is given as

$$
\begin{aligned}
y(n) &= b_0[x(n) + x(n - L + 1)] + b_1[x(n - 1) + x(n - L + 2)] \\
&\quad + \cdots + b_{L/2-1}[x(n - L/2 + 1) + x(n - L/2)] \\
&= \sum_{l=0}^{L/2-1} b_l[x(n - l) + x(n - L + 1 + l)].
\end{aligned}
\quad (5.2.23)
$$

A realization of $H(z)$ defined in (5.2.22) is illustrated in Figure 5.8. For an anti-symmetric FIR filter, the addition of two signals is replaced by subtraction. That is,

$$y(n) = \sum_{l=0}^{L/2-1} b_l[x(n - l) - x(n - L + 1 + l)]. \quad (5.2.24)$$

As shown in (5.2.23) and Figure 5.8, the number of multiplications is cut in half by adding the pair of samples, then multiplying the sum by the corresponding coefficient.

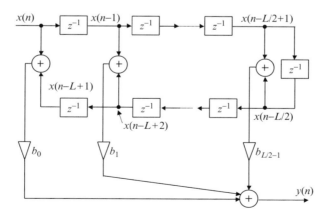

Figure 5.8 Signal flow diagram of symmetric FIR filter, *L* is even

The trade-off is that instead of accessing data linearly through the same buffer with a single pointer, we need two address pointers that point at both ends for $x(n - l)$ and $x(n - L + 1 + l)$. The TMS320C55x provides two special instructions for implementing the symmetric and anti-symmetric FIR filters efficiently. In Section 5.6, we will demonstrate how to use the symmetric FIR instructions for experiments.

There are applications where data is already collected and stored for later processing, i.e., the processing is not done in real time. In these cases, the 'current' time n can be located arbitrarily as the data is processed, so that the current output of the filter may depend on past, current, and future input values. Such a filter is 'non-realizable' in real time, but is easy to implement for the stored data. The non-causal filter has the I/O equation

$$y(n) = \sum_{l=-L_1}^{L_2} b_l x(n - l) \tag{5.2.25}$$

and the transfer function

$$H(z) = \sum_{l=-L_1}^{L_2} b_l z^{-l}. \tag{5.2.26}$$

Some typical applications of non-causal filters are the smoothing filters, the interpolation filters, and the inverse filters. A simple example of a non-causal filter is a Hanning filter with coefficients $\{0.25, 0.5, 0.25\}$ for smoothing estimated pitch in speech processing.

5.2.4   Realization of FIR Filters

An FIR filter can be realized to operate either on a block basis or a sample-by-sample basis. In the block processing case, the input is segmented into multiple blocks. Filtering is performed on one block at a time, and the resulting output blocks are recombined to form the overall output. The filtering of each block can be implemented using the linear convolution technique discussed in Section 5.2.1, or fast convolution using FFT, which will be introduced in Chapter 7. The implementation of block-FIR filter with the TMS320C55x will be introduced in Section 5.6. In the sample-by-sample case, the input samples are processed at every sampling period after the current input $x(n)$ is available. This approach is useful in real-time applications and will be discussed in this section.

Once the coefficients, b_l, $l = 0, 1, \ldots, L - 1$, have been determined, the next step is to decide on the structure (form) of the filter. The direct form implementation of digital FIR (transversal) filter is shown in Figure 3.6 and is described by the difference equation (5.2.1). The transfer function of the FIR filter given in (5.2.2) can be factored as

$$H(z) = b_0(1 - z_1 z^{-1})(1 - z_2 z^{-1}) \cdots (1 - z_{L-1} z^{-1}), \tag{5.2.27}$$

where the zeros $z_l, l = 1, 2, \cdots, L - 1$ must occur in complex-conjugate pairs for a real-valued filter. The factorization of $H(z)$ can be carried out in MATLAB using the function `roots`. If we pair two zeros and multiply out the expressions, we obtain a cascade of second-order sections as

$$H(z) = b_0(1 + b_{11}z^{-1} + b_{12}z^{-2})(1 + b_{21}z^{-1} + b_{22}z^{-2}) \cdots (1 + b_{M1}z^{-1} + b_{M2}z^{-2})$$

$$= b_0 \prod_{m=1}^{M} (1 + b_{m1}z^{-1} + b_{m2}z^{-2})$$

$$= b_0 H_1(z) H_2(z) \cdots H_M(z), \tag{5.2.28}$$

where $M = (L - 1)/2$ if L is odd and $M = L/2$ if L is even. Thus the higher order $H(z)$ given in (5.2.2) is broken up and can be implemented in cascade form as illustrated in Figure 5.9. Splitting the filter in this manner reduces roundoff errors, which may be critical for some applications. However, the direct form is more efficient for implementation on most commercially available DSP processors such as the TMS320C55x.

The output $y(n)$ is a linear combination of a finite number of inputs $\{x(n), x(n - 1), \ldots, x(n - L + 1)\}$ and L coefficients $\{b_l, l = 0, 1, \ldots, L - 1\}$, which can be represented as tables illustrated in Figure 5.10. In order to compute the output at any time, we simply have to multiply the corresponding values in each table and sum the results. That is,

$$y(n) = b_0 x(n) + b_1 x(n - 1) + \cdots + b_{L-1}x(n - L + 1). \tag{5.2.29}$$

In FIR filtering, the coefficient values are constant, but the data in the signal buffer changes every sampling period, T. That is, the $x(n)$ value at time n becomes $x(n - 1)$ in the next sampling period, then $x(n - 2)$, etc., until it simply drops off the end of the delay chain.

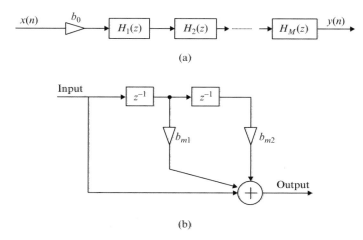

(a)

(b)

Figure 5.9 A cascade structure of FIR filter: (a) overall structure, and (b) flow diagram of second-order FIR section

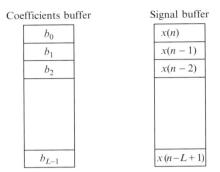

Figure 5.10   Tables of coefficient vector and signal buffer

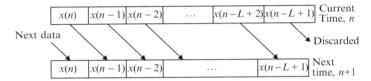

Figure 5.11   Refreshing the signal buffer for FIR filtering

The signal buffer is refreshed in every sampling period in the fashion illustrated in Figure 5.11, where the oldest sample $x(n - L + 1)$ is discarded and other signals are shifted one location to the right in the buffer. A new sample (from ADC in real-time application) is inserted to the memory location labeled as $x(n)$. The FIR filtering operation that computes $y(n)$ using (5.2.29) is then performed. The process of refreshing the signal buffer shown in Figure 5.11 requires intensive processing time if the operation is not implemented by the DSP hardware.

The most efficient method for handling a signal buffer is to load the signal samples into a circular buffer, as illustrated in Figure 5.12(a). Instead of shifting the data forward while holding the buffer addresses fixed, the data is kept fixed and the addresses are shifted backwards (counterclockwise) in the circular buffer. The beginning of the signal sample, $x(n)$, is pointed at with a pointer and the previous samples are loaded sequentially from that point in a clockwise direction. As we receive a new sample, it is placed at the position $x(n)$ and our filtering operation defined in (5.2.29) is performed. After calculating the output $y(n)$, the pointer is moved counterclockwise one position to the point at $x(n - L + 1)$ and we wait for the next input signal. The next input at time $n + 1$ is written to the $x(n - L + 1)$ position, and is referred to as $x(n)$ for the next iteration. This is permitted because the old $x(n - L + 1)$ signal dropped off the end of our delay chain after the previous calculation as shown in Figure 5.11. The circular buffer implementation of a signal buffer, or a tapped-delay-line is very efficient. The update is carried out by adjusting the address pointer without physically shifting any data in memory. It is especially useful in implementing a comb filter when L is large, since we only need to access two adjacent samples $x(n)$ and $x(n - L)$ in the circular

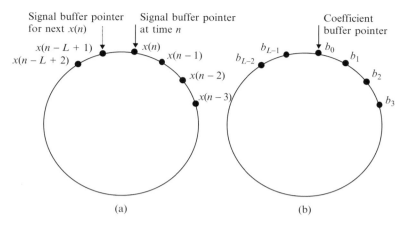

Figure 5.12 Circular buffers for FIR filter: (a) circular buffer for holding the signals for FIR filtering. The pointer to $x(n)$ is updated in the counterclockwise direction, and (b) circular buffer for FIR filter coefficients, the pointer always pointing to b_0 at the beginning of filtering

buffer. It is also used in sinewave generators and wavetable sound synthesis, where a stored waveform can be generated periodically by cycling over the circular buffer.

Figure 5.12(b) shows a circular buffer for FIR filter coefficients. Circular buffer allows the coefficient pointer to wrap around when it reaches to the end of the coefficient buffer. That is, the pointer moves from b_{L-1} to b_0 such that the filtering will always start with the first coefficient.

5.3 Design of FIR Filters

The objective of FIR filter design is to determine a set of filter coefficients $\{b_l, l = 0, 1, \ldots, L - 1\}$ such that the filter performance is close to the given specifications. A variety of techniques have been proposed for the design of FIR filters. In this section, we discuss two direct design methods. The first method is based on truncating the Fourier series representation of the desired frequency response. The second method is based on specifying equally spaced frequency samples of the frequency response of the desired filter.

5.3.1 Filter Design Procedure

The design of digital FIR filters involves five steps:

1. Specification of filter requirements.

2. Calculation and optimization of filter coefficients.

3. Realization of the filter by a suitable structure.

4.   Analysis of finite wordlength effects on filter performance.

5.   Implementation of filter in software and/or hardware.

These five steps are not necessarily independent, and they may be conducted in a different order. Specification of filter characteristics and realization of desired filters were discussed in Section 5.2. In this section, we focus on designing FIR filters for given specifications.

There are several methods for designing FIR filters. The methods discussed in this section are the Fourier series (window) method and the frequency-sampling method. The Fourier series method offers a very simple and flexible way of computing FIR filter coefficients, but it does not allow the designer adequate control over the filter parameters. The main attraction of the frequency-sampling method is that it allows recursive realization of FIR filters, which can be computationally efficient. However, it lacks flexibility in specifying or controlling filter parameters.

With the availability of an efficient and easy-to-use filter design program such as MATLAB, the Park–McClellan algorithm is now widely used in industry for FIR filter design. The Park–McClellan algorithm should be the method of first choice for most practical applications.

5.3.2   Fourier Series Method

The basic idea of Fourier series method is to design an FIR filter that approximates the desired frequency response of filter by calculating its impulse response. This method utilizes the fact that the frequency response $H(\omega)$ of a digital filter is a periodic function with period 2π. Thus it can be expanded in a Fourier series as

$$H(\omega) = \sum_{n=-\infty}^{\infty} h(n)e^{-j\omega n}, \tag{5.3.1}$$

where

$$h(n) = \frac{1}{2\pi} \int_{-\pi}^{\pi} H(\omega)e^{j\omega n} \, d\omega, \quad -\infty \le n \le \infty. \tag{5.3.2}$$

This equation shows that the impulse response $h(n)$ is double-sided and has infinite length. If $H(\omega)$ is an even function in the interval $|\omega| \le \pi$, we can show that (see exercise problem)

$$h(n) = \frac{1}{\pi} \int_0^{\pi} H(\omega) \cos(\omega n) d\omega, \quad n \ge 0 \tag{5.3.3}$$

and the impulse response is symmetric about $n = 0$. That is,

$$h(-n) = h(n), \quad n \ge 0. \tag{5.3.4}$$

For a given desired frequency response $H(\omega)$, the corresponding impulse response (filter coefficients) $h(n)$ can be calculated for a non-recursive filter if the integral (5.3.2) or (5.3.3) can be evaluated. However, in practice there are two problems with this simple design technique. First, the impulse response for a filter with any sharpness to its frequency response is infinitely long. Working with an infinite number of coefficients is not practical. Second, with negative values of n, the resulting filter is non-causal, thus is non-realizable for real-time applications.

A finite-duration impulse response $\{h'(n)\}$ of length $L = 2M + 1$ that is the best approximation (minimum mean-square error) to the ideal infinite-length impulse response can be simply obtained by truncation. That is,

$$h'(n) = \begin{cases} h(n), & -M \le n \le M \\ 0, & \text{otherwise.} \end{cases} \qquad (5.3.5)$$

Note that in this definition, we assume L to be an odd number otherwise M will not be an integer. On the unit circle, we have $z = e^{j\omega}$ and the system transfer function is expressed as

$$H'(z) = \sum_{n=-M}^{M} h'(n)z^{-n}. \qquad (5.3.6)$$

It is clear that this filter is not physically realizable in real time since the filter must produce an output that is advanced in time with respect to the input.

A causal FIR filter can be derived by delaying the $h'(n)$ sequence by M samples. That is, by shifting the time origin to the left of the vector and re-indexing the coefficients as

$$b'_l = h'(l - M), \quad l = 0, 1, \ldots, 2M. \qquad (5.3.7)$$

The transfer function of this causal FIR filter is

$$B'(z) = \sum_{l=0}^{L-1} b'_l z^{-l}. \qquad (5.3.8)$$

This FIR filter has $L \ (= 2M + 1)$ coefficients b'_l, $l = 0, 1, \ldots, L - 1$. The impulse response is symmetric about b'_M due to the fact that $h(-n) = h(n)$ given in (5.3.4). The duration of the impulse response is $2MT$ where T is the sampling period.

From (5.3.6) and (5.3.8), we can show that

$$B'(z) = z^{-M} H'(z) \qquad (5.3.9)$$

and

$$B'(\omega) = e^{-j\omega M} H'(\omega). \qquad (5.3.10)$$

Since $|e^{-j\omega M}| = 1$, we have

$$|B'(\omega)| = |H'(\omega)|. \tag{5.3.11}$$

This causal filter has the same magnitude response as that of the non-causal filter. If $h(n)$ is real, then $H'(\omega)$ is a real function of ω (see exercise problem). As discussed in Section 5.2.3, if $H'(\omega) \geq 0$, then the phase of $B'(\omega)$ is equal to $-M\omega$. If $H'(\omega) < 0$, then the phase of $B'(\omega)$ is equal to $\pi - M\omega$. Therefore the phase of $B'(\omega)$ is a linear function of ω and thus the transfer function $B'(z)$ has a constant group delay.

Example 5.5: The ideal lowpass filter of Figure 5.1(a) has frequency response

$$H(\omega) = \begin{cases} 1, & |\omega| \leq \omega_c \\ 0, & \text{otherwise.} \end{cases} \tag{5.3.12}$$

The corresponding impulse response can be computed using (5.3.2) as

$$h(n) = \frac{1}{2\pi} \int_{-\pi}^{\pi} H(\omega)e^{j\omega n} d\omega = \frac{1}{2\pi} \int_{-\omega_c}^{\omega_c} e^{j\omega n} d\omega$$

$$= \frac{1}{2\pi} \left[\frac{e^{j\omega n}}{jn} \right]_{-\omega_c}^{\omega_c} = \frac{1}{2\pi} \left[\frac{e^{j\omega_c n} - e^{-j\omega_c n}}{jn} \right]$$

$$= \frac{\sin(\omega_c n)}{\pi n} = \frac{\omega_c}{\pi} \operatorname{sinc}\left(\frac{\omega_c n}{\pi}\right), \tag{5.3.13a}$$

is referred to as sinc function, where a commonly used precise form for the sinc function is defined as

$$\operatorname{sinc}(x) = \frac{\sin(\pi x)}{\pi x}. \tag{5.3.13b}$$

Taking the limit as $n \to 0$, we have

$$h(0) = \omega_c/\pi. \tag{5.3.14}$$

By setting all impulse response coefficients outside the range $-M \leq n \leq M$ to zero, we obtain an FIR filter with the symmetry property $h'(n) = h'(-n)$, $n = 0, 1, \ldots, M$. For $M = 7$, we have $\{\sqrt{2}/6\pi, 1/2\pi, \sqrt{2}/2\pi, 1/4, \sqrt{2}/2\pi, 1/2\pi, \sqrt{2}/6\pi\}$. By shifting M units to the right, we obtain a causal FIR filter of finite-length $L = 2M + 1$ with coefficients

$$b_l' = \begin{cases} \dfrac{\omega_c}{\pi} \operatorname{sinc}\left[\dfrac{\omega_c(l - M)}{\pi}\right], & 0 \leq l \leq L - 1 \\ 0, & \text{otherwise.} \end{cases} \tag{5.3.15}$$

Example 5.6: Design a lowpass FIR filter with the frequency response

$$H(f) = \begin{cases} 1, & 0 \leq f \leq 1 \text{ kHz} \\ 0, & 1 \text{ kHz} < f \leq 4 \text{ kHz} \end{cases}$$

when the sampling rate is 8 kHz. The duration of the impulse response is limited to 2.5 msec.

Since $2MT = 0.0025$ seconds and $T = 0.000125$ seconds, we obtain $M = 10$. Thus the actual filter has 21 coefficients. From Table 3.1, 1 kHz corresponds to $\omega_c = 0.25\pi$. From (5.3.13), we have

$$h(n) = 0.25 \operatorname{sinc}\left(\frac{0.25\pi n}{\pi}\right), \quad n = 0, 1, \ldots, 10.$$

Since $h(-n) = h(n)$, for $n = 0, 1, \ldots, 10$, we can obtain, $b'_l = h(l - 10)$, $l = 0, 1, \ldots, 20$. The transfer function of the designed causal filter is

$$B'(z) = \sum_{l=0}^{20} b'_l z^{-1}.$$

Example 5.7: Design a lowpass filter of cut-off frequency $\omega_c = 0.4\pi$ with filter length $L = 41$ and $L = 61$.

When $L = 41$, $M = (L - 1)/2 = 20$. From (5.3.15), the designed impulse response is given by

$$b'_l = 0.4 \operatorname{sinc}\left(\frac{0.4\pi(l - 20)}{\pi}\right), \quad l = 0, 1, \ldots, 40.$$

When $L = 61$, $M = (L - 1)/2 = 30$. The impulse response becomes

$$b'_l = 0.4 \operatorname{sinc}\left(\frac{0.4\pi(l - 30)}{\pi}\right), \quad l = 0, 1, \ldots, 60.$$

The magnitude responses are computed and plotted in Figure 5.13 using the MATLAB script exam5_7.m given in the software package.

5.3.3 Gibbs Phenomenon

As shown in Figure 5.13, the causal FIR filter obtained by simply truncating the impulse response coefficients of the desired filter exhibits an oscillatory behavior (or ripples) in its magnitude response. As the length of the filter is increased, the number of ripples in both passband and stopband increases, and the width of the ripples decrease. The ripple becomes narrower, but its height remains almost constant. The largest ripple occurs near the transition discontinuity and their amplitude is independent of L. This undesired effect is called the Gibbs phenomenon. This is an unavoidable consequence of having an abrupt discontinuity (or truncation) of impulse response in time domain.

The truncation operation described in (5.3.5) can be considered as multiplication of the infinite-length sequence $\{h(n)\}$ by the rectangular sequence $\{w(n)\}$. That is,

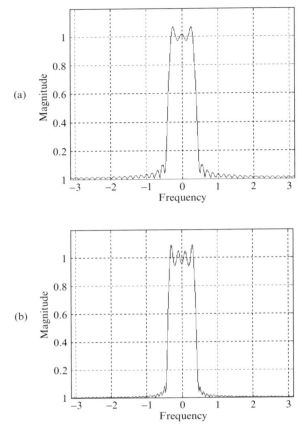

Figure 5.13 Magnitude responses of lowpass filters designed by Fourier series method: (a) $L = 41$, and (b) $L = 61$

$$h'(n) = h(n)w(n), \qquad -\infty \leq n \leq \infty, \qquad (5.3.16)$$

where the rectangular window $w(n)$ is defined as

$$w(n) = \begin{cases} 1, & -M \leq n \leq M \\ 0, & \text{otherwise.} \end{cases} \qquad (5.3.17)$$

The discrete-time Fourier transform (DTFT) of $h'(n)$ defined in (5.3.16) can be expressed as

$$H'(\omega) = H(\omega) * W(\omega) = \frac{1}{2\pi} \int_{-\pi}^{\pi} H(\varphi) W(\omega - \varphi) \, d\varphi, \qquad (5.3.18)$$

where $W(\omega)$ is the DTFT of $w(n)$ defined in (5.3.17). Thus the designed filter $H'(\omega)$ will be a smeared version of the desired filter $H(\omega)$.

Equation (5.3.18) shows that $H'(\omega)$ is obtained by the convolution of the desired frequency response $H(\omega)$ with the rectangular window's frequency response $W(\omega)$. If

$$W(\omega - \varphi) = 2\pi\delta(\omega - \varphi), \tag{5.3.19}$$

we have the desired result $H'(\omega) = H(\omega)$. Equation (5.3.19) implies that if $W(\omega)$ is a very narrow pulse centered at $\omega = 0$ such as a delta function $W(\omega) = 2\pi\delta(\omega)$, then $H'(\omega)$ will approximate $H(\omega)$ very closely. From Table 4.1, this condition requires the optimum window

$$w(n) = 1, \quad |n| < \infty, \tag{5.3.20}$$

which has infinite length.

In practice, the length of the window should be as small as possible in order to reduce the computational complexity of the FIR filter. Therefore we have to use sub-optimum windows that have the following properties:

1. They are even functions about $n = 0$.

2. They are zero in the range $|n| > M$.

3. Their frequency responses $W(\omega)$ have a narrow mainlobe and small sidelobes as suggested by (5.3.19).

The oscillatory behavior of a truncated Fourier series representation of FIR filter, observed in Figure 5.13, can be explained by the frequency response of rectangular window defined in (5.3.17). It can be expressed as

$$W(\omega) = \sum_{n=-M}^{M} e^{-j\omega n} \tag{5.3.21a}$$

$$= \frac{\sin[(2M + 1)\omega/2]}{\sin(\omega/2)}. \tag{5.3.21b}$$

A plot of $W(\omega)$ is illustrated in Figure 5.14 for $M = 8$ and 20. The MATLAB script `fig5_14.m` that generated this figure is available in the software package. The frequency response $W(\omega)$ has a mainlobe centered at $\omega = 0$. All the other ripples in the frequency response are called the sidelobes. The magnitude response $|W(\omega)|$ has the first zero at $(2M + 1)\omega/2 = \pi$. That is, $\omega = 2\pi/(2M + 1)$. Therefore the width of the mainlobe is $4\pi/(2M + 1)$. From (5.3.21a), it is easy to show that the magnitude of mainlobe is $|W(0)| = 2M + 1$. The first sidelobe is approximately at frequency $\omega_1 = 3\pi/(2M + 1)$ with magnitude $|W(\omega_1)| \approx 2(2M + 1)/3\pi$ for $M \gg 1$. The ratio of the mainlobe magnitude to the first sidelobe magnitude is

$$\left| \frac{W(0)}{W(\omega_1)} \right| \approx \frac{3\pi}{2} = 13.5 \, \text{dB}. \tag{5.3.22}$$

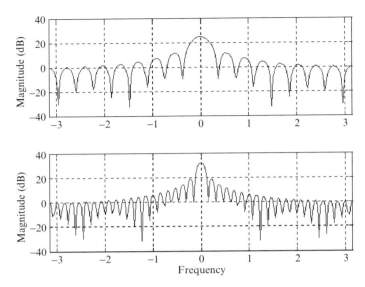

Figure 5.14   Frequency response of the rectangular window for $M = 8$ (top) and 20 (bottom)

As ω increases to the Nyquist frequency, π, the denominator grows larger. This attenuates the higher-frequency numerator terms, resulting in the damped sinusoidal function shown in Figure 5.14.

As M increases, the width of the mainlobe decreases as desired. However, the area under each lobe remains constant, while the width of each lobe decreases with an increase in M. This implies that with increasing M, ripples in $H'(\omega)$ around the point of discontinuity occur more closely but with no decrease in amplitude.

The rectangular window has an abrupt transition to 0 outside the range $-M \leq n \leq M$, which causes the Gibbs phenomenon in the magnitude response of the windowed filter's impulse response. The Gibbs phenomenon can be reduced by either using a window that tapers smoothly to 0 at each end, or by providing a smooth transition from the passband to the stopband. A tapered window causes the height of the sidelobes to diminish and increases in the mainlobe width, resulting in a wider transition at the discontinuity. This phenomenon is often referred to as leakage or smearing.

5.3.4   Window Functions

A large number of tapered windows have been developed and optimized for different applications. In this section, we restrict our discussion to four commonly used windows of length $L = 2M + 1$. That is, $w(n)$, $n = 0, 1, \ldots, L-1$ and is symmetric about its middle, $n = M$. Two parameters that predict the performance of the window in FIR filter design are its mainlobe width and the relative sidelobe level. To ensure a fast transition from the passband to the stopband, the window should have a small mainlobe width. On the other hand, to reduce the passband and stopband ripples, the area under the sidelobes should be small. Unfortunately, there is a trade-off between these two requirements.

The Hann (Hanning) window function is one period of the raised cosine function defined as

$$w(n) = 0.5\left[1 - \cos\left(\frac{2\pi n}{L - 1}\right)\right], \quad n = 0, 1, \ldots, L - 1. \tag{5.3.23}$$

Note that the Hanning window has an actual length of $L - 2$ since the two end values given by (5.3.23) are zero. The window coefficients can be generated by the MATLAB built-in function

```
w = hanning(L);
```

which returns the L-point Hanning window function in array w. Note that the MATLAB window functions generate coefficients $w(n)$, $n = 1, \ldots, L$, and is shown in Figure 5.15 (top). The magnitude response of the Hanning window is shown in the bottom of Figure 5.15. The MATLAB script han.m is included in the software package. For a large L, the peak-to-sidelobe ratio is approximately 31 dB, an improvement of 17.5 dB over the rectangular window. However, since the width of the transition band corresponds roughly to the mainlobe width, it is more than twice that resulting from the rectangular window shown in Figure 5.14.

The Hamming window function is defined as

$$w(n) = 0.54 - 0.46\cos\left(\frac{2\pi n}{L - 1}\right), \quad n = 0, 1, \ldots, L - 1, \tag{5.3.24}$$

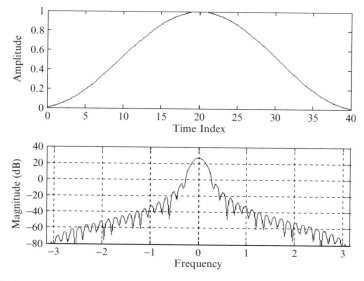

Figure 5.15 Hanning window function (top) and its magnitude response (bottom), $L = 41$

which also corresponds to a raised cosine, but with different weights for the constant and cosine terms. The Hamming function does not taper the end values to 0, but rather to 0.08. MATLAB provides the Hamming window function as

```
w = hamming(L);
```

This window function and its magnitude response are shown in Figure 5.16, and the MATLAB script ham.m is given in the software package.

The mainlobe width is about the same as for the Hanning window, but has an additional 10 dB of stopband attenuation (41 dB). In designing a lowpass filter, the Hamming window provides low ripple over the passband and good stopband attenuation is usually more appropriate for FIR filter design than the Hanning window.

Example 5.8: Design a lowpass filter of cut-off frequency $w_c = 0.4\pi$ and order $L = 61$ using the Hamming window. Using MATLAB script (exam5_8.m in the software package) similar to the one used in Example 5.7, we plot the magnitude response in Figure 5.17. Compared with Figure 5.13(b), we observe that the ripples produced by rectangular window design are virtually eliminated from the Hamming window design. The trade-off for eliminating the ripples is loss of resolution, which is shown by increasing transition width.

The Blackman window function is defined as

$$w(n) = 0.42 - 0.5 \cos\left(\frac{2\pi n}{L-1}\right) + 0.08 \cos\left(\frac{4\pi n}{L-1}\right), \tag{5.3.25}$$

$n = 0, 1, \ldots, L-1.$

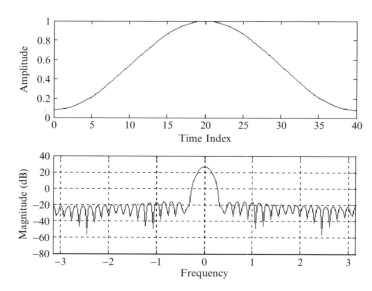

Figure 5.16   Hamming window function (top) and its magnitude response (bottom), $L = 41$

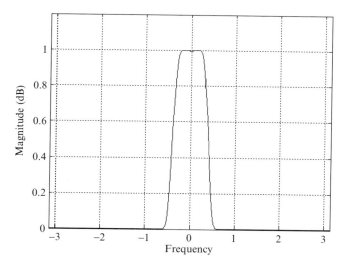

Figure 5.17 Magnitude response of lowpass filter using Hamming window, $L = 61$

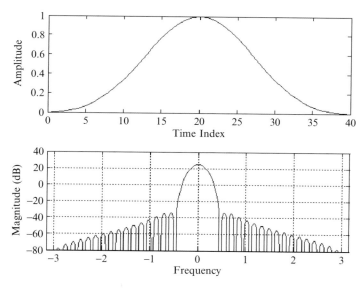

Figure 5.18 Blackman window function (top) and its magnitude response (bottom), $L = 41$

This function is also supported by the MATLAB function

```
w = blackman(L);
```

This window and its magnitude response are shown in Figure 5.18 using the MATLAB script bmw.m given in the software package.

The addition of the second cosine term in (5.3.25) has the effect of increasing the width of the mainlobe (50 percent), but at the same time improving the peak-to-sidelobe ratio to about 57 dB. The Blackman window provides 74 dB of stopband attenuation, but with a transition width six times that of the rectangular window.

The Kaiser window function is defined as

$$w(n) = \frac{I_0\left[\beta\sqrt{1 - (n - M)^2/M^2}\right]}{I_0(\beta)}, \quad n = 0, 1, \ldots, L - 1, \tag{5.3.26a}$$

where β is an adjustable (shape) parameter and

$$I_0(\beta) = \sum_{k=0}^{\infty}\left[\frac{(\beta/2)^k}{k!}\right]^2 \tag{5.3.26b}$$

is the zero-order modified Bessel function of the first kind. In practice, it is sufficient to keep only the first 25 terms in the summation of (5.3.26b). Because $I_0(0) = 1$, the Kaiser window has the value $1/I_0(\beta)$ at the end points $n = 0$ and $n = L - 1$, and is symmetric about its middle $n = M$. This is a useful and very flexible family of window functions.

MATLAB provides Kaiser window function as

```
kaiser(L, beta);
```

The window function and its magnitude response are shown in Figure 5.19 for $L = 41$ and $\beta = 8$ using the MATLAB script ksw.m given in the software package. The Kaiser window is nearly optimum in the sense of having the most energy in the mainlobe for a given peak sidelobe level. Providing a large mainlobe width for the given stopband attenuation implies the sharpness transition width. This window can provide different transition widths for the same L by choosing the parameter β to determine the trade-off between the mainlobe width and the peak sidelobe level.

As shown in (5.3.26), the Kaiser window is more complicated to generate, but the window function coefficients are generated only once during filter design. Since a window is applied to each filter coefficient when designing a filter, windowing will not affect the run-time complexity of the designed FIR filter.

Although δ_p and δ_s can be specified independently as given in (5.1.8) and (5.1.9), FIR filters designed by all windows will have equal passband and stopband ripples. Therefore we must design the filter based on the smaller of the two ripples expressed as

$$\delta = \min(\delta_p, \delta_s). \tag{5.3.27}$$

The designed filter will have passband and stopband ripples equal to δ. The value of δ can be expressed in dB scale as

$$A = -20\log_{10}\delta \text{ dB.} \tag{5.3.28}$$

In practice, the design is usually based on the stopband ripple, i.e., $\delta = \delta_s$. This is because any reasonably good choices for the passband and stopband attenuation (such as $A_p = 0.1$ dB and $A_s = 60$ dB) will result in $\delta_s < \delta_p$.

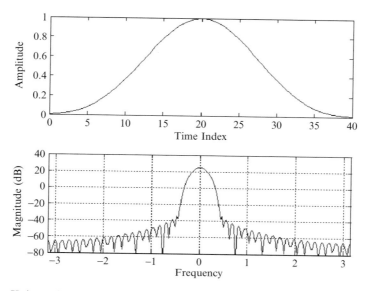

Figure 5.19 Kaiser window function (top) and its magnitude response (bottom), $L = 41$ and $\beta = 8$

The main limitation of Hanning, Hamming, and Blackman windows is that they produce a fixed value of δ. They limit the achievable passband and stopband attenuation to only certain specific values. However, the Kaiser window does not suffer from the above limitation because it depends on two parameters, L and β, to achieve any desired value of ripple δ or attenuation A. For most practical applications, $A \geq 50$ dB. The Kaiser window parameters are determined in terms of the filter specifications δ and transition width Δf as follows [5]:

$$\beta = 0.1102(A - 8.7) \qquad (5.3.29)$$

and

$$L = \frac{(A - 7.59)f_s}{14.36\Delta f} + 1, \qquad (5.3.30)$$

where $\Delta f = f_{stop} - f_{pass}$.

Example 5.9: Design a lowpass filter using the Kaiser window with the following specifications: $f_s = 10\,\text{kHz}$, $f_{pass} = 2\,\text{kHz}$, $f_{stop} = 2.5\,\text{kHz}$, $A_p = 0.1$ dB, and $A_s = 80$ dB.

From (5.1.8), $\delta_p = \dfrac{10^{0.1/20} - 1}{10^{0.1/20} + 1} = 0.0058$. From (5.1.9), $\delta_s = 10^{-80/20} = 0.0001$.

Thus we choose $\delta = \delta_s = 0.0001$ from (5.3.27) and $A = -20\log_{10}\delta = 80 = A_s$ from (5.3.28). The shaping parameter β is computed as

$$\beta = 0.1102(80 - 8.7) = 7.875.$$

The required window length L is computed as

$$L = \frac{(80 - 7.59)10}{14.36(2.5 - 2)} + 1 \approx 101.85.$$

Thus we choose the filter order $L = 103$.

The procedures of designing FIR filters using windows are summarized as follows:

1. Determine the window type that will satisfy the stopband attenuation requirements.

2. Determine the window size L based on the given transition width.

3. Calculate the window coefficients $w(n)$, $n = 0, 1, \ldots, L-1$.

4. Generate the ideal impulse response $h(n)$ using (5.3.3) for the desired filter.

5. Truncate the ideal impulse response of infinite length using (5.3.5) to obtain $h'(n)$, $-M \le n \le M$.

6. Make the filter causal by shifting the result M units to the right using (5.3.7) to obtain b'_l, $l = 0, 1, \ldots, L-1$.

7. Multiply the window coefficients obtained in step 3 and the impulse response coefficients obtained in step 6 sample-by-sample. That is,

$$b_l = b'_l \cdot w(l), \quad l = 0, 1, \ldots, L-1. \tag{5.3.31}$$

Applying a window to an FIR filter's impulse response has the effect of smoothing the resulting filter's magnitude response. A symmetric window will preserve a symmetric FIR filter's linear-phase response.

The advantage of the Fourier series method with windowing is its simplicity. It does not require sophisticated mathematics and can be carried out in a straightforward manner. However, there is no simple rule for choosing M so that the resulting filter will meet exactly the specified cut-off frequencies. This is due to the lack of an exact relation between M and its effect on leakage.

5.3.5 Frequency Sampling Method

The frequency-sampling method is based on sampling a desired amplitude spectrum and obtaining filter coefficients. In this approach, the desired frequency response $H(\omega)$ is first uniformly sampled at L equally spaced points $\omega_k = \frac{2\pi k}{L}$, $k = 0, 1, \ldots, L-1$. The frequency-sampling technique is particularly useful when only a small percentage of the frequency samples are non-zero, or when several bandpass functions are desired simultaneously. A unique attraction of the frequency-sampling method is that it also allows recursive implementation of FIR filters, leading to computationally efficient filters. However, the disadvantage of this method is that the actual magnitude response of the filter will match the desired filter only at the points that were samples.

For a given frequency response $H(\omega)$, we take L samples at frequencies of kf_s/L, $k = 0, 1, \ldots, L - 1$ to obtain $H(k)$, $k = 0, 1, \ldots L - 1$. The filter coefficients b_l can be obtained as the inverse DFT of these frequency samples. That is,

$$b_l = \frac{1}{L} \sum_{k=0}^{L-1} H(k) e^{j(2\pi/L)lk}, \quad l = 0, 1, \ldots, L - 1. \tag{5.3.32}$$

The resulting filter will have a frequency response that is exactly the same as the original response at the sampling instants. However, the response may be significantly different between the samples. To obtain a good approximation to the desired frequency response, we have to use a sufficiently large number of frequency samples. That is, we have to use a large L.

Let $\{B_k\}$ be the DFT of $\{b_l\}$ so that

$$B_k = \sum_{l=0}^{L-1} b_l e^{-j(2\pi/L)lk}, \quad k = 0, 1, \ldots, L - 1 \tag{5.3.33}$$

and

$$b_l = \frac{1}{L} \sum_{k=0}^{L-1} B_k e^{j(2\pi/L)lk}, \quad l = 0, 1, \ldots, L - 1. \tag{5.3.34}$$

Using the geometric series $\sum_{l=0}^{L-1} x^l = \frac{1 - x^L}{1 - x}$ (see Appendix A.2), the desired filter's transfer function can be obtained as

$$H(z) = \sum_{l=0}^{L-1} b_l z^{-l} = \sum_{l=0}^{L-1} \left(\frac{1}{L} \sum_{k=0}^{L-1} B_k e^{j(2\pi lk/L)} \right) z^{-l}$$

$$= \frac{1}{L}(1 - z^{-L}) \sum_{k=0}^{L-1} \frac{B_k}{1 - e^{j(2\pi k/L)} z^{-1}}. \tag{5.3.35}$$

This equation changes the transfer function into a recursive form, and $H(z)$ can be viewed as a cascade of two filters: a comb filter, $(1 - z^{-L})/L$ as discussed in Section 5.2.2, and a sum of L first-order all-pole filters.

The problem is now to relate $\{B_k\}$ to the desired sample set $\{H(k)\}$ used in (5.3.32). In general, the frequency samples $H(k)$ are complex. Thus a direct implementation of (5.3.35) would require complex arithmetic. To avoid this complication, we use the symmetry inherent in the frequency response of any FIR filter with real impulse response $h(n)$. Suppose that the desired amplitude response $|H(\omega)|$ is sampled such that B_k and $H(k)$ are equal in amplitude. For a linear-phase filter (assume positive symmetry), we have

$$H_k = |H(k)| = |H(L - k)| = |B_k|, \quad k \le \frac{L}{2}. \tag{5.3.36}$$

The remaining problem is to adjust the relative phases of the B_k so that H_k will provide a smooth approximation to $|H(\omega)|$ between the samples. The phases of two adjacent contributions to $|H(\omega)|$ are in phase except between the sample points, which are 180 degrees out of phase [10]. Thus the two adjacent terms should be subtracted to provide a smooth reconstruction between sample points. Therefore B_k should be equal to H_k with alternating sign. That is,

$$B_k = (-1)^k H_k, \quad k \le \frac{L}{2}. \tag{5.3.37}$$

This is valid for L being odd or even, although it is convenient to assume that B_0 is zero and $B_{L/2}$ is 0 if L is even.

With these assumptions, the transfer function given in (5.3.35) can be further expressed as

$$H(z) = \frac{1}{L}(1 - z^{-L}) \sum_{k \le L/2} (-1)^k H_k \left[\frac{1}{1 - e^{j(2\pi k/L)}z^{-1}} + \frac{1}{1 - e^{j[2\pi(L-k)/L]}z^{-1}} \right]$$

$$= \frac{2}{L}(1 - z^{-L}) \sum_{k \le L/2} (-1)^k H_k \frac{1 - \cos(2\pi k/L)z^{-1}}{1 - 2\cos(2\pi k/L)z^{-1} + z^{-2}}. \tag{5.3.38}$$

This equation shows that $H(z)$ has poles at $e^{\pm j(2\pi k/L)}$ on the unit circle in the z-plane. The comb filter $1 - z^{-L}$ provides L zeros at $z_k = e^{j(2\pi k/L)}$, $k = 0, 1, \ldots, L - 1$, equally spaced around the unit circle. Each non-zero sample H_k brings to the digital filter a conjugate pair of poles, which cancels a corresponding pair of zeros at $e^{\pm j(2\pi k/L)}$ on the unit circle. Therefore the filter defined in (5.3.38) is recursive, but does not have poles in an essential sense. Thus it still has a finite impulse response.

Although the pole–zero cancellation is exact in theory, it will not be in practice due to finite wordlength effects in the digital implementation. The $\cos(2\pi k/L)$ terms in (5.3.38) will in general be accurate only if more bits are used. An effective solution to this problem is to move the poles and zeros slightly inside the unit circle by moving them into radius r, where $r < 1$.

Therefore (5.3.38) is modified to

$$H(z) = \frac{2}{L}(1 - r^L z^{-L}) \sum_{k \le L/2} (-1)^k H_k \frac{1 - r\cos(2\pi k/L)z^{-1}}{1 - 2r\cos(2\pi k/L)z^{-1} + r^2 z^{-2}}. \tag{5.3.39}$$

The frequency-sampling filter defined in (5.3.39) can be realized by means of a comb filter

$$C(z) = \frac{2}{L}(1 - r^L z^{-L})$$
(5.3.40)

and a bank of resonators

$$R_k(z) = \frac{1 - r\cos(2\pi k/L)z^{-1}}{1 - 2r\cos(2\pi k/L)z^{-1} + r^2 z^{-2}}, \quad 0 \le k \le L/2,$$
(5.3.41)

as illustrated in Figure 5.20. They effectively act as narrowband filters, each passing only those frequencies centered at and close to resonant frequencies $2\pi k/L$ and excluding others outside these bands. Banks of these filters weighted by frequency samples H_k can then be used to synthesize a desired frequency response.

The difference equation representing the comb filter $C(z)$ can be written in terms of variables in Figure 5.20 as

$$u(n) = \frac{2}{L}[x(n) - r^L x(n - L)].$$
(5.3.42)

The block diagram of this modified comb filter is illustrated in Figure 5.21. An effective technique to implement this comb filter is to use the circular buffer, which is available in most modern DSP processors such as the TMS320C55x. The comb filter output $u(n)$ is common to all the resonators $R_k(z)$ connected in parallel.

The resonator with $u(n)$ as common input can be computed as

$$f_k(n) = u(n) - r\cos\left(\frac{2\pi k}{L}\right)u(n - 1) + 2r\cos\left(\frac{2\pi k}{L}\right)f_k(n - 1) - r^2 f_k(n - 2)$$

$$= u(n) + r\cos\left(\frac{2\pi k}{L}\right)[2f_k(n - 1) - u(n - 1)] - r^2 f_k(n - 2), \quad 0 \le k \le L/2 \quad (5.3.43)$$

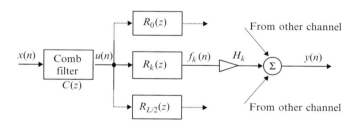

Figure 5.20 Block diagram of frequency sampling filter structure, channel k

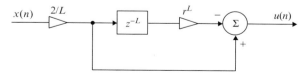

Figure 5.21 Detailed block diagram of comb filter

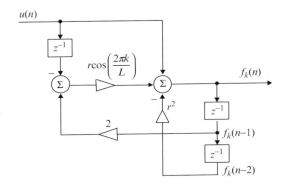

Figure 5.22 Detailed block diagram of resonator, $R_k(z)$

where $f_k(n)$ is the output from the kth resonator. The detailed flow diagram of the resonator is illustrated in Figure 5.22. Note that only one coefficient $r\cos\left(\frac{2\pi k}{L}\right)$ is needed for each resonator. This significantly reduces the memory requirements when compared to other second-order IIR bandpass filters.

Finally, the output of each resonator $f_k(n)$ is weighted by H_k and combined into the overall output

$$y(n) = \sum_{k \le L/2} (-1)^k H_k f_k(n). \qquad (5.3.44)$$

Example 5.10: Design a bandpass filter of sampling rate 10 kHz, passband 1–1.25 kHz, power gain at ends of passband is 0.5 (-3 dB), and duration of impulse response is 20 ms.

Since the impulse response duration is 20 ms,

$$L = 0.02/T = 200.$$

The frequency samples must be $10\,000/200 = 50$ Hz apart. The passband is sampled at frequencies 1, 1.05, 1.1, 1.15, 1.2, and 1.25 kHz that correspond to $k = 20, 21, 22, 23, 24$, and 25.

Therefore the frequency-sampling filter consists of six channels (resonators) as shown in Figure 5.20 with

$$H_{20} = H_{25} = \sqrt{0.5} = 0.707\ (-3\text{dB}) \quad \text{and} \quad H_{21} = H_{22} = H_{23} = H_{24} = 1.$$

The designed frequency-sampling filter is implemented in C (`fsf.c` in the software package) using the circular buffer. In the program, we assume the input data file is using binary floating-point format, and save the output using the same format.

5.4 Design of FIR Filters Using MATLAB

The filter design methods described in Section 5.3 can be easily realized using a computer. A number of filter design algorithms are based on iterative optimization techniques that are used to minimize the error between the desired and actual frequency responses. The most widely used algorithm is the Parks–McClellan algorithm for designing the optimum linear-phase FIR filter. The algorithm spreads out the error within a frequency band, which produces ripples of equal magnitude. The user can specify the relative importance of each band. For example, there would be less ripples in the passband than in the stopband. A variety of software packages are commercially available that have made digital filter design rather simple on a computer. In this section, we consider only the MATLAB application for designing FIR filters.

As discussed in Section 5.1.3, the filter specifications ω_p, ω_s, δ_p, and δ_s are given. The filter order L can be estimated using the simple formula

$$L = 1 + \frac{-20 \log_{10} \sqrt{\delta_p \delta_s} - 13}{14.6 \Delta f}, \tag{5.4.1}$$

where

$$\Delta f = \frac{\omega_s - \omega_p}{2\pi}. \tag{5.4.2}$$

A highly efficient procedure, the Remez algorithm, is developed to design the optimum linear-phase FIR filters based on the Parks–McClellan algorithm. The algorithm uses the Remez exchange and Chebyshev approximation theory to design a filter with an optimum fit between the desired and actual frequency responses. This algorithm is implemented as an M-file function `remez` that is available in the *Signal Processing Toolbox* of MATLAB. There are various versions of this function:

```
b = remez(N, f, m);
b = remez(N, f, m, w);
b = remez(N, f, m, 'ftype');
b = remez(N, f, m, w, 'ftype');
```

The function returns row vector b containing the $N + 1$ coefficients of the FIR filter of order $L = (N + 1)$. The vector f specifies bandedge frequencies in the range between 0 and 1, where 1 corresponds to the Nyquist frequency $f_N = f_s/2$. The frequencies must be in an increasing order with the first element being 0 and the last element being 1. The desired values of the FIR filter magnitude response at the specified bandedge frequencies in f are given in the vector m, with the elements given in equal-valued pairs. The vector f and m must be the same length with the length being an even number.

Example 5.11: Design a linear-phase FIR bandpass filter of order 18 with a passband from 0.4 to 0.6. Plot the desired and actual frequency responses using the following MATLAB script (`exam5_11.m` in the software package):

```
f = [0  0.3  0.4  0.6  0.7  1];
m = [0  0  1  1  0  0];
```

```
b = remez(17, f, m);
[h, omega] = freqz(b, 1, 512);
plot(f, m, omega/pi, abs(h));
```

The graph is shown in Figure 5.23.

The desired magnitude response in the passband and the stopband can be weighted by an additional vector w. The length of w is half of the length of f and m. As shown in Figure 5.7, there are four types of linear-phase FIR filters. Types III (*L* even) and IV (*L* odd) are used for specialized filter designs: the Hilbert transformer and differentiator. To design these two types of FIR filters, the arguments 'hilbert' and 'differentiator' are used for 'ftype' in the last two versions of remez.

Similar to remez, MATLAB also provides firls function to design linear-phase FIR filters which minimize the weighted, integrated squared error between the ideal filter and the actual filter's magnitude response over a set of desired frequency bands. The synopsis of function firls is identical to the function remez.

Two additional functions available in the MATLAB *Signal Processing Toolbox*, fir1 and fir2, can be used in the design of FIR filters using windowed Fourier series method. The function fir1 designs windowed linear-phase lowpass, highpass, bandpass, and bandstop FIR filters with the following forms:

```
b = fir1(N, Wn);
b = fir1(N, Wn, 'filtertype');
b = fir1(N, Wn, window);
b = fir1(N, Wn, 'filtertype', window);
```

The basic form, b = fir1(N, Wn), generates the length $L = N + 1$ vector b containing the coefficients of a lowpass filter with a normalized cut-off frequency Wn between 0 and

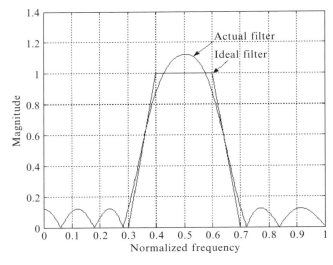

Figure 5.23   Magnitude responses of the desired and actual FIR filters

1 using the Hamming window. If `Wn` is a two-element vector, `Wn = [w1 w2]`, `fir1` generates a bandpass filter with passband $w1 \leq \omega \leq w2$. The argument 'filtertype' specifies a filter type, where 'high' for a highpass filter with cut-off frequency `Wn`, and 'stop' for a bandstop filter with cut-off frequency `Wn = [w1 w2]`. The vector `window` must have L elements and is generated by

```
window = hamming(L);
window = hanning(L);
window = blackman(L);
window = kaiser(L, beta);
```

If no window is specified, the Hamming window will be used as the default for filter design.

The function `fir2` is used to design an FIR filter with arbitrarily shaped magnitude response. The various forms of this function are:

```
b = fir2(N, f, m);
b = fir2(N, f, m, window);
b = fir2(N, f, m, npt);
b = fir2(N, f, m, npt, window);
b = fir2(N, f, m, npt, lap);
b = fir2(N, f, m, npt, lap, window);
```

The basic form, b = fir2(N, f, m), is used to design an FIR filter of order $L = N + 1$. The vector m defines the magnitude response sampled at the specified frequency points given by the vector f. The argument `npt` specifies the number of points for the grid onto which `fir2` interpolates the frequency response. The argument `lap` specifies the size of the region which `fir2` inserts around duplicated frequency points. The details are given in the *Signal Processing Toolbox* manual.

5.5 Implementation Considerations

Discrete-time FIR filters designed in the previous section can be implemented in the following forms: hardware, firmware, and software. Digital filters are realized by digital computers or DSP hardware uses quantized coefficients to process quantized signals. In this section, we discuss the software implementation of digital FIR filters using MATLAB and C to illustrate the main issues. We will consider finite wordlength effects in this section. The DSP chip implementation using the TMS320C55x will be presented in the next section.

5.5.1 Software Implementations

The software implementation of digital filtering algorithms on a PC is often carried out first to verify whether or not the chosen algorithm meets the goals of the application. MATLAB and C implementation may be adequate if the application does not require real-time signal processing. For simulation purposes, it is convenient to use a powerful software package such as MATLAB for signal processing.

MATLAB provides the built-in function `filter` for **FIR** and **IIR** filtering. The basic form of this function is

```
y = filter(b, a, x)
```

For FIR filtering, a $= 1$ and filter coefficients b_l are contained in the vector b. The input vector is x while the output vector generated by the filter is y.

Example 5.12: The following C function `fir.c` implements the linear convolution (FIR filtering, inner product, or dot product) operation given in (5.2.1). The arrays x and h are declared to proper dimension in the main program `firfltr.c` given in Appendix C.

```
/****************************************************************
*  FIR – This function performs FIR filtering (linear convolution)
*            ntap-1
*     y(n) = sum hi*x(n-i)
*            i=0
****************************************************************/
float fir(float *x, float *h, int ntap)
{
   float yn = 0.0;              /* Output of FIR filter */
   int i;                       /* Loop index          */
   for(i = 0; i > ntap; i++)
   {
     yn += h[i] * x[i];         /* Convolution of x(n) with h(n) */
   }
   return(yn);                  /* Return y(n) to main function  */
}
```

The signal buffer x is refreshed in every sampling period as shown in Figure 5.11, and is implemented in the following C function called `shift.c`. At each iteration, the oldest sample $x(n - L + 1)$ discarded and other signals are shifted one location to the right. The most recent sample $x(n)$ is inserted to memory location x[0] .

```
/****************************************************************
*   SHIFT – This function updates signal vector of order ntap
*           data stored as [x(n) x(n-1) ... x(n-ntap+1)]
****************************************************************/
void  shift(float *x, int ntap,  float in)
{
   int i;                       /* Loop index            */
   for(i = ntap-1; i > 0; i--)
   {
     x[i] = x[i-1];             /* Shift old data x(n-i) */
   }
   x[0] = in;                   /* Insert new data x(n)  */
   return;
}
```

The FIR filtering defined in (5.2.1) can be implemented using DSP chips or special purpose ASIC devices. Modern programmable DSP chips, such as the TMS320C55x, have architecture that is optimized for the repetitive nature of multiplications and accumulations. They are also optimized for performing the memory moves required in updating the contents of the signal buffer, or realizing the circular buffers. The implementation of FIR filters using the TMS320C55x will be discussed in Section 5.6.

5.5.2 Quantization Effects in FIR Filters

Consider an FIR filter transfer function given in (5.2.2). The filter coefficients, b_l, are determined by a filter design package such as MATLAB for given specifications. These coefficients are usually represented by double-precision floating-point numbers and have to be quantized using a finite number of bits for a given fixed-point processor such as 16-bit for the TMS320C55x. The filter coefficients are only quantized once in the design process, and those values remain constant in the filter implementation. We must check the quantized design. If it no longer meets the given specifications, we can optimize, redesign, restructure, and/or use more bits to satisfy the specifications. It is especially important to consider quantization effects when designing filters for implementation using fixed-point arithmetic.

Let b'_l denote the quantized values corresponding to b_l. As discussed in Chapter 3, the nonlinear quantization can be modeled as a linear operation expressed as

$$b'_l = Q[b_l] = b_l + e(l), \tag{5.5.1}$$

where $e(l)$ is the quantization error and can be assumed as a uniformly distributed random noise of zero mean and variance defined in (3.5.6).

Quantization of the filter coefficients results in a new filter transfer function

$$H'(z) = \sum_{l=0}^{L-1} b'_l z^{-l} = \sum_{l=0}^{L-1} [b_l + e(l)] z^{-l} = H(z) + E(z), \tag{5.5.2}$$

where

$$E(z) = \sum_{l=0}^{L-1} e(l) z^{-l} \tag{5.5.3}$$

is the FIR filter representing the error in the transfer function due to coefficient quantization. The FIR filter with quantized coefficients can be modeled as a parallel connection of two FIR filters as illustrated in Figure 5.24.

The frequency response of the actual FIR filter with quantized coefficients b'_l can be expressed as

$$H'(\omega) = H(\omega) + E(\omega), \tag{5.5.4}$$

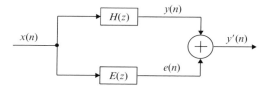

Figure 5.24 Model of the FIR filter with quantized coefficients

where

$$E(\omega) = \sum_{l=0}^{L-1} e(l)e^{-j\omega l} \qquad (5.5.5)$$

represents the error in the desired frequency response $H(\omega)$. The error is bounded by

$$|E(\omega)| = \left| \sum_{l=0}^{L-1} e(l)e^{-j\omega l} \right| \leq \sum_{l=0}^{L-1} |e(l)||e^{-j\omega l}| \leq \sum_{l=0}^{L-1} |e(l)|. \qquad (5.5.6)$$

As shown in (3.5.3),

$$|e(l)| \leq \frac{\Delta}{2} = 2^{-B}. \qquad (5.5.7)$$

Thus Equation (5.5.6) becomes

$$|E(\omega)| \leq L \cdot 2^{-B}. \qquad (5.5.8)$$

This bound is too conservative because it can only be reached if all errors $e(l)$ are of the same sign and have the maximum value in the range. A more realistic bound can be derived assuming $e(l)$ are statistically independent random variables. The variance of $E(\omega)$ can be obtained as

$$\sigma_E^2(\omega) \leq 2^{-B+1} \sqrt{\frac{2L-1}{12}}. \qquad (5.5.9)$$

This bound can be used to estimate the wordlength of the FIR coefficients required to meet the given filter specifications.

As discussed in Section 3.6.3, the most effective technique in preventing overflow is scaling down the magnitude of signals. The scaling factor used to prevent overflow in computing the sum of products defined in (5.2.1) is given in (3.6.4) or (3.6.5).

As discussed in Section 5.2.3, most FIR filters are linear-phase and the coefficients are constrained to satisfy the symmetry condition (5.2.15) or the anti-symmetry property (5.2.18). Quantizing both sides of (5.2.15) or (5.2.18) has the same quantized value for each l, which implies that the filter still has linear phase after quantization. Only the magnitude response of the filter is changed. This constraint greatly reduces the sensitivity of the direct-form FIR filter implementation given in (5.2.1). There is no need to use the cascade form shown in Figure 5.9 for FIR filters, unlike the IIR filters that require cascade form. This issue will be discussed in the next chapter.

5.6 Experiments Using the TMS320C55x

FIR filters are widely used in a variety of areas such as audio, video, wireless communications, and medical devices. For many practical applications such as wireless communications (CDMA/TDMA), streamed video (MPEG/JPEG), and voice over internet protocol (VoIP), the digital samples are usually grouped in frames with time duration from a few milliseconds to several hundred milliseconds. It is more efficient for the C55x to process samples in frames (blocks). The FIR filter program `fir.c` given in Section 5.5.1 are designed for processing signals sample-by-sample. It can be easily modified to handle a block of samples. We call the filter that processes signals block-by-block a block filter. Example 5.13 is an example of a block-FIR filter written in C.

Example 5.13: The following C function, `block_fir.c`, implements an L-tap FIR filter that processes a block of M input signals at a time.

```
/*****************************************************************
*   BLOCK_FIR - This function performs block FIR filter of size M
****************************************************************** /
void fir(float *in, int M, float *h, int L, float *out, float *x)
{
  float yn;                    /* Output of FIR filter */
  int i,j;                     /* Loop index          */
  for(j = 0; j < M; j++)
  {
    x[0] = in[j];              /* Insert new data x(n) */
    /*****************************************************************
    *   FIR filtering (linear convolution)
    *              L-1
    *       y(n) = sum hi* x (n-i)
    *              i=0
    ****************************************************************** /
    for(yn = 0.0, i = 0; i < L; i++)
    {
      yn += h[i] * x[i];       /* Convolution of x(n) with h(n) */
    }
    out[j] = yn;               /* Store y(n) to output buffer   */
    /*****************************************************************
    * Updates signal buffer, as [x(n) x(n-1) ...x(n-L+1)]
    ****************************************************************** /
```

```
       for(i = L−1; i > 0; i−−)
       {
          x[i] = x[i−1] ;        /* Shift old data x(n−i)                   */
       }
    }
    return;
}
```

Simulation and emulation methods are commonly used in DSP software development. They are particularly useful for the study and analysis of DSP algorithms. By using a software signal generator, we can produce the exact same signals repeatedly during the debug and analysis processes. Table 5.1 lists the example of sinusoid signal generator, signal_gen.c that is used to generate the experimental data input5.dat for experiments in this section.

Table 5.1   List of sinusoid signal generator for the experiments

```
/*
     signal_gen.c — Generate sinewaves as testing data in Q15 format

     Prototype:   void signal_gen(int *, int)
                  arg0: − data buffer pointer for output
                  arg1: − number of samples
*/
#include <math.h>

#define T 0.000125                      /* 8000 Hz sampling frequency */
#define f1 800                          /* 800 Hz frequency           */
#define f2 1800                         /* 1800 Hz frequency          */
#define f3 3300                         /* 3300 Hz frequency          */
#define PI 3.1415926
#define two_pi_f1_T(2* PI* f1* T)  /* 2*pi* f1/Fs                    */
#define two_pi_f2_T(2* PI* f2* T)  /* 2*pi* f2/Fs                    */
#define two_pi_f3_T(2* PI* f3* T)  /* 2*pi* f3/Fs                    */
#define a1 0.333                        /* Magnitude for wave 1       */
#define a2 0.333                        /* Magnitude for wave 2       */
#define a3 0.333                        /* Magnitude for wave 3       */

static unsigned int n = 0;
void signal_gen (int *x, int N)
{
   float temp;
   int i;
     for(i = 0; i < N; i++)
     {
         temp = a1* cos((double) two_pi_f1_T* n);
         temp += a2* cos((double) two_pi_f2_T* n);
         temp += a3* cos((double) two_pi_f3_T* n);
```

Table 5.1 (*continued*)

```
        n++;
        x[i] = (int)((0x7fff*temp)+0.5);
     }
  }
```

5.6.1 Experiment 5A – Implementation of Block FIR Filter

The difference equation (5.2.1) and Example 5.13 show that FIR filtering includes two different processes: (1) It performs a summation of products generated by multiplying the incoming signals with the filter coefficients. (2) The entire signal buffer is updated to include a new sample. For an FIR filter of L coefficients, L multiplications, $(L - 1)$ additions, and additional data memory move operations are required for the complete filtering operations. Refreshing the signal buffer in Example 5.13 uses the memory shift shown in Figure 5.11. To move $(L - 1)$ samples in the signal buffer to the next memory location requires additional instruction cycles. These extensive operations make FIR filtering a computation-intensive task for general-purpose microprocessors.

The TMS320C55x has three important features to support FIR filtering. It has multiply–accumulate instructions, the circular addressing modes, and the zero-overhead nested loops. Using multiply–accumulate instructions, the C55x can perform both multiplication and addition with rounding options in one cycle. That is, the C55x can complete the computation of one filter tap at each cycle. In Example 5.13, the updating of the signal buffers by shifting data in memory requires many data-move operations. In practice, we can use the circular buffers as shown in Figure 5.12. The FIR filtering in Example 5.13 can be tightly placed into the loops. To reduce the overhead of loop control, the loop counters in the TMS320C55x are handled using hardware, which can support three levels of zero-overhead nested loops using BRC0, BRC1, and CSR registers.

The block FIR filter in Example 5.13 can be implemented with circular buffers using the following TMS320C55x assembly code:

```
      mov    # M−1,BRC0
      mov    # L−3,CSR
   || rptblocal sample_loop-1  ; Start the outer loop
      mov    *AR0+,*AR3         ; Put the new sample to signal buffer
      mpym   *AR3+,*AR1+,AC0    ; Do the 1st operation
   || rpt    CSR                ; Start the inner loop
      macm   *AR3+,*AR1+,AC0
      macmr  *AR3,*AR1+,AC0     ; Do the last operation
      mov    hi(AC0),*AR2+      ; Save result in Q15 format
  sample_loop
```

Four auxiliary registers, AR0–AR3, are used as pointers in this example. AR0 points to the input buffer in[]. The signal buffer x[] containing the current input $x(n)$ and the $L - 1$ old samples is pointed at by AR3. The filter coefficients in the array h[] are pointed at by AR1. For each iteration, a new sample is placed into the signal buffer and the inner loop repeats the multiply–accumulate instructions. Finally, the filter output $y(n)$ is rounded and stored in the output buffer out[] that is pointed at by AR2. Both AR1 and AR3 use circular addressing mode. At the end of the computation, the coefficient pointer AR1 will be wrapped around, thus pointing at the first coefficient again. The signal buffer pointer AR3 will point at the oldest sample, $x(n - L + 1)$. In the next iteration, AR1 will start from the first tap, while the oldest sample in the signal buffer will be replaced with the new input sample as shown in Figure 5.12.

In the assembly code, we initialize the repeat counter CSR with the value $L-3$ for $L-2$ iterations. This is because we use a multiplication instruction before the loop and a multiply–accumulate-rounding instruction after the repeat loop. Moving instructions outside the repeat loops is called loop unrolling. It is clear from using the loop unrolling technique, the FIR filter must have at least three coefficients. The complete assembly program fir.asm is given in the experimental software package.

The C program exp5a.c listed in Table 5.2 will be used for Experiment 5A. It uses the data file input5.dat as input, and calls fir() to perform lowpass filtering. Since we use circular buffers, we define a global variable index as the signal buffer index for tracking the starting position of the signal buffer for each sample block. The C55x compiler supports several pragma directives. We apply the two most frequently used directives, CODE_SECTION and DATA_SECTION, to allocate the C functions' program and data variables for experiments. For a complete list of C pragma directives that the compiler supports, please refer to the TMS320C55x Optimizing C Compiler User's Guide [11].

Table 5.2   List of the C program for Experiment 5A

```
/*
      exp5a.c — Block FIR filter experiment using input data file
*/
#define M 128           /* Input sample size  */
#define L    48         /* FIR filter order   */
#define SN   L          /* Signal buffer size */
extern unsigned int fir(int *, unsigned int, int *, unsigned int,
                        int *, int *, unsigned int);

/* Define DSP system memory map */
#pragma DATA_SECTION(LP_h, "fir_coef");
#pragma DATA_SECTION(x, "fir_data");
#pragma DATA_SECTION(index, "fir_data");
#pragma DATA_SECTION(out, "output");
#pragma DATA_SECTION(in, "input");
#pragma DATA_SECTION(input, "input");
#pragma CODE_SECTION(main, "fir_code");

/* Input data */
#include "input5.dat"
```

Table 5.2   (*continued*)

```
/* Low-pass FIR filter coefficients */
int LP_h[L] =
{-6,28,46,27,-35,-100,-93,26,191,240,52,-291,-497,-278,
337,888,773,-210,-1486,-1895,-442,2870,6793,9445,
9445,6793,2870,-442,-1895,-1486,-210,773,888,337,
-278,-497,-291,52,240,191,26,-93,-100,-35,27,46,28,-6};

int x[SN];              /* Signal buffer       */
unsigned int index;     /* Signal buffer index */
int out[M];             /* Output buffer       */
int in[M];              /* Input buffer        */

void main(void)
{
    unsigned int i,j;

    /* Initialize filter signal buffer      */
    for(i = 0; i < SN;i++)
         x[i] = 0;
    index = 0;
    /* Processing samples using a block FIR filter          */
    j = 0;
    for(;;)
    {
      for(i = 0; i < M; i++)
      {
         in[i] = input[j++]; /* Get a buffer of samples     */
         if(j == 160)
            j = 0;
      }
      index = fir(in,M,LP_h,L,out,x,index);  /* FIR filtering */
    }
}
```

The `CODE_SECTION` is a pragma directive that allocates code objects to a named code section. The syntax of the pragma is given as:

```
#pragma CODE_SECTION(func_name, "section_name");
```

where the `func_name` is the name of the C function that will be allocated into the program memory section defined by the `section_name`. The linker command file uses this section to define and allocate a specific memory section for the C function.

The `DATA_SECTION` is the pragma directive that allocates data objects to a named data section. The syntax of the pragma is given as:

```
#pragma DATA_SECTION(var_name, "section_name");
```

where the var_name is the variable name contained in the C function that will be allocated into the data section defined by the section_name. The linker command file uses this name for data section allocation to system memory.

Go through the following steps for Experiment 5A:

1. Copy the assembly program fir.asm, the C function epx5a.c, the linker command file exp5.cmd, and the experimental data input5.dat from the software package to the working directory.

2. Create the project exp5a and add the files fir.asm, epx5a.c, and exp5.cmd to the project.

   The prototype of the FIR routine is defined as:

   ```
   unsigned int fir(int *in, unsigned int M, int *h,
                    unsigned int L, int *out, int *x, int index);
   ```

   where in is the pointer to the input data buffer in[], M defines the number of samples in the input data buffer in[], h is the pointer to the filter coefficient buffer LP_h[], L is the number of FIR filter coefficients, out is the pointer to the output buffer out[], x is the pointer to the signal buffer x[], and index is the index for signal buffer.

3. Build, debug, and run the project exp5a. The lowpass filter will attenuate two higher frequency components at 1800 and 3300 Hz and pass the 800 Hz sinewave. Figure 5.25 shows the time-domain and frequency-domain plots of the input and output signals.

4. Use the CCS animation capability to view the FIR filtering process frame by frame. Profile the FIR filter to record the memory usage and the average C55x cycles used for processing one block of 128 samples.

5.6.2 Experiment 5B – Implementation of Symmetric FIR Filter

As shown in Figure 5.7, a symmetric FIR filter has the characteristics of symmetric impulse responses (or coefficients) about its center index. Type I FIR filters have an even number of symmetric coefficients, while Type II filters have an odd number. An even symmetric FIR filter shown in Figure 5.8 indicates that only the first half of the filter coefficients are necessary for computing the filter result.

The TMS320C55x has two special instructions, firsadd and firssub, for implementing symmetric and anti-symmetric FIR filters. The former can be used to compute symmetric FIR filters given in (5.2.23), while the latter can be used for anti-symmetric FIR filters defined in (5.2.24). The syntax for symmetric and anti-symmetric filter instructions are

```
firsadd Xmem,Ymem,Cmem,ACx,ACy  ; Symmetric FIR filter
firssub Xmem,Ymem,Cmem,ACx,ACy  ; Anti-symmetric FIR filter
```

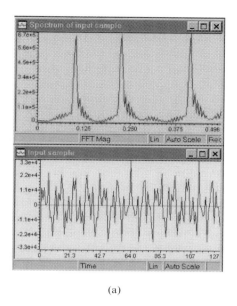

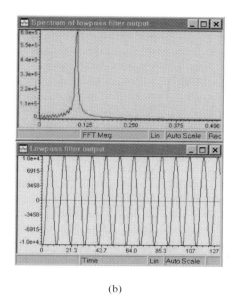

(a)   (b)

Figure 5.25   Input and output signals of Experiment 5A: (a) input signals in the frequency (top) and time (bottom) domains, and (b) output signals in the frequency (top) and time (bottom) domains

where Xmem and Ymem are the signal buffers for $\{x(n), x(n-1), \ldots x(n-L/2+1)\}$ and $\{x(n-L/2), \ldots x(n-L+1)\}$, and Cmem is the coefficient buffer.

For a symmetric FIR filter, the firsadd instruction is equivalent to performing the following parallel instructions in one cycle:

```
        macm  *CDP+,ACx,ACy       ; b_l[x(n−1) + x(n + l − L +1)]
     || add   *ARx+,*ARy+,ACx     ; x(n − l +1) + x(n + l − L + 2)
```

While the macm instruction carries out the multiply–accumulate portion of the symmetric filter operation, the add instruction adds up a pair of samples for the next iteration. This parallel arrangement effectively improves the computation of symmetric FIR filters. The following assembly program shows an implementation of symmetric FIR filter using the TMS320C55x:

```
        mov       #M−1,BRC0          ; Outer loop counter for execution
        mov       #(L/2−3),CSR       ; Inner loop for (L/2−2) iteration
        mov       #L/2,T0            ; Set up pointer offset for AR1
        sub       #(L/2−2),T1        ; Set up pointer offset for AR3
     || rptblocal sample_loop-1      ; To prevent overflow in addition
        mov       *AR0+,AC1          ; Get new sample
        mov       #0,AC0             ; Input is scaled to Q14 format
     || mov       AC1≪#−1,*AR3       ; Put input to signal buffer
        add       *AR3+,*AR1−,AC1    ; AC1 = [x(n)+x(n−L+1)] ≪ 16
     || rpt       CSR                ; Do L/2−2 iterations
```

```
firsadd *AR3+,*AR1-,*CDP+,AC1,AC0
firsadd *(AR3-T0),*(AR1+T1),*CDP+,AC1,AC0
macm    *CDP+,AC1,AC0          ; Finish the last macm instruction
mov     rnd(hi(AC0≪1)),*AR2+; Save rounded & scaled result
sample_loop
```

Although the assembly program of the symmetric FIR filter is similar to the regular
FIR filter, there are several differences when we implement it using the `firsadd`
instruction: (1) We only need to store the first half of the symmetric FIR filter coefficients. (2) The inner-repeat loop is set to $L/2-2$ since each multiply–accumulate
operation accounts for a pair of samples. (3) In order to use `firsadd` instructions
inside a repeat loop, we add the first pair of filter samples using a dual-memory add
instruction, `add *AR1+,*AR3-, AC1`. We also place the last instruction, `macmr
*CDP+, AC1, AC0`, outside the repeat loop for the final calculation. (4) We use
two data pointers, AR1 and AR3, to address the signal buffer. AR3 points at the
newest sample in the buffer, and AR1 points at the oldest sample in the buffer.
Temporary registers, T1 and T0, are used as the offsets for updating circular buffer
pointers. The offsets are initialized to $T0 = L/2$ and $T1 = L/2-2$. After AR3 and AR1
are updated, they will point to the newest and the oldest samples again. Figure 5.26
illustrates this two-pointer circular buffer for a symmetric FIR filtering. The `firsadd`
instruction accesses three data buses simultaneously (Xmem and Ymem for signal samples
and Cmem for filter coefficient). The coefficient pointer CDP is set as the circular
pointer for coefficients. The input and output samples are pointed at by AR0 and AR2.

Two implementation issues should be considered: (1) The symmetric FIR filtering
instruction `firsadd` adds two corresponding samples, and then performs multiplication. The addition may cause an undesired overflow. (2) The `firsadd` instruction
accesses three read operations in the same cycle. This may cause data memory bus
contention. The first problem can be resolved by scaling the new sample to Q14 format

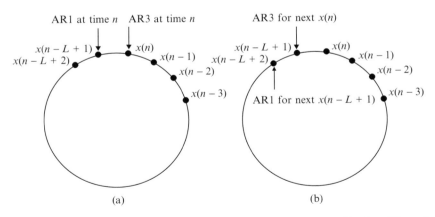

Figure 5.26 Circular buffer for accessing signals for a symmetric FIR filtering. The pointers to
$x(n)$ and $x(n - L + 1)$ are updated at the counter-clockwise direction: (a) circular buffer for a
symmetric FIR filter at time n, and (b) circular buffer for a symmetric FIR filter at time $n + 1$

prior to saving it to the signal buffer. The filter result needs to be scaled back before it can be stored into the output buffer. The second problem can be resolved by placing the filter coefficient buffer and the signal buffer into different memory blocks.

Go through the following steps for Experiment 5B:

1. Copy the assembly program `firsymm.asm`, the C function `epx5b.c`, and the linker command file `exp5.cmd` from the software package into the working directory.

2. Create the project `exp5b` and add the files `firsymm.asm`, `epx5b.c`, and `exp5.cmd` into the project.

3. The symmetric FIR routine is defined as:

    ```
    unsigned int firsymm(int *in, unsigned int M, int *h, unsigned int L,
                         int *out, int *x, unsigned int index);
    ```

    where all the arguments are the same as those defined by `fir()` in Experiment 5A.

4. Build and run the project `exp5b`. Compare the results with Figure 5.25.

5. Profile the symmetric filter performance and record the memory usage. How many instruction cycles were reduced as a result of using the symmetric FIR filter implementation? How many memory locations have been saved?

5.6.3 Experiment 5C – Implementation of FIR Filter Using Dual-MAC

As introduced in Chapter 2, the TMS320C55x has two multiply–accumulate (MAC) units and four accumulators. We can take advantage of the dual-MAC architecture to improve the execution speed of FIR filters. Unlike the symmetric FIR filter implementations given in the previous experiment, FIR filters implemented using dual-MAC can be symmetric, anti-symmetric, or any other type. The basic idea of using dual-MAC to improve the processing speed is to generate two outputs in parallel. That is, we use one MAC unit to compute $y(n)$ and the other to generate $y(n+1)$ at the same time. For example, the following parallel instructions can be used for dual-MAC filtering operations:

```
   rpt  CSR
   mac  *ARx+,*CDP+,ACx   ; ACx += bl*x(n)
:: mac  *ARy+,*CDP+,ACy   ; ACy += bl*x(n+1)
```

In this example, ARx and ARy are data pointers to $x(n)$ and $x(n + 1)$, respectively, and CDP is the coefficient pointer. The repeat loop produces two filter outputs, $y(n)$ and $y(n + 1)$. After execution, the addresses of the data pointers ARx and ARy are increased by one. The coefficient pointer CDP is also incremented by one, although the coefficient pointer CDP is set for auto-increment mode in both instructions. This is because when

CDP pointer is used in parallel instructions, it can be incremented only once. Figure 5.27 shows the C55x dual-MAC architecture for FIR filtering. The CDP uses B-bus to fetch filter coefficients, while ARx and ARy use C-bus and D-bus to get data from the signal buffer. The dual-MAC filtering results are temporarily stored in the accumulators ACx and ACy.

The following example shows the C55x assembly implementation using the dual-MAC and circular buffer for a block-FIR filter:

```
       mov    #M−1,BRC0     ; Outer loop counter
       mov    #(L/2−3),CSR  ; Inner loop counter as L/2−2
    || rptblocal sample_loop-1
       mov    *AR0+,*AR1     ; Put new sample to signal buffer x[n]
       mov    *AR0+,*AR3     ; Put next new sample to location x[n+1]
       mpy    *AR1+,*CDP+,AC0          ; First operation
    :: mpy    *AR3+,*CDP+,AC1
    || rpt    CSR
       mac    *AR1+,*CDP+,AC0          ; Rest MAC iterations
    :: mac    *AR3+,*CDP+,AC1
       macr   *AR1,*CDP+,AC0
    :: macr   *AR3,*CDP+,AC1           ; Last MAC operation
       mov    pair(hi(AC0)),dbl(*AR2+) ; Store two output data
    sample_loop
```

There are three implementation issues to be considered: (1) In order to use dual-MAC units, we need to increase the length of the signal buffer by one in order to accommodate an extra memory location required for computing two output signals. With an additional space in the buffer, we can form two sample sequences in the signal buffer, one pointed at by AR1 and the other by AR3. (2) The dual-MAC implementation of the FIR filter also makes three memory reads simultaneously. Two memory reads are used to get data samples from the signal buffer into MAC units, and the third one is used to fetch the filter coefficient. To avoid memory bus contention, we shall place the coefficients in a different memory block. (3) We place the convolution sums in two accumulators when we use dual-MAC units. To store both filter results, it requires two memory store instructions. It will be more efficient if we can use the dual-memory-store instruction, mov pair (hi (AC0)), dbl (*AR2+), to save both outputs $y(n)$ and $y(n+1)$ to the data memory in the same cycle. However, this requires the data memory to be

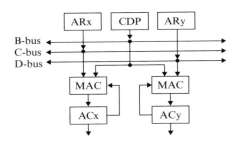

Figure 5.27   Diagram of TMS320C55x dual-MAC architecture

aligned on an even word boundary. This alignment can be done using the linker command file with the key word, align 4, see the linker command file. We use the DATA_SECTION pragma directive to tell the linker where to place the output sequence.

Go through the following steps for Experiment 5C:

1. Copy the assembly program firs2macs.asm, the C function epx5c.c, and the linker command file exp5.cmd from the software package to the working directory. The dual-MAC FIR routine is defined as:

```
unsigned int fir2macs(int * in, unsigned int M, int * h, unsigned int L,
                      int * out, int * x, unsigned int index);
```

where all the arguments are the same as Experiment 5A.

2. Create the project exp5c and add files fir2macs.asm, epx5c.c, and exp5.cmd into the project.

3. Build and run the project. Compare the results with the results from the two previous experiments.

4. Profile the filter performance and record the memory usage. How many instruction cycles were reduced by using the dual-MAC implementation? Why is the dual-MAC implementation more efficient than the symmetric FIR implementation?

References

[1] N. Ahmed and T. Natarajan, *Discrete-Time Signals and Systems*, Englewood Cliffs, NJ: Prentice-Hall, 1983.
[2] V. K. Ingle and J. G. Proakis, *Digital Signal Processing Using MATLAB V.4*, Boston: PWS Publishing, 1997.
[3] *Signal Processing Toolbox for Use with MATLAB*, Math Works, 1994.
[4] A. V. Oppenheim and R. W. Schafer, *Discrete-Time Signal Processing*, Englewood Cliffs, NJ: Prentice-Hall, 1989.
[5] S. J. Orfanidis, *Introduction to Signal Processing*, Englewood Cliffs, NJ: Prentice-Hall, 1996.
[6] J. G. Proakis and D. G. Manolakis, *Digital Signal Processing – Principles, Algorithms, and Applications*, 3rd Ed., Englewood Cliffs, NJ: Prentice-Hall, 1996.
[7] S. K. Mitra, *Digital Signal Processing: A Computer-Based Approach*, 2nd Ed., New York, NY: McGraw Hill, 1998.
[8] D. Grover and J. R. Deller, *Digital Signal Processing and the Microcontroller*, Englewood Cliffs, NJ: Prentice-Hall, 1999.
[9] F. Taylor and J. Mellott, *Hands-On Digital Signal Processing*, New York, NY: McGraw Hill, 1998.
[10] S. D. Stearns and D. R. Hush, *Digital Signal Analysis*, 2nd Ed., Englewood Cliffs, NJ: Prentice-Hall, 1990.
[11] Texas Instruments, Inc., *TMS320C55x Optimizing C Compiler User's Guide*, Literature no. SPRU281, 2000.

Exercises

Part A

1. Consider the moving-average filter given in Example 5.4. What is the 3-dB bandwidth of this filter if the sampling rate is 8 kHz?

2. Consider the FIR filter with the impulse response $h(n) = \{1, 1, 1\}$. Calculate the magnitude and phase responses and show that the filter has linear phase.

3. Given a linear time-invariant filter with an exponential impulse response

$$h(n) = a^n u(n),$$

show that the output due to a unit-step input, $u(n)$, is

$$y(n) = \frac{1 - a^{n+1}}{1 - a}, \quad n \geq 0.$$

4. A rectangular function of length L can be expressed as

$$x(n) = u(n) - u(n - L) = \begin{cases} 1, & 0 \leq n \leq L - 1 \\ 0, & \text{elsewhere,} \end{cases}$$

show that

(a) $r(n) = u(n) * u(n)$, where $*$ denotes linear convolution and $r(n) = (n + 1)u(n)$ is called the unit-ramp sequence.

(b) $t(n) = x(n) * x(n) = r(n) - 2r(n - L) + r(n - 2L)$ is the triangular pulse.

5. Using the graphical interpretation of linear convolution given in Figure 5.4 to compute the linear convolution of $h(n) = \{1, 2, 1\}$ and $x(n)$, $n = 0, 1, 2$ defined as follows:

(a) $x(n) = \{1, -1, 2\}$,

(b) $x(n) = \{1, 2, -1\}$, and

(c) $x(n) = \{1, 3, 1\}$.

6. The comb filter also can be described as

$$y(n) = x(n) + x(n - L),$$

find the transfer function, zeros, and the magnitude response of this filter and compare the results with Figure 5.6.

7. Show that at $\omega = 0$, the magnitude of the rectangular window function is $2M + 1$.

8. Assuming $h(n)$ has the symmetry property $h(n) = h(-n)$ for $n = 0, 1, \ldots M$, show that $H(\omega)$ can be expressed as

$$H(\omega) = h(0) + \sum_{n=1}^{M} 2h(n) \cos(\omega n).$$

9. The simplest digital approximation to a continuous-time differentiator is the first-order operation defined as

$$y(n) = \frac{1}{T}[x(n) - x(n-1)].$$

Find the transfer function $H(z)$, the frequency response $H(\omega)$, and the phase response of the differentiator.

10. Redraw the signal-flow diagram shown in Figure 5.8 and modify equations (5.2.23) and (5.2.24) in the case that L is an odd number.

11. Consider the rectangular window $w(n)$ of length $L = 2M + 1$ defined in (5.3.17). Show that the convolution of $w(n)$ with itself and then divided by L yields the triangular window.

12. Assuming that $H(\omega)$ given in (5.3.2) is an even function in the interval $|\omega| < \pi$, show that

$$h(n) = \frac{1}{\pi} \int_0^\pi H(\omega) \cos(\omega n) d\omega, \quad n \leq 0$$

and $h(-n) = h(n)$.

13. Design a lowpass FIR filter of length $L = 5$ with a linear phase to approximate the ideal lowpass filter of cut-off frequency $\omega_c = 1$. Use the Hamming window to eliminate the ripples in the magnitude response.

14. The ideal highpass filter of Figure 5.1(b) has frequency response

$$H(\omega) = \begin{cases} 0, & |\omega| < \omega_c \\ 1, & \omega_c \leq |\omega| \leq \pi. \end{cases}$$

Compute the coefficients of a causal FIR filter of length $L = 2M + 1$ obtained by truncating and shifting the impulse response of the ideal bandpass filter.

15. Design a bandpass filter

$$H(f) = \begin{cases} 1, & 1.6\,\text{kHz} \leq f \leq 2\,\text{kHz} \\ 0, & \text{otherwise} \end{cases}$$

with the sampling rate 8 kHz and the duration of impulse response be 50 msec using Fourier series method.

Part B

16. Consider the FIR filters with the following impulse responses:

(a) $h(n) = \{-4, 1, -1, -2, 5, 0, -5, 2, 1, -1, 4\}$

(b) $h(n) = \{-4, 1, -1, -2, 5, 6, 5, -2, -1, 1, -4\}$

Using MATLAB to plot magnitude responses, phase responses, and locations of zeros of the FIR filter's transfer function $H(z)$.

17. Show the frequency response of the lowpass filter given in (5.2.10) for $L = 8$ and compare the result with Figure 5.6.

18. Plot the magnitude response of a linear-phase FIR highpass filter of cut-off frequency $\omega_c = 0.6\pi$ by truncating the impulse response of the ideal highpass filter to length $L = 2M + 1$ for $M = 32$ and 64.

19. Repeat problem 18 using Hamming and Blackman window functions. Show that oscillatory behavior is reduced using the windowed Fourier series method.

20. Write C (or MATLAB) program that implement a comb filter of $L = 8$. The program must have the input/output capability as introduced in Appendix C. Test the filter using the sinusoidal signals of frequencies and $\omega_1 = \pi/4$ and $\omega_2 = 3\pi/8$. Explain the results based on the distribution of the zeros of the filter.

21. Rewrite the above program using the circular buffer.

22. Rewrite the program `firfltr.c` given in Appendix C using circular buffer. Implement the circular pointer updated in a new C function to replace the function `shift.c`.

Part C

23. Based on the assembly routines given in Experiments 5A, 5B, and 5C, what is the minimum number of the FIR filter coefficients if the FIR filter is

    (a) symmetric and L is even,

    (b) symmetric and L is odd,

    (c) anti-symmetric and L is even,

    (d) anti-symmetric and L is odd.

    Do we need to modify these routines if the FIR filter has odd number of taps?

24. Design a 24th-order bandpass FIR filter using MATLAB. The filter will attenuate the 800 Hz and 3.3 kHz frequency components of the signal generated by the signal generator `signal_gen()`. Implement this filter using the C55x assembly routines `fir.asm`, `firsymm.asm`, and `fir2macs.asm`. Plot the filter results in both the time domain and the frequency domain.

25. When design highpass or bandstop FIR filters using MATLAB, the number of filter coefficients is an odd number. This ensures the unit gain at the half-sampling frequency. Design a highpass FIR filter, such that it will pass the 3.3 kHz frequency components of the input signal. Implement this filter using the dual-MAC block FIR filter. Plot the results in

both the time domain and the frequency domain (Hint: modify the assembly routine `fir2macs.asm` to handle the odd number coefficients).

26. Design an anti-symmetric bandpass FIR filter to allow only the frequency component at 1.8 kHz to pass. Using `firssub` instruction to implement the FIR filter and plot the filter results in both the time domain and the frequency domain.

27. Experiment 5B demonstrates a symmetric FIR filter implementation. This filter can also be implemented efficiently using the C55x dual-MAC architecture. Modify the dual-MAC FIR filter assembly routine `fir2macs.asm` to implement the Experiment 5B based on the Equation (5.2.23). Compare the profiling results with Experiment 5B that uses the symmetric FIR filter `firsadd` instruction.

28. Use TMS320C55x EVM (or DSK) for collecting real-time signal from an analog signal generator.

    - set the TMS320C55x EVM or DSK to 8 kHz sampling rate
    - connect the signal generator output to the audio input of the EVM/DSK
    - write an interrupt service routine (ISR) to handle input samples
    - process the samples at 128 samples per block
    - verify your result using an oscilloscope or spectrum analyzer

6

Design and Implementation of IIR Filters

We have discussed the design and implementation of digital FIR filters in the previous chapter. In this chapter, our attention will be focused on the design, realization, and implementation of digital IIR filters. The design of IIR filters is to determine the transfer function $H(z)$ that satisfies the given specifications. We will discuss the basic characteristics of digital IIR filters, and familiarize ourselves with the fundamental techniques used for the design and implementation of these filters. IIR filters have the best roll-off and lower sidelobes in the stopband for the smallest number of coefficients.

Digital IIR filters can be easily obtained by beginning with the design of an analog filter, and then using mapping technique to transform it from the s-plane into the z-plane. The Laplace transform will be introduced in Section 6.1 and the analog filter will be discussed in Section 6.2. The impulse-invariant and bilinear-transform methods for designing digital IIR filters will be introduced in Section 6.3, and realization of IIR filters using direct, cascade, and parallel forms will be introduced in Section 6.4. The filter design using MATLAB will be described in Section 6.5, and the implementation considerations are given in Section 6.6. The software development and experiments using the TMS320C55x will be given in Section 6.7.

6.1 Laplace Transform

As discussed in Chapter 4, the Laplace transform is the most powerful technique used to describe, represent, and analyze analog signals and systems. In order to introduce analog filters in the next section, a brief review of the Laplace transform is given in this section.

6.1.1 Introduction to the Laplace Transform

Many practical aperiodic functions such as a unit step function $u(t)$, a unit ramp $tu(t)$, or an impulse train $\sum_{k=-\infty}^{\infty} \delta(t - kT)$ do not satisfy the integrable condition given in (4.1.11), which is a sufficient condition for a function $x(t)$ that possesses a Fourier transform. Given a positive-time function, $x(t) = 0$, for $t < 0$, a simple way to find the Fourier transform is to multiply $x(t)$ by a convergence factor $e^{-\sigma t}$, where σ is a positive number such that

$$\int_0^\infty \left| x(t) e^{-\sigma t} \right| dt < \infty. \tag{6.1.1}$$

Taking the Fourier transform defined in (4.1.10) on the composite function $x(t) e^{-\sigma t}$, we have

$$X(s) = \int_0^\infty x(t) e^{-\sigma t} e^{-j\Omega t} dt = \int_0^\infty x(t) e^{-(\sigma + j\Omega)t} dt$$

$$= \int_0^\infty x(t) e^{-st} dt, \tag{6.1.2}$$

where

$$s = \sigma + j\Omega \tag{6.1.3}$$

is a complex variable. This is called the one-sided Laplace transform of $x(t)$ and is denoted by $X(s) = \mathrm{LT}[x(t)]$. Table 6.1 lists the Laplace transforms of some simple time functions.

Example 6.1: Find the Laplace transform of signal

$$x(t) = a + b e^{-ct}, \quad t \geq 0.$$

From Table 6.1, we have the transform pairs

$$a \leftrightarrow \frac{a}{s} \quad \text{and} \quad e^{-ct} \leftrightarrow \frac{1}{s+c}.$$

Using the linear property, we have

$$X(s) = \frac{a}{s} + \frac{b}{s+c}.$$

The inverse Laplace transform can be expressed as

$$x(t) = \frac{1}{2\pi j} \int_{\sigma - j\infty}^{\sigma + j\infty} X(s) e^{st} ds. \tag{6.1.4}$$

The integral is evaluated along the straight line $\sigma + j\Omega$ in the complex plane from $\Omega = -\infty$ to $\Omega = \infty$, which is parallel to the imaginary axis $j\Omega$ at a distance σ from it.

Table 6.1 Basic Laplace transform pairs

$x(t), \; t \geq 0$	$X(s)$
$\delta(t)$	1
$u(t)$	$\dfrac{1}{s}$
c	$\dfrac{c}{s}$
ct	$\dfrac{c}{s^2}$
ct^{n-1}	$\dfrac{c(n-1)!}{s^n}$
e^{-at}	$\dfrac{1}{s+a}$
$\sin \Omega_0 t$	$\dfrac{\Omega_0}{s^2 + \Omega_0^2}$
$\cos \Omega_0 t$	$\dfrac{s}{s^2 + \Omega_0^2}$
$x(t)\cos \Omega_0 t$	$\dfrac{1}{2}[X(s+j\Omega_0) + X(s-j\Omega_0)]$
$x(t)\sin \Omega_0 t$	$\dfrac{j}{2}[X(s+j\Omega_0) - X(s-j\Omega_0)]$
$e^{\pm at}x(t)$	$X(s \mp a)$
$x(at)$	$\dfrac{1}{a}X\left(\dfrac{s}{a}\right)$

Equation (6.1.2) clearly shows that the Laplace transform is actually the Fourier transform of the function $x(t)e^{-\sigma t}$, $t > 0$. From (6.1.3), we can think of a complex s-plane with a real axis σ and an imaginary axis $j\Omega$. For values of s along the $j\Omega$ axis, i.e., $\sigma = 0$, we have

$$X(s)|_{s=j\Omega} = \int_0^\infty x(t)e^{-j\Omega t}dt, \tag{6.1.5}$$

which is the Fourier transform of the causal signal $x(t)$. Given a function $X(s)$, we can find its frequency characteristics by setting $s = j\Omega$.

There are convolution properties associated with the Laplace transform. If

$$y(t) = x(t) * h(t) = \int_0^\infty x(\tau)h(t-\tau)d\tau = \int_0^\infty h(\tau)x(t-\tau)d\tau, \tag{6.1.6}$$

then

$$Y(s) = X(s)H(s), \tag{6.1.7}$$

where $Y(s)$, $H(s)$, and $X(s)$ are the Laplace transforms of $y(t)$, $h(t)$, and $x(t)$, respectively. Thus convolution in the time domain is equivalent to multiplication in the Laplace (or frequency) domain.

In (6.1.7), $H(s)$ is the transfer function of the system defined as

$$H(s) = \frac{Y(s)}{X(s)} = \int_0^\infty h(t)e^{-st}dt, \tag{6.1.8}$$

where $h(t)$ is the impulse response of the system. The general form of a transfer function is expressed as

$$H(s) = \frac{b_0 + b_1 s + \cdots + b_{L-1}s^{L-1}}{a_0 + a_1 s + \cdots + a_M s^M} = \frac{N(s)}{D(s)}. \tag{6.1.9}$$

The roots of $N(s)$ are the zeros of the transfer function $H(s)$, while the roots of $D(s)$ are the poles.

Example 6.2: The input signal $x(t) = e^{-2t}u(t)$ is applied to an LTI system, and the output of the system is given as

$$y(t) = (e^{-t} + e^{-2t} - e^{-3t})u(t).$$

Find the system's transfer function $H(s)$ and the impulse response $h(t)$.

From Table 6.1, we have

$$X(s) = \frac{1}{s+2} \quad \text{and} \quad Y(s) = \frac{1}{s+1} + \frac{1}{s+2} - \frac{1}{s+3}.$$

From (6.1.8), we obtain

$$H(s) = \frac{Y(s)}{X(s)} = 1 + \frac{s+2}{s+1} - \frac{s+2}{s+3}.$$

This transfer function can be written as

$$H(s) = \frac{s^2 + 6s + 7}{(s+1)(s+3)} = 1 + \frac{1}{s+1} + \frac{1}{s+3}.$$

From Table 6.1, we have

$$h(t) = \delta(t) + (e^{-t} + e^{-3t})u(t).$$

The stability condition for a system can be represented in terms of its impulse response $h(t)$ or its transfer function $H(s)$. A system is stable if

$$\lim_{t \to \infty} h(t) = 0. \tag{6.1.10}$$

This condition is equivalent to requiring that all the poles of $H(s)$ must be in the left-half of the s-plane, i.e., $\sigma < 0$.

Example 6.3: Consider the impulse response

$$h(t) = e^{-at}u(t).$$

This function satisfies (6.1.10) for $a > 0$. From Table 6.1, the transfer function

$$H(s) = \frac{1}{s + a}, \quad a > 0$$

has the pole at $s = -a$, which is located at the left-half s-plane. Thus the system is stable.

If $\lim_{t \to \infty} h(t) \to \infty$, the system is unstable. This condition is equivalent to the system that has one or more poles in the right-half s-plane, or has multiple-order pole(s) on the $j\Omega$ axis. The system is marginally stable if $h(t)$ approaches a non-zero value or a bounded oscillation as t approaches infinity. If the system is stable, then the natural response goes to zero as $t \to \infty$. In this case, the natural response is also called the transient response. If the input signal is periodic, then the corresponding forced response is called the steady-state response. When the input signal is the sinusoidal signal in the form of $\sin \Omega t$, $\cos \Omega t$, or $e^{j\Omega t}$, the steady-state output is called the sinusoidal steady-state response.

6.1.2 Relationships between the Laplace and z-Transforms

An analog signal $x(t)$ can be converted into a train of narrow pulses $x(nT)$ as

$$x(nT) = x(t)\delta_T(t), \tag{6.1.11}$$

where

$$\delta_T(t) = \sum_{n=-\infty}^{\infty} \delta(t - nT) \tag{6.1.12}$$

represents a unit impulse train and is called a sampling function. Clearly, $\delta_T(t)$ is not a signal that we could generate physically, but it is a useful mathematical abstraction when dealing with discrete-time signals. Assuming that $x(t) = 0$ for $t < 0$, we have

$$x(nT) = x(t) \sum_{n=-\infty}^{\infty} \delta(t - nT) = \sum_{n=0}^{\infty} x(nT)\delta(t - nT). \tag{6.1.13}$$

To obtain the frequency characteristics of the sampled signal, take the Laplace transform of $x(nT)$ given in (6.1.13). Integrating term-by-term and using the property of the impulse function $\int_{-\infty}^{\infty} x(t)\delta(t - \tau)dt = x(\tau)$, we obtain

$$X(s) = \int_{-\infty}^{\infty} \left[\sum_{n=0}^{\infty} x(nT)\delta(t - nT) \right] e^{st} dt = \sum_{n=0}^{\infty} x(nT)e^{-nsT}. \tag{6.1.14}$$

When defining a complex variable

$$z = e^{sT}, \tag{6.1.15}$$

Equation (6.1.14) can be expressed as

$$X(z) = X(s)|_{z=e^{sT}} = \sum_{n=0}^{\infty} x(nT)z^{-n}, \tag{6.1.16}$$

where $X(z)$ is the z-transform of the discrete-time signal $x(nT)$. Thus the z-transform can be viewed as the Laplace transform of the sampled function $x(t)$ with the change of variable $z = e^{sT}$.

As discussed in Chapter 4, the Fourier transform of a sequence $x(nT)$ can be obtained from the z-transform by replacing z with $e^{j\omega}$. That is, by evaluating the z-transform on the unit circle of $|z| = 1$. The whole procedure can be summarized in Figure 6.1.

6.1.3 Mapping Properties

The relationship $z = e^{sT}$ defined in (6.1.15) represents the mapping of a region in the s-plane to the z-plane since both s and z are complex variables. Since $s = \sigma + j\Omega$, we have

$$z = e^{sT} = e^{\sigma T}e^{j\Omega T} = |z|e^{j\omega}, \tag{6.1.17}$$

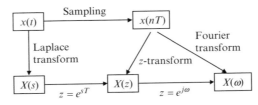

Figure 6.1   Relationships between the Laplace, Fourier, and z-transforms

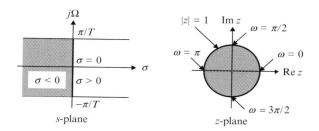

Figure 6.2 Mapping between the s-plane and z-plane

where the magnitude

$$|z| = e^{\sigma T} \tag{6.1.18a}$$

and the angle

$$\omega = \Omega T. \tag{6.1.18b}$$

When $\sigma = 0$, the amplitude given in (6.1.18a) is $|z| = 1$, and Equation (6.1.17) is simplified to $z = e^{j\Omega T}$. It is apparent that the portion of the $j\Omega$-axis between $\Omega = -\pi/T$ and $\Omega = \pi/T$ in the s-plane is mapped onto the unit circle in the z-plane from $-\pi$ to π as illustrated in Figure 6.2. As Ω increases from π/T to $3\pi/T$ in the s-plane, another counterclockwise encirclement of the unit circle results in the z-plane. Thus as Ω varies from 0 to ∞, there are an infinite number of encirclements of the unit circle in the counterclockwise direction. Similarly, there are an infinite numbers of encirclements of the unit circle in the clockwise direction as Ω varies from 0 to $-\infty$.

From (6.1.18a), $|z| < 1$ when $\sigma < 0$. Thus each strip of width $2\pi/T$ in the left-half of the s-plane is mapped onto the unit circle. This mapping occurs in the form of concentric circles in the z-plane as σ varies from 0 to $-\infty$. Equation (6.1.18a) also implies that $|z| > 1$ if $\sigma > 0$. Thus each strip of width $2\pi/T$ in the right-half of the s-plane is mapped outside of the unit circle. This mapping also occurs in concentric circles in the z-plane as σ varies from 0 to ∞.

In conclusion, the mapping from the s-plane to the z-plane is not one-to-one, since there is more than one point in the s-plane that corresponds to a single point in the z-plane. This issue will be discussed later when we design a digital filter from a given analog filter.

6.2 Analog Filters

Analog filter design is a well-developed technique. Many of the techniques employed in studying digital filters are analogous to those used in studying analog filters. The most systematic approach to designing IIR filters is based on obtaining a suitable analog filter function and then transforming it into the discrete-time domain. This is not possible when designing FIR filters as they have no analog counterpart.

6.2.1 Introduction to Analog Filters

In this section, we briefly introduce some basic concepts of analog filters. Knowledge of analog filter transfer functions is readily available since analog filters have already been investigated in great detail. In Section 6.3, we will introduce a conventional powerful bilinear-transform method to design digital IIR filters utilizing analog filters.

From basic circuit theory, capacitors and inductors have an impedance (X) that depends on frequency. It can be expressed as

$$X_C = \frac{1}{j\Omega C} \tag{6.2.1}$$

and

$$X_L = j\Omega L, \tag{6.2.2}$$

where C is the capacitance with units in Farads (F), and L is the inductance with units in Henrys (H). When either component is combined with a resistor, we can build frequency-dependent voltage dividers. In general, capacitors and resistors are used to design analog filters since inductors are bulky, more expensive, and do not perform as well as capacitors.

Example 6.4: Consider a circuit containing a resistor and a capacitor as shown in Figure 6.3. Applying Ohm's law to this circuit, we have

$$V_{\text{in}} = I(R + X_C) \quad \text{and} \quad V_{\text{out}} = IR.$$

From (6.2.1), the transfer function of the circuit is

$$H(\Omega) = \frac{V_{\text{out}}}{V_{\text{in}}} = \frac{R}{R + \dfrac{1}{j\Omega C}} = \frac{j\Omega RC}{1 + j\Omega RC}. \tag{6.2.3}$$

The magnitude response of circuit can be expressed as

$$|H(\Omega)| = \frac{R}{\sqrt{R^2 + \dfrac{1}{\Omega^2 C^2}}}.$$

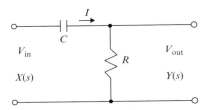

Figure 6.3 An analog filter with a capacitor and resistor

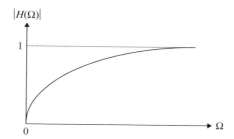

Figure 6.4   Amplitude response of analog circuit shown in Figure 6.3

The plot of the magnitude response $|H(\Omega)|$ vs. the frequency Ω is shown in Figure 6.4. For a constant input voltage, the output is approximately equal to the input at high frequencies, and the output approaches zero at low frequencies. Therefore the circuit shown in Figure 6.3 is called a highpass filter since it only allows high frequencies to pass without attenuation.

The transfer function of the circuit shown in Figure 6.3 is given by

$$H(s) = \frac{Y(s)}{X(s)} = \frac{R}{R + 1/Cs} = \frac{RCs}{1 + RCs}. \tag{6.2.4}$$

To design an analog filter, we can use computer programs to calculate the correct values of the resistor and the capacitor for desired magnitude and phase responses. Unfortunately, the characteristics of the components drift with temperature and time. It is sometimes necessary to re-tune the circuit while it is being used.

6.2.2   Characteristics of Analog Filters

In this section, we briefly describe some important characteristics of commonly used analog filters based on lowpass filters. We will discuss frequency transformations for converting a lowpass filter to highpass, bandpass, and bandstop filters in the next section. Approximations to the ideal lowpass prototype are obtained by first finding a polynomial approximation to the desired squared magnitude $|H(\Omega)|^2$, and then converting this polynomial into a rational function. An error criterion is selected to measure how close the function obtained is to the desired function. These approximations to the ideal prototype will be discussed briefly based on Butterworth filters, Chebyshev filters type I and II, elliptic filters, and Bessel filters.

The lowpass Butterworth filter is an all-pole approximation to the ideal filter, which is characterized by the squared magnitude response

$$|H(\Omega)|^2 = \frac{1}{1 + \left(\Omega/\Omega_p\right)^{2L}}, \tag{6.2.5}$$

where L is the order of the filter. It is shown that $|H(0)| = 1$ and $|H(\Omega_p)| = 1/\sqrt{2}$ or, equivalently, $20 \log_{10} |H(\Omega_p)| = -3\,\text{dB}$ for all values of L. Thus Ω_p is called the 3-dB

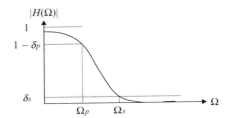

Figure 6.5 Magnitude response of Butterworth lowpass filter

cut-off frequency. The magnitude response of a typical Butterworth lowpass filter is illustrated in Figure 6.5. This figure shows that the magnitude is monotonically decreasing in both the passband and the stopband. The Butterworth filter has a completely flat magnitude response over the passband and the stopband. It is often referred to as the 'maximally flat' filter. This flat passband is achieved at the expense of the transition region from Ω_p to Ω_s, which has a very slow roll-off. The phase response is nonlinear around the cut-off frequency.

From the monotonic nature of the magnitude response, it is clear that the specifications are satisfied if we choose

$$1 \geq |H(\Omega_p)| \geq 1 - \delta_p, \quad |\Omega| \leq \Omega_p \tag{6.2.6a}$$

in the passband, and

$$|H(\Omega_s)| \leq \delta_s, \quad |\Omega| \geq \Omega_s \tag{6.2.6b}$$

in the stopband. The order of the filter required to satisfy an attenuation, δ_s, at a specified frequency, Ω_s, can be determined by substituting $\Omega = \Omega_s$ into (6.2.5), resulting in

$$L = \frac{\log_{10}[(1 - \delta_s^2) - 1]}{2\log_{10}(\Omega_s/\Omega_p)}. \tag{6.2.7}$$

The parameter L determines how closely the Butterworth characteristic approximates the ideal filter.

If we increase the order of the filter, the flat region of the passband gets closer to the cut-off frequency before it drops away and we have the opportunity to improve the roll-off. Although the Butterworth filter is very easy to design, the rate at which its magnitude decreases in the frequency range $\Omega \geq \Omega_p$ is rather slow for a small L. Therefore for a given transition band, the order of the Butterworth filter required is often higher than that of other types of filters. In addition, for a large L, the overshoot of the step response of a Butterworth filter is rather large.

To obtain the filter transfer function $H(s)$, we use $H(s)H(-s)|_{s=j\Omega} = |H(\Omega)|^2$. From (6.2.5), the poles of the Butterworth filter are defined by

$$1 + (-s^2)^L = 0. \tag{6.2.8}$$

By solving this equation, we obtain the poles

$$s_k = e^{j(2k+L-1)\pi/2L}, \quad k = 0, 1, \ldots, 2L - 1. \tag{6.2.9}$$

These poles are located uniformly on a unit circle in the s-plane at intervals of π/L radians. The pole locations are symmetrical with respect to both the real and imaginary axes. Since $2L - 1$ cannot be an even number, it is clear that as there are no poles on the $j\Omega$ axis, there are exactly L poles in each of the left- and right-half planes.

To obtain a stable Lth-order IIR filter, we choose only the poles in the left-half s-plane. That is, we choose

$$s_k = e^{j(2k+L-1)\pi/2L}, \quad k = 1, 2, \ldots, L. \tag{6.2.10}$$

Therefore the transfer function of Butterworth filter is defined as

$$H(s) = \frac{1}{(s - s_1)(s - s_2)\ldots(s - s_L)} = \frac{1}{s^L + a_{L-1}s^{L-1} + \ldots + a_1 s + 1}. \tag{6.2.11}$$

The coefficients a_k are real numbers because the poles s_k are symmetrical with respect to the imaginary axis. Table 6.2 lists the denominator of the Butterworth filter transfer function $H(s)$ in factored form for values of L ranging from $L = 1$ to $L = 4$.

Example 6.5: Obtain the transfer function of a lowpass Butterworth filter for $L = 3$. From (6.2.9), the poles are located at

$$s_0 = e^{j\pi/3}, \quad s_1 = e^{j2\pi/3}, \quad s_2 = e^{j\pi}, \quad s_3 = e^{j4\pi/3}, \quad s_4 = e^{j5\pi/3}, \quad \text{and} \quad s_5 = 0.$$

These poles are shown in Figure 6.6. To obtain a stable IIR filter, we choose the poles in the left-half plane to get

Table 6.2 Analog Butterworth lowpass filter transfer functions

L	$H(s)$
1	$\dfrac{1}{s + 1}$
2	$\dfrac{1}{s^2 + \sqrt{2}s + 1}$
3	$\dfrac{1}{(s + 1)(s^2 + s + 1)}$
4	$\dfrac{1}{(s^2 + 0.7653s + 1)(s^2 + 1.8477s + 1)}$

$$H(s) = \frac{1}{(s - s_1)(s - s_2)(s - s_3)} = \frac{1}{(s - e^{j2\pi/3})(s - e^{j\pi})(s - e^{j4\pi/3})}$$

$$= \frac{1}{(s + 1)(s^2 + s + 1)}.$$

Chebyshev filters permit a certain amount of ripples in the passband, but have a much steeper roll-off near the cut-off frequency than what the Butterworth design can achieve. The Chebyshev filter is called the equiripple filter because the ripples are always of equal size throughout the passband. Even if we place very tight limits on the passband ripple, the improvement in roll-off is considerable when compared with the Butterworth filter. There are two types of Chebyshev filters. Type I Chebyshev filters are all-pole filters that exhibit equiripple behavior in the passband and a monotonic characteristic in the stopband (see Figure 6.7a). The family of type II Chebyshev filters contains both poles and zeros, and exhibit a monotonic behavior in the passband and an equiripple behavior in the stopband, as shown in Figure 6.7(b). In general, the Chebyshev filter meets the specifications with a fewer number of poles than the corresponding Butterworth filter. Although the Chebyshev filter is an improvement over the Butterworth filter with respect to the roll-off, it has a poorer phase response.

The sharpest transition from passband to stopband for any given δ_p, δ_s, and L can be achieved using the elliptic design. In fact, the elliptic filter is the optimum design in this sense. As shown in Figure 6.8, elliptic filters exhibit equiripple behavior in both the

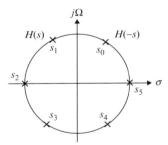

Figure 6.6   Poles of the Butterworth polynomial for $L = 3$

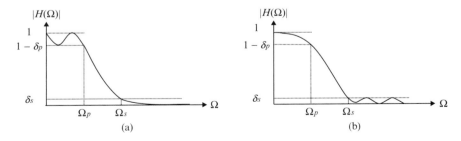

Figure 6.7   Magnitude responses of Chebyshev lowpass filters: (a) type I, and (b) type II

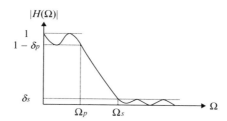

Figure 6.8 Magnitude response of elliptic lowpass filter

passband and the stopband. In addition, the phase response of elliptic filter is extremely nonlinear in the passband (especially near cut-off frequency), so we can only use the design where the phase is not an important design parameter.

Butterworth, Chebyshev, and elliptic filters approximate an ideal rectangular bandwidth. The Butterworth filter has a monotonic magnitude response. By allowing ripples in the passband for type I and in the stopband for type II, the Chebyshev filter can achieve sharper cutoff with the same number of poles. An elliptic filter has even sharper cutoffs than the Chebyshev filter for the same complexity, but it results in both passband and stopband ripples. The design of these filters strives to achieve the ideal magnitude response with trade-offs in phase response.

Bessel filters are a class of all-pole filters that approximate linear phase in the sense of maximally flat group delay in the passband. However, we must sacrifice steepness in the transition region. In addition, acceptable Bessel IIR designs are derived by transformation only for a relatively limited range of specifications such as sufficiently low cut-off frequency Ω_p.

6.2.3 Frequency Transforms

We have discussed the design of prototype analog lowpass filters with a cut-off frequency Ω_p. Although the same procedure can be applied to designing highpass, bandpass, or bandstop filters, it is much easier to obtain these filters from the desired lowpass filter using frequency transformations. In addition, most classical filter design tables only generate lowpass filters and must be converted using spectral transformation into highpass, bandpass, or bandstop filters. Filter design packages such as MATLAB often incorporate and perform the frequency transformations directly.

Butterworth highpass filter's transfer function $H_{\mathrm{hp}}(s)$ can be obtained from the corresponding lowpass filter's transfer function $H(s)$ by using the relationship

$$H_{\mathrm{hp}}(s) = H(s)|_{s=\frac{1}{s}} = H\left(\frac{1}{s}\right). \qquad (6.2.12)$$

For example, consider $L = 1$. From Table 6.2, we have $H(s) = 1/(s+1)$. From (6.2.12), we obtain

$$H_{\text{hp}}(s) = \frac{1}{s+1}\bigg|_{s=\frac{1}{s}} = \frac{s}{s+1}. \tag{6.2.13}$$

Similarly, we can calculate $H_{\text{hp}}(s)$ for higher order filters. We can show that the denominator polynomials of $H(s)$ and $H_{\text{hp}}(s)$ are the same, but the numerator becomes s^L for the Lth-order highpass filters. Thus $H_{\text{hp}}(s)$ has an additional Lth-order zero at the origin, and has identical poles s_k as given in (6.2.10).

Transfer functions of bandpass filters can be obtained from the corresponding lowpass filters by replacing s with $(s^2 + \Omega_m^2)/\text{BW}$. That is,

$$H_{\text{bp}}(s) = H(s)\bigg|_{s = \frac{s^2+\Omega_m^2}{\text{BW}}}, \tag{6.2.14}$$

where Ω_m is the center frequency of the bandpass filter and BW is its bandwidth. As illustrated in Figure 5.3 and defined in (5.1.10) and (5.1.11), the center frequency is defined as

$$\Omega_m = \sqrt{\Omega_a \Omega_b}, \tag{6.2.15}$$

where Ω_a and Ω_b are the lower and upper cut-off frequencies. The filter bandwidth is defined by

$$\text{BW} = \Omega_b - \Omega_a. \tag{6.2.16}$$

Note that for an Lth-order lowpass filter, we obtain a $2L$th-order bandpass filter transfer function.

For example, consider $L = 1$. From Table 6.2 and (6.2.14), we have

$$H_{\text{bp}}(s) = \frac{1}{s+1}\bigg|_{s = \frac{s^2+\Omega_m^2}{\text{BW}}} = \frac{\text{BW}s}{s^2 + \text{BW}s + \Omega_m^2}. \tag{6.2.17}$$

In general, $H_{\text{bp}}(s)$ has L zeros at the origin and L pole-pairs.

Bandstop filter transfer functions can be obtained from the corresponding highpass filters by replacing s in the highpass filter transfer function with $(s^2 + \Omega_m^2)/\text{BW}s$. That is,

$$H_{\text{bs}}(s) = H_{\text{hp}}(s)\bigg|_{s = \frac{s^2+\Omega_m^2}{\text{BW}s}}, \tag{6.2.18}$$

where Ω_m is the center frequency defined in (6.2.15) and BW is the bandwidth defined in (6.2.16).

6.3 Design of IIR Filters

In this section, we discuss the design of digital filters that have an infinite impulse response. In designing IIR filters, the usual starting point will be an analog filter transfer function $H(s)$. Because analog filter design is a mature and well-developed field, it is not surprising that we begin the design of digital IIR filters in the analog domain and then convert the design into the digital domain. The problem is to determine a digital filter $H(z)$ which will approximate the performance of the desired analog filter $H(s)$. There are two methods, the impulse-invariant method and the bilinear transform, for designing digital IIR filters based on existing analog IIR filters. Instead of designing the digital IIR filter directly, these methods map the digital filter into an equivalent analog filter, which can be designed by one of the well-developed analog filter design methods. The designed analog filter is then mapped back into the desired digital filter.

The impulse-invariant method preserves the impulse response of the original analog filter by digitizing the impulse response of analog filter, but not its frequency (magnitude) response. Because of inherent aliasing, this method is inappropriate for highpass or bandstop filters. The bilinear-transform method yields very efficient filters, and is well suited for the design of frequency selective filters. Digital filters resulting from the bilinear transform will preserve the magnitude response characteristics of the analog filters, but not the time domain properties. In general, the impulse-invariant method is good for simulating analog filters, but the bilinear-transform method is better for designing frequency selective IIR filters.

6.3.1 Review of IIR Filters

As discussed in Chapters 3 and 4, an IIR filter can be specified by its impulse response $\{h(n), n = 0, 1, \ldots, \infty\}$, I/O difference equation, or transfer function. The general form of the IIR filter transfer function is defined in (4.3.10) as

$$H(z) = \frac{\sum_{l=0}^{L-1} b_l z^{-l}}{1 + \sum_{m=1}^{M} a_m z^{-m}}. \tag{6.3.1}$$

The design problem is to find the coefficients b_l and a_m so that $H(z)$ satisfies the given specifications. This IIR filter can be realized by the I/O difference equation

$$y(n) = \sum_{l=0}^{L-1} b_l x(n-l) - \sum_{m=1}^{M} a_m y(n-m). \tag{6.3.2}$$

The impulse response $h(n)$ of the IIR filter is the output that results when the input is the unit impulse response defined in (3.1.1). Given the impulse response, the filter output $y(n)$ can also be obtained by linear convolution expressed as

$$y(n) = x(n) * h(n) = \sum_{k=0}^{\infty} h(k)x(n-k). \tag{6.3.3}$$

However, Equation (6.3.3) is not computationally feasible because it uses an infinite number of coefficients. Therefore we restrict our attention to IIR filters that are described by the linear difference equation given in (6.3.2).

By factoring the numerator and denominator polynomials of $H(z)$ given in (6.3.1) and assuming $M = L - 1$, the transfer function can be expressed in (4.3.12) as

$$H(z) = b_0 \frac{\displaystyle\prod_{m=1}^{M}(z - z_m)}{\displaystyle\prod_{m=1}^{M}(z - p_m)}, \tag{6.3.4}$$

where z_m and p_m are the mth zero and pole, respectively. For a system to be stable, it is necessary that all its poles lie strictly inside the unit circle on the z-plane.

6.3.2   Impulse-Invariant Method

The design technique for an impulse-invariant digital filter is illustrated in Figure 6.9. Assuming the impulse function $\delta(t)$ is used as a signal source, the output of the analog filter will be the impulse response $h(t)$. Sampling this continuous-time impulse response yields the sample values $h(nT)$. In the second signal path, the impulse function $\delta(t)$ is sampled first to yield the discrete-time impulse sequence $\delta(n)$. Filtering this signal by $H(z)$ yields the impulse response $h(n)$ of the digital filter. If the coefficients of $H(z)$ are adjusted so that the impulse response coefficients are identical to the previous specified $h(nT)$, that is,

$$h(n) = h(nT), \tag{6.3.5}$$

the digital filter $H(z)$ is the impulse invariant equivalent of the analog filter $H(s)$. An analog filter $H(s)$ and a digital filter $H(z)$ are impulse invariant if the impulse response of $H(z)$ is the same as the sampled impulse response of $H(s)$. Thus in effect, we sample the continuous-time impulse response to produce the discrete-time filter as described by (6.3.5).

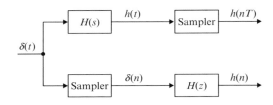

Figure 6.9   The concept of impulse-invariant design

The impulse-invariant design is usually not performed directly in the form of (6.3.5). In practice, the transfer function of an analog filter $H(s)$ is first expanded into a partial-fraction form

$$H(s) = \sum_{i=1}^{P} \frac{c_i}{s + s_i},$$

(6.3.6)

where $s = -s_i$ is the pole of $H(s)$, and c_i is the residue of the pole at $-s_i$. Note that we have assumed there are no multiple poles. Taking the inverse Laplace transform of (6.3.6) yields

$$h(t) = \sum_{i=1}^{P} c_i e^{-s_i t}, \quad t \geq 0,$$

(6.3.7)

which is the impulse response of the analog filter $H(s)$.

The impulse response samples are obtained by setting t equal to nT. From (6.3.5) and (6.3.7), we have

$$h(n) = \sum_{i=1}^{P} c_i e^{-s_i n T}, \quad n \geq 0,$$

(6.3.8)

The z-transform of the sampled impulse response is given by

$$H(z) = \sum_{n=0}^{\infty} h(n) z^{-n} = \sum_{i=1}^{P} c_i \sum_{n=0}^{\infty} (e^{-s_i T} z^{-1})^n = \sum_{i=1}^{P} \frac{c_i}{1 - e^{-s_i T} z^{-1}}.$$

(6.3.9)

The impulse response of $H(z)$ is obtained by taking the inverse z-transform of (6.3.9). Therefore the filter described in (6.3.9) has an impulse response equivalent to the sampled impulse response of the analog filter $H(s)$ defined in (6.3.6). Comparing (6.3.6) with (6.3.9), the parameters of $H(z)$ may be obtained directly from $H(s)$ without bothering to evaluate $h(t)$ or $h(n)$.

The magnitude response of the digital filter will be scaled by $f_s (= 1/T)$ due to the sampling operation. Scaling the magnitude response of the digital filter to approximate magnitude response of the analog filter requires the multiplication of $H(z)$ by T. The transfer function of the impulse-invariant digital filter given in (6.3.9) is modified as

$$H(z) = T \sum_{i=1}^{P} \frac{c_i}{1 - e^{-s_i T} z^{-1}}.$$

(6.3.10)

The frequency variable ω for the digital filter bears a linear relationship to that for the analog filter within the operating range of the digital filter. This means that when ω varies from 0 to π around the unit circle in the z-plane, Ω varies from 0 to π/T along the $j\Omega$-axis in the s-plane. Recall that $\omega = \Omega T$ as given in (3.1.7). Thus critical frequencies such as cutoff and bandwidth frequencies specified for the digital filter can be used directly in the design of the analog filter.

Example 6.6: Consider the analog filter expressed as

$$H(s) = \frac{0.5(s+4)}{(s+1)(s+2)} = \frac{1.5}{s+1} - \frac{1}{s+2}.$$

The impulse response of the filter is

$$h(t) = 1.5e^{-t} - e^{-2t}.$$

Taking the *z*-transform and scaling by *T* yields

$$H(z) = \frac{1.5T}{1 - e^{-T}z^{-1}} - \frac{T}{1 - e^{-2T}z^{-1}}.$$

It is interesting to compare the frequency response of the two filters given in Example 6.6. For the analog filter, the frequency response is

$$H(\Omega) = \frac{0.5(4+j\Omega)}{(1+j\Omega)(2+j\Omega)}.$$

For the digital filter, we have

$$H(\omega) = \frac{1.5T}{1 - e^{-T}e^{-j\omega T}} - \frac{T}{1 - e^{-2T}e^{-j\omega T}}.$$

The DC response of the analog filter is given by

$$H(0) = 1, \tag{6.3.11}$$

and

$$H(0) = \frac{1.5T}{1 - e^{-T}} - \frac{T}{1 - e^{-2T}} \tag{6.3.12}$$

for the digital filter. Thus the responses are different due to aliasing at DC. For a high sampling rate, *T* is small and the approximations $e^{-T} \approx 1 - T$ and $e^{-2T} \approx 1 - 2T$ are valid. Thus Equation (6.3.12) can be approximated with

$$H(0) \approx \frac{1.5T}{1 - (1-T)} - \frac{T}{1 - (1-2T)} = 1. \tag{6.3.13}$$

Therefore by using a high sampling rate, the aliasing effect becomes negligible and the DC gain is one as shown in (6.3.13).

While the impulse-invariant method is straightforward to use, it suffers from obtaining a discrete-time system from a continuous-time system by the process of sampling. Recall that sampling introduces aliasing, and that the frequency response corresponding to the sequence $h(nT)$ is obtained from (4.4.18) as

$$H(\omega) = \frac{1}{T} \sum_{k=-\infty}^{\infty} H\left(\Omega - \frac{2\pi k}{T}\right). \tag{6.3.14}$$

This is not a one-to-one transformation from the s-plane to the z-plane. Therefore $H(\omega) = \frac{1}{T} H(\Omega)$ is true only if $H(\Omega) = 0$ for $|\Omega| \geq \pi/T$. As shown in (6.3.14), $H(\omega)$ is the aliased version of $H(\Omega)$. Hence the stopband characteristics are maintained adequately if the aliased tails of $H(\Omega)$ are sufficiently small. The passband is also affected, but this effect is usually less serious. Thus the resulting digital filter does not exactly meet the original design specifications.

In a bandlimited filter, the magnitude response of the analog filter is negligibly small at frequencies exceeding half the sampling frequency in order to reduce the aliasing effect. Thus we must have

$$|H(\omega)| \rightarrow 0, \quad \text{for } \omega \geq \pi. \tag{6.3.15}$$

This condition can hold for lowpass and bandpass filters, but not for highpass and bandstop filters.

MATLAB supports the design of impulse invariant digital filters through the function `impinvar` in the *Signal Processing Toolbox*. The s-domain transfer function is first defined along with the sampling frequency. The function `impinvar` determines the numerator and denominator of the z-domain transfer function. The MATLAB command is expressed as

```
[bz, az] = impinvar(b, a, Fs)
```

where `bz` and `az` are the numerator and denominator coefficients of a digital filter, `Fs` is the sampling rate, and `b` and `a` represent coefficients of the analog filter.

6.3.3   Bilinear Transform

As discussed in the previous section, the time-domain impulse-invariant method of filter design is simple, but has an undesired aliasing effect. This is because the impulse-invariant method uses the transformation $\omega = \Omega T$ or equivalently, $z = e^{sT}$. As discussed in Section 6.1.3, such mapping leads to aliasing problems. In this section, we discuss the most commonly used technique for designing IIR filters with prescribed magnitude response specifications – the bilinear transform. The procedure of designing digital filters using bilinear transform is illustrated in Figure 6.10. Instead of designing the digital filter directly, this method maps the digital filter specifications to an equivalent analog filter, which can be designed by using analog filter design methods introduced in Section 6.2. The designed analog filter is then mapped back to the desired digital filter.

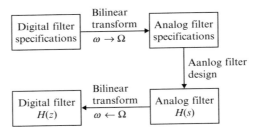

Figure 6.10 Digital IIR filter design using the bilinear transform

The bilinear transform is a mapping or transformation that relates points on the s- and z-planes. It is defined as

$$s = \frac{2}{T}\left(\frac{z-1}{z+1}\right) = \frac{2}{T}\left(\frac{1-z^{-1}}{1+z^{-1}}\right), \tag{6.3.16}$$

or equivalently,

$$z = \frac{1+(T/2)s}{1-(T/2)s}. \tag{6.3.17}$$

This is called the bilinear transform because of the linear functions of z in both the numerator and denominator of (6.3.16).

As discussed in Section 6.1.2, the $j\Omega$-axis of the s-plane ($\sigma = 0$) maps onto the unit circle in the z-plane. The left ($\sigma < 0$) and right ($\sigma > 0$) halves of the s-plane map into the inside and outside of the unit circle, respectively. Because the $j\Omega$-axis maps onto the unit circle ($|z| = 1$), there is a direct relationship between the s-plane frequency Ω and the z-plane frequency ω. Substituting $s = j\Omega$ and $z = e^{j\omega}$ into (6.3.16), we have

$$j\Omega = \frac{2}{T}\left(\frac{e^{j\omega}-1}{e^{j\omega}+1}\right). \tag{6.3.18}$$

It can be easily shown that the corresponding mapping of frequencies is obtained as

$$\Omega = \frac{2}{T}\tan\left(\frac{\omega}{2}\right), \tag{6.3.19}$$

or equivalently,

$$\omega = 2\tan^{-1}\left(\frac{\Omega T}{2}\right). \tag{6.3.20}$$

Thus the entire $j\Omega$-axis is compressed into the interval $[-\pi/T, \pi/T]$ for ω in a one-to-one manner. The range $0 \to \infty$ portion in the s-plane is mapped onto the $0 \to \pi$ portion of the unit circle in the z-plane, while the $0 \to -\infty$ portion in the s-plane is mapped onto

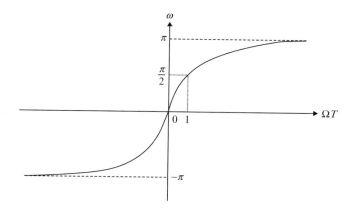

Figure 6.11 Plot of transformation given in (6.3.20)

the $0 \rightarrow -\pi$ portion of the unit circle in the z-plane. Each point in the s-plane is uniquely mapped onto the z-plane. This fundamental relation enables us to locate a point Ω on the $j\Omega$-axis for a given point on the unit circle.

The relationship in (6.3.20) between the frequency variables Ω and ω is illustrated in Figure 6.11. The bilinear transform provides a one-to-one mapping of the points along the $j\Omega$-axis onto the unit circle, i.e., the entire $j\Omega$ axis is mapped uniquely onto the unit circle, or onto the Nyquist band $|\omega| \leq \pi$. However, the mapping is highly nonlinear. The point $\Omega = 0$ is mapped to $\omega = 0$ (or $z = 1$), and the point $\Omega = \infty$ is mapped to $\omega = \pi$ (or $z = -1$). The entire band $\Omega T \geq 1$ is compressed onto $\pi/2 \leq \omega \leq \pi$. This frequency compression effect associated with the bilinear transform is known as frequency warping due to the nonlinearity of the arctangent function given in (6.3.20). This nonlinear frequency-warping phenomenon must be taken into consideration when designing digital filters using the bilinear transform. This can be done by pre-warping the critical frequencies and using frequency scaling.

The bilinear transform guarantees that

$$H(s)\big|_{s=j\Omega} = H(z)\big|_{z \, = \, e^{j\omega}}, \qquad (6.3.21)$$

where $H(z)$ is the transfer function of the digital filter, and $H(s)$ is the transfer function of an analog filter with the desired frequency characteristics.

6.3.4 Filter Design Using Bilinear Transform

The bilinear transform of an analog filter function $H(s)$ is obtained by simply replacing s with z using Equation (6.3.16). The filter specifications will be in terms of the critical frequencies of the digital filter. For example, the critical frequency ω for a lowpass filter is the bandwidth of the filter, and for a notch filter, it is the notch frequency. If we use the same critical frequencies for the analog design and then apply the bilinear transform, the digital filter frequencies would be in error because of the frequency wrapping given in (6.3.20). Therefore we have to pre-wrap the critical frequencies of the analog filter.

There are three steps involved in the bilinear design procedure. These steps are summarized as follows:

1. Pre-wrap the critical frequency ω_c of the digital filter using (6.3.19) to obtain the corresponding analog filter's frequency Ω_c.

2. Frequency scale of the designed analog filter $H(s)$ with Ω_c to obtain

$$\hat{H}(s) = \hat{H}(s)|_{s=s/\Omega_c} = H\left(\frac{s}{\Omega_c}\right), \qquad (6.3.22)$$

where $\hat{H}(s)$ is the scaled transfer function corresponding to $H(s)$.

3. Replace s in $\hat{H}(s)$ by $2(z-1)/(z+1)T$ to obtain desired digital filter $H(z)$. That is

$$H(z) = \hat{H}(s)|_{s=2(z-1)/(z+1)T}, \qquad (6.3.23)$$

where $H(z)$ is the desired digital filter.

Example 6.7: Consider the transfer function of the simple analog lowpass filter given as

$$H(s) = \frac{1}{1+s}.$$

Use this $H(s)$ and the bilinear transform method to design the corresponding digital lowpass filter whose bandwidth is 1000 Hz and the sampling frequency is 8000 Hz.
    The critical frequency for the lowpass filter is the filter bandwidth $\omega_c = 2\pi(1000/8000)$ radians/sample and $T = 1/8000$ second.

Step 1:

$$\Omega_c = \frac{2}{T}\tan\left(\frac{\omega_c}{2}\right) = \frac{2}{T}\tan\left(\frac{2000\pi}{16\,000}\right) = \frac{2}{T}\tan\left(\frac{\pi}{8}\right) = \frac{0.8284}{T}.$$

Step 2: We use frequency scaling to obtain

$$\hat{H}(s) = H(s)|_{s=s/(0.8284/T)} = \frac{0.8284}{sT + 0.8284}.$$

Step 3: The bilinear transform in (6.3.12) yields the desired transfer function

$$H(z) = \hat{H}(s)|_{s=2(z-1)/(z+1)T} = 0.2929\,\frac{1+z^{-1}}{1-0.4142z^{-1}}.$$

MATLAB provides the function `bilinear` to design digital filters using the bilinear transform. The transfer function for the analog prototype is first determined. The numerator and denominator polynomials of the analog prototype are then mapped to the polynomials for the digital filter using the bilinear transform. For example, the following MATLAB script can be used for design a lowpass filter using bilinear transform:

```
Fs = 2000;                  % Sampling frequency
Wn = 2*pi*500;              % Edge frequency
n = 2;                      % Order of analog filter
[b, a] = butter(n, Wn, 's' );   % Design analog filter
[bz, az] = bilinear(b, a, Fs);  % Determine digital filter
```

6.4 Realization of IIR Filters

As discussed earlier, a digital IIR filter can be described by the linear convolution (6.3.3), the transfer function (6.3.1), or the I/O difference equation (6.3.2). These equations are equivalent mathematically, but may be different in realization. In DSP implementation, we have to consider the required operations, memory storage, and the finite wordlength effects. A given transfer function $H(z)$ can be realized in several forms or configurations. In this section, we will discuss direct-form I, direct-form II, cascade, and parallel realizations. Many additional structures such as wave digital filters, ladder structures, and lattice structures can be found in the reference book [7].

6.4.1 Direct Forms

Given an IIR filter described by (6.3.1), the direct-form I realization is defined by the I/O Equation (6.3.2). It has $L + M$ coefficients and needs $L + M + 1$ memory locations to store $\{x(n - l),\ l = 0, 1, \ldots, L - 1\}$ and $\{y(n - m),\ m = 0, 1, \ldots, M\}$. It also requires $L + M$ multiplications and $L + M - 1$ additions for implementation on a DSP system. The detailed signal-flow diagram for $L = M + 1$ is illustrated in Figure 4.6.

Example 6.8: Given a second-order IIR filter transfer function

$$H(z) = \frac{b_0 + b_1 z^{-1} + b_2 z^{-2}}{1 + a_1 z^{-1} + a_2 z^{-2}}, \tag{6.4.1}$$

the I/O difference equation of direct-form I realization is described as

$$y(n) = b_0 x(n) + b_1 x(n - 1) + b_2 x(n - 2) - a_1 y(n - 1) - a_2 y(n - 2). \tag{6.4.2}$$

The signal-flow diagram is illustrated in Figure 6.12.

As shown in Figure 6.12, the IIR filter can be interpreted as the cascade of two transfer functions $H_1(z)$ and $H_2(z)$. That is,

$$H(z) = H_1(z)H_2(z). \tag{6.4.3}$$

where $H_1(z) = b_0 + b_1 z^{-1} + b_2 z^{-2}$ and $H_2(z) = 1/(1 + a_1 z^{-1} + a_2 z^{-2})$. Since multiplication is commutative, we have

$$H(z) = H_2(z)H_1(z). \tag{6.4.4}$$

Therefore Figure 6.12 can be redrawn as Figure 6.13.

Note that in Figure 6.13, the intermediate signal $w(n)$ is common to both signal buffers of $H_1(z)$ and $H_2(z)$. There is no need to use two separate buffers, thus these two signal buffers can be combined into one, shared by both filters as illustrated in Figure 6.14. We observe that this realization requires three memory locations to realize the second-order IIR filter, as opposed to six memory locations required for the direct-form I realization given in Figure 6.12. Therefore the direct-form II realization is called

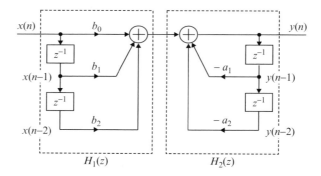

Figure 6.12   Direct-form I realization of second-order IIR filter

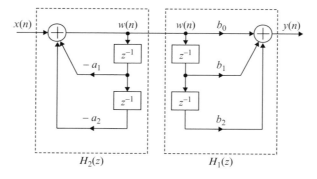

Figure 6.13   Signal-flow diagram of $H(z) = H_2(z)H_1(z)$

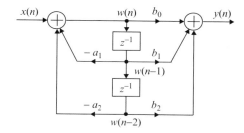

Figure 6.14 Direct-form II realization of second-order IIR filter

the canonical form since it realizes the given transfer function with the smallest possible numbers of delays, adders, and multipliers.

It is worthwhile verifying that the direct-form II realization does indeed implement the second-order IIR filter. From Figure 6.14, we have

$$y(n) = b_0 w(n) + b_1 w(n-1) + b_2 w(n-2),$$ (6.4.5)

where

$$w(n) = x(n) - a_1 w(n-1) - a_2 w(n-2).$$ (6.4.6)

Taking the z-transform of both sides of these two equations and re-arranging terms, we obtain

$$Y(z) = W(z)\left(b_0 + b_1 z^{-1} + b_2 z^{-2}\right)$$ (6.4.7)

and

$$X(z) = W(z)\left(1 + a_1 z^{-1} + a_2 z^{-2}\right).$$ (6.4.8)

The overall transfer function equals to

$$H(z) = \frac{Y(z)}{X(z)} = \frac{b_0 + b_1 z^{-1} + b_2 z^{-2}}{1 + a_1 z^{-1} + a_2 z^{-2}}$$

which is identical to (6.4.1). Thus the direct-form II realization described by (6.4.5) and (6.4.6) is identical to the direct-form I realization described in (6.4.2).

Figure 6.14 can be expanded as Figure 6.15 to realize the general IIR filter defined in (6.3.1) using the direct-form II structure. The block diagram realization of this system assumes $M = L - 1$. If $M \neq L - 1$, one must draw the maximum number of common delays. Although direct-form II still satisfies the difference Equation (6.3.2), it does not implement this difference equation directly. Similar to (6.4.5) and (6.4.6), it is a direct implementation of a pair of I/O equations:

$$w(n) = x(n) - \sum_{m=1}^{M} a_m w(n-m)$$ (6.4.9)

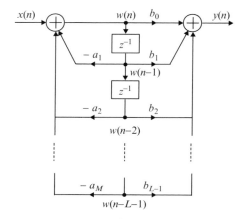

Figure 6.15   Direct-form II realization of general IIR filter, $L = M + 1$

and

$$y(n) = \sum_{l=0}^{L-1} b_l w(n - l). \tag{6.4.10}$$

The computed value of $w(n)$ from the first equation is passed into the second equation to compute the final output $y(n)$.

6.4.2   Cascade Form

The cascade realization of an IIR filter assumes that the transfer function is the product of first-order and/or second-order IIR sections. By factoring the numerator and the denominator polynomials of the transfer function $H(z)$ as a product of lower order polynomials, an IIR filter can be realized as a cascade of low-order filter sections. Consider the transfer function $H(z)$ given in (6.3.4), it can be expressed as

$$H(z) = b_0 H_1(z) H_2(z) \cdots H_K(z) = b_0 \prod_{k=1}^{K} H_k(z), \tag{6.4.11}$$

where each $H_k(z)$ is a first- or second-order IIR filter and K is the total number of sections. That is

$$H_k(z) = \frac{z - z_i}{z - p_j} = \frac{1 + b_{1k}z^{-1}}{1 + a_{1k}z^{-1}}, \tag{6.4.12}$$

or

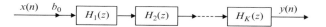

Figure 6.16 Cascade realization of digital filter

$$H_k(z) = \frac{(z - z_i)(z - z_j)}{(z - p_l)(z - p_m)} = \frac{1 + b_{1k}z^{-1} + b_{2k}z^{-2}}{1 + a_{1k}z^{-1} + a_{2k}z^{-2}}. \tag{6.4.13}$$

The realization of Equation (6.4.11) in cascade form is illustrated in Figure 6.16. In this form, any real-valued roots can be left as they are or combined into pairs, and the complex-conjugate roots must be grouped into the same section to guarantee that all the coefficients of $H_k(z)$ are real numbers. Assuming that every $H_k(z)$ is the second-order IIR filter described by (6.4.13), the I/O equations describing the time-domain operations of the cascade realization are expressed as

$$w_k(n) = x_k(n) - a_{1k}w_k(n - 1) - a_{2k}w_k(n - 2), \tag{6.4.14a}$$

$$y_k(n) = w_k(n) + b_{1k}w_k(n - 1) + b_{2k}w_k(n - 2), \tag{6.4.14b}$$

$$x_{k+1}(n) = y_k(n), \tag{6.4.14c}$$

for $k = 1, 2, \ldots, K$ and

$$x_1(n) = b_0 x(n), \tag{6.4.15a}$$

$$y(n) = y_k(n). \tag{6.4.15b}$$

By different ordering and different pairing, it is possible to obtain many different cascade realizations for the same transfer function $H(z)$. Ordering means the order of connecting $H_k(z)$, and pairing means the grouping of poles and zeros of $H(z)$ to form a section. These different cascade realizations are mathematically equivalent. In practice, each cascade realization behaves differently from others due to the finite-wordlength effects. In DSP hardware implementation, the internal multiplications in each section will generate a certain amount of roundoff error, which is then propagated into the next section. The total roundoff noise at the final output will depend on the particular pairing/ordering. The best ordering is the one that generates the minimum overall roundoff noise. It is not a simple task to determine the best realization among all possible cascade realizations. However, the complex-conjugate roots should be paired together, and we may pair the poles and zeros that are closest to each other in each section.

In the direct-form realization shown in Figure 6.15, the variation of one parameter will affect the locations of all the poles of $H(z)$. In a cascade realization, the variation of one parameter will affect only pole(s) in that section. Therefore the cascade realization is less sensitive to parameter variation (due to coefficient quantization, etc.) than the direct-form structure. In practical implementations of digital IIR filters, the cascade form is preferred.

Example 6.9: Given the second-order IIR filter

$$H(z) = \frac{0.5(z^2 - 0.36)}{z^2 + 0.1z - 0.72},$$

realize it using cascade form in terms of first-order sections.

By factoring the numerator and denominator polynomials of $H(z)$, we obtain

$$H(z) = \frac{0.5(1 + 0.6z^{-1})(1 - 0.6z^{-1})}{(1 + 0.9z^{-1})(1 - 0.8z^{-1})}.$$

By different pairings of poles and zeros, there are four different realizations of $H(z)$. For example, we choose

$$H_1(z) = \frac{1 + 0.6z^{-1}}{1 + 0.9z^{-1}} \quad \text{and} \quad H_2(z) = \frac{1 - 0.6z^{-1}}{1 - 0.8z^{-1}}.$$

The IIR filter can be realized by the cascade form expressed as

$$H(z) = 0.5H_1(z)H_2(z).$$

6.4.3 Parallel Form

The expression of $H(z)$ in a partial-fraction expansion leads to another canonical structure called the parallel form. It is expressed as

$$H(z) = c + H_1(z) + H_2(z) + \cdots + H_K(z), \tag{6.4.16}$$

where c is a constant, K is a positive integer, and $H_k(z)$ are transfer functions of first- or second-order IIR filters with real coefficients. That is,

$$H_k(z) = \frac{b_{0k}}{1 + a_{1k}z^{-1}}, \tag{6.4.17}$$

or

$$H_k(z) = \frac{b_{0k} + b_{1k}z^{-1}}{1 + a_{1k}z^{-1} + a_{2k}z^{-2}}. \tag{6.4.18}$$

The realization of Equation (6.4.16) in parallel form is illustrated in Figure 6.17. In order to produce real-valued coefficients in the filter structure, the terms in the partial-fraction-expansion corresponding to complex-conjugate pole pairs must be combined into second-order terms. Each second-order section can be implemented as direct-form II as shown in Figure 6.14, or direct-form I as shown in Figure 6.12.

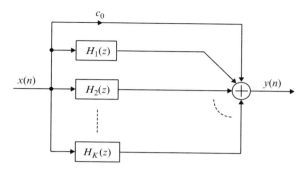

Figure 6.17 A parallel realization of digital IIR filter

The variation of parameters in a parallel form affects only the poles of the $H_k(z)$ associated with the parameters. The variation of any parameter in the direct-form realization will affect all the poles of $H(z)$. Therefore the pole sensitivity of a parallel realization is less than that of the direct form.

Example 6.10: Consider the transfer function $H(z)$ as given in Example 6.9, we can express

$$H'(z) = \frac{H(z)}{z} = \frac{0.5(1 + 0.6z^{-1})(1 - 0.6z^{-1})}{z(1 + 0.9z^{-1})(1 - 0.8z^{-1})} = \frac{A}{z} + \frac{B}{z + 0.9} + \frac{C}{z - 0.8},$$

where

$$A = zH'(z)|_{z=0} = 0.25,$$
$$B = (z + 0.9)H'(z)|_{z=-0.9} = 0.147, \text{ and}$$
$$C = (z - 0.8)H'(z)|_{z=0.8} = 0.103.$$

We can obtain

$$H(z) = 0.25 + \frac{0.147}{1 + 0.9z^{-1}} + \frac{0.103}{1 - 0.8z^{-1}}.$$

6.4.4 Realization Using MATLAB

The cascade realization of an IIR transfer function $H(z)$ involves its factorization in the form of (6.3.4). This can be done in MATLAB using the function `roots`. For example, the statement

```
r = roots(b);
```

will return the roots of the numerator vector b containing the coefficients of polynomial in z^{-1} in ascending power of z^{-1} in the output vector r. Similarly, we can use

```
d = roots(a);
```

to obtain the roots of the denominator vector a in the output vector d. From the computed roots, the coefficients of each section can be determined by pole–zero pairings.

A much simpler approach is to use the function tf2zp in the *Signal Processing Toolbox*, which finds the zeros, poles, and gains of systems in transfer functions of single-input or multiple-output form. For example, the statement

```
[z, p, c] = tf2zp(b, a);
```

will return the zero locations in the columns of matrix z, the pole locations in the column vector p, and the gains for each numerator transfer function in vector c. Vector a specifies the coefficients of the denominator in descending powers of z^{-1}, and the matrix b indicates the numerator coefficients with as many rows as there are outputs.

Example 6.11: The zeros, poles, and gain of the system

$$H(z) = \frac{2z^{-1} + 3z^{-2}}{1 + 0.4z^{-1} + z^{-2}}$$

can be obtained using the MATLAB script as follows:

```
b = [2, 3];
a = [1, 0.4, 1];
[z, p, c] = tf2zp(b,a);
```

MATLAB also provides a useful function zp2sos in the *Signal Processing Toolbox* to convert a zero-pole-gain representation of a given system to an equivalent representation of second-order sections. The function

```
[sos, G] = zp2sos(z, p, c);
```

finds the overall gain G and a matrix sos containing the coefficients of each second-order section of the equivalent transfer function $H(z)$ determined from its zero–pole form. The zeros and poles must be real or in complex-conjugate pairs. The matrix sos is a $K \times 6$ matrix

$$sos = \begin{bmatrix} b_{01} & b_{11} & b_{21} & a_{01} & a_{11} & a_{21} \\ b_{02} & b_{12} & b_{22} & a_{02} & a_{12} & a_{22} \\ \vdots & \vdots & \vdots & \vdots & \vdots & \vdots \\ b_{0K} & b_{1K} & b_{2K} & a_{0K} & a_{1K} & a_{2K} \end{bmatrix}, \tag{6.4.19}$$

whose rows contain the numerator and denominator coefficients, b_{ik} and a_{ik}, $i = 0, 1, 2$ of the kth second-order section $H_k(z)$. The overall transfer function is expressed as

$$H(z) = \prod_{k=1}^{K} H_k(z) = \prod_{k=1}^{K} \frac{b_{0k} + b_{1k}z^{-1} + b_{2k}z^{-2}}{a_{0k} + a_{1k}z^{-1} + a_{2k}z^{-2}}. \tag{6.4.20}$$

The parallel realizations discussed in Section 6.4.3 can be developed in MATLAB using the function `residuez` in the *Signal Processing Toolbox*. This function converts the transfer function expressed as (6.3.1) to the partial-fraction-expansion (or residue) form as (6.4.16). The function

```
[r, p, c] = residuez(b, a);
```

returns that the column vector `r` contains the residues, `p` contains the pole locations, and `c` contains the direct terms.

6.5   Design of IIR Filters Using MATLAB

As discussed in Chapter 5, digital filter design is a process of determining the values of the filter coefficients for given specifications. This is not an easy task and is generally better to be performed using computer software packages. A filter design package can be used to evaluate the filter design methods, to calculate the filter coefficients, and to simulate the filter's magnitude and phase responses. MATLAB is capable of designing Butterworth, Chebyshev I, Chebyshev II, and elliptic IIR filters in four different types of filters: lowpass, highpass, bandpass, and bandstop.

The *Signal Processing Toolbox* provides a variety of M-files for designing IIR filters. The IIR filter design using MATLAB requires two processes. First, the filter order N and the frequency-scaling factor Wn are determined from the given specifications. The coefficients of the filter are then determined using these two parameters. In the first step, the MATLAB functions to be used for estimating filter order are

```
[N, Wn] = buttord(Wp, Ws, Rp, Rs);
[N, Wn] = cheb1ord(Wp, Ws, Rp, Rs);
[N, Wn] = cheb2ord(Wp, Ws, Rp, Rs);
[N, Wn] = ellip(Wp, Ws, Rp, Rs);
```

for Butterworth, Chebyshev type I, Chebyshev type II, and elliptic filters, respectively. The parameters Wp and Ws are the normalized passband and stopband edge frequencies, respectively. The range of Wp and Ws are between 0 and 1, where 1 corresponds to the Nyquist frequency $(f_N = f_s/2)$. The parameters Rp and Rs are the passband ripple and the minimum stopband attenuation specified in dB, respectively. These four functions return the order N and the frequency scaling factor Wn. These two parameters are needed in the second step of IIR filter design using MATLAB.

For lowpass filters, the normalized frequency range of passband is $0 < F < Wp$, the stopband is $Ws < F < 1$, and $Wp < Ws$. For highpass filters, the normalized frequency range of stopband is $0 < F < Ws$, the passband is $Wp < F < 1$, and $Wp > Ws$. For bandpass and bandstop filters, Wp and Ws are two-element vectors that specify the transition bandages, with the lower-frequency edge being the first element of the vector, and N is half of the order of the filter to be designed.

In the second step of designing IIR filters based on the bilinear transformation, the *Signal Processing Toolbox* provides the following functions:

```
[b, a] = butter(N, Wn);
[b, a] = cheby1(N, Rp, Wn);
```

```
[b, a] = cheby2(N, Rs, Wn);
[b, a] = ellip(N, Rp, Rs, Wn);
```

The input parameters N and Wn are determined in the order estimation stage. These functions return the filter coefficients in length N+1 row vectors b and a with coefficients in descending powers of z^{-1}. The form of the transfer function obtained is given by

$$H(z) = \frac{b(1) + b(2)z^{-1} + \cdots + b(N+1)z^{-N}}{1 + a(2)z^{-1} + \cdots + a(N+1)z^{-N}}. \tag{6.5.1}$$

As introduced in Chapter 4, the frequency response of digital filters can be computed using the function freqz, which returns the frequency response $H(\omega)$ of a given numerator and denominator coefficients in vectors b and a, respectively.

Example 6.12: Design a lowpass Butterworth filter with less than 1.0 dB of ripple from 0 to 800 Hz, and at least 20 dB of stopband attenuation from 1600 Hz to the Nyquist frequency 4000 Hz.

The MATLAB script (exam6_12.m in the software package) for designing the specified filter is listed as follows:

```
Wp = 800/4000; Ws = 1600/4000;
Rp = 1.0; Rs = 20.0;
[N, Wn] = buttord(Wp, Ws, Rp, Rs);
[b, a] = butter(N, Wn);
freqz(b, a, 512, 8000);
```

The Butterworth filter coefficients are returned via vectors b and a by MATLAB function butter(N, Wn). The magnitude and phase responses of the designed fourth-order IIR filter are shown in Figure 6.18. This filter will be used for the IIR filter experiments in Sections 6.7.

Example 6.13: Design a bandpass filter with passband of 100 Hz to 200 Hz and the sampling rate is 1 kHz. The passband ripple is less than 3 dB and the stopband attenuation is at least 30 dB by 50 Hz out on both sides of the passband.

The MATLAB script (exam6_13.m in the software package) for designing the specified bandpass filter is listed as follows:

```
Wp = [100 200] /500; Ws = [50 250] /500;
Rp = 3; Rs = 30;
[N, Wn] = buttord(Wp, Ws, Rp, Rs);
[b, a] = butter(N, Wn);
freqz(b, a, 128, 1000);
```

The magnitude and phase responses of the designed bandpass filter are shown in Figure 6.19.

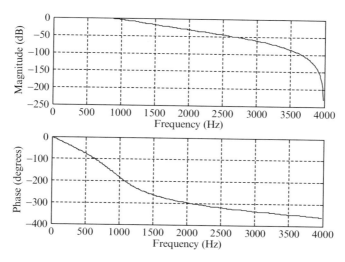

Figure 6.18 Frequency response of fourth-order Butterworth lowpass filter

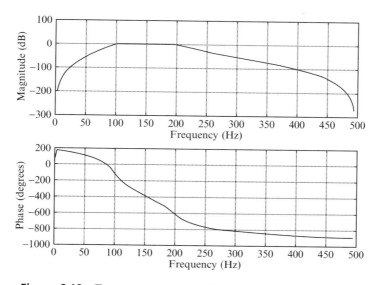

Figure 6.19 Frequency response of the designed bandpass filter

6.6 Implementation Considerations

As discussed in Section 6.4, the common IIR filter structures are direct, parallel, and cascade forms. The cascade forms are most often employed in practical applications for the reasons concerning quantization effects and DSP implementation. These issues will be discussed in this section.

6.6.1 Stability

The IIR filter described by the transfer function given in (6.3.4) is stable if all the poles lie within the unit circle. That is,

$$|p_m| < 1, \quad m = 1, 2, \ldots, M. \tag{6.6.1}$$

In this case, we can show that

$$\lim_{n \to \infty} h(n) = 0. \tag{6.6.2}$$

If $|p_m| > 1$ for any $0 \le m \le M$, then the IIR filter defined in (6.3.4) is unstable since

$$\lim_{n \to \infty} h(n) \to \infty. \tag{6.6.3}$$

In addition, an IIR filter is unstable if $H(z)$ has multiple-order pole(s) on the unit circle. For example, if $H(z) = z/(z-1)^2$, there is a second-order pole at $z = 1$. The impulse response of the system is $h(n) = n$, which is an unstable system as defined in (6.6.3).

Example 6.14: Given the transfer function

$$H(z) = \frac{1}{1 - az^{-1}},$$

the impulse response of the system is

$$h(n) = a^n, \quad n \ge 0.$$

If the pole is inside the unit circle, that is, $|a| < 1$, the impulse response

$$\lim_{n \to \infty} h(n) = \lim_{n \to \infty} a^n = 0.$$

Thus the IIR filter is stable. However,

$$\lim_{n \to \infty} h(n) = \lim_{n \to \infty} a^n \to \infty \quad \text{if } |a| > 1.$$

Thus the IIR filter is unstable for $|a| > 1$.

An IIR filter is marginally stable (or oscillatory bounded) if

$$\lim_{n \to \infty} h(n) = c, \tag{6.6.4}$$

where c is a non-zero constant. For example, $H(z) = 1/1 + z^{-1}$. There is a first-order pole on the unit circle. It is easy to show that the impulse response oscillates between ± 1 since $h(n) = (-1)^n, n \ge 0$.

Consider the second-order IIR filter defined by equation (6.4.1). The denominator can be factored as

$$1 + a_1 z^{-1} + a_2 z^{-2} = (1 - p_1 z^{-1})(1 - p_2 z^{-1}), \qquad (6.6.5)$$

where

$$a_1 = -(p_1 + p_2) \qquad (6.6.6)$$

and

$$a_2 = p_1 p_2. \qquad (6.6.7)$$

The poles must lie inside the unit circle for stability, that is $|p_1| < 1$ and $|p_2| < 1$. From (6.6.7), we obtain

$$|a_2| = |p_1 p_2| < 1. \qquad (6.6.8)$$

The corresponding condition on a_1 can be derived from the Schur–Cohn stability test and is given by

$$|a_1| < 1 + a_2. \qquad (6.6.9)$$

Stability conditions (6.6.8) and (6.6.9) are illustrated in Figure 6.20, which shows the resulting stability triangle in the $a_1 - a_2$ plane. That is, the second-order IIR filter is stable if and only if the coefficients define a point (a_1, a_2) that lies inside the stability triangle.

6.6.2   Finite-Precision Effects and Solutions

As discussed in Chapter 3, there are four types of quantization effects in digital filters – input quantization, coefficient quantization, roundoff errors, and overflow. In practice, the digital filter coefficients obtained from a filter design package are quantized to a finite number of bits so that the filter can be implemented using DSP hardware. The filter coefficients, b_l and a_m, of the discrete-time filter defined by (6.3.1) and (6.3.2) are determined by the filter design techniques introduced in Section 6.3, or by a filter design

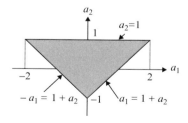

Figure 6.20   Region of coefficient values for a stable second-order IIR filter

package such as MATLAB that uses double-precision floating-point format to represent filter coefficients. Let b'_l and a'_m denote the quantized values corresponding to b_l and a_m, respectively. The I/O equation that can be actually implemented is given by Equation (3.5.10), and its transfer function is expressed as

$$H'(z) = \frac{\sum\limits_{l=0}^{L-1} b'_l z^{-l}}{1 + \sum\limits_{m=1}^{M} a'_m z^{-m}}. \tag{6.6.10}$$

Similar to the concept of input quantization discussed in Section 3.5, the nonlinear operation of coefficient quantization can be modeled as a linear process that introduces a quantization noise expressed as

$$b'_l = Q[b_l] = b_l + e(l) \tag{6.6.11}$$

and

$$a'_m = Q[a_m] = a_m + e(m), \tag{6.6.12}$$

where the coefficient quantization errors $e(l)$ and $e(m)$ can be assumed to be a random noise that has zero mean and variance as defined in (3.5.6).

If the wordlength is not large enough, some undesirable effects occur. For example, the frequency characteristics such as magnitude and phase responses of $H'(z)$ may be different from those of $H(z)$. In addition, for high-order filters whose poles are closely clustered in the z-plane, small changes in the denominator coefficients can cause large shifts in the location of the poles. If the poles of $H(z)$ are close to the unit circle, the pole(s) of $H'(z)$ may move outside the unit circle after coefficient quantization, resulting in an unstable implementation. These undesired effects are more serious when higher-order filters are implemented using the direct-form I and II realizations discussed in Section 6.4. Therefore the cascade and parallel realizations are preferred in practical DSP implementations with each $H_k(z)$ in a first- or second-order section.

Example 6.15: Given the IIR filter with transfer function

$$H(z) = \frac{1}{1 - 0.9z^{-1} + 0.2z^{-2}},$$

the poles are located at $z = 0.4$ and $z = 0.5$. This filter can be realized in the cascade form as

$$H(z) = H_1(z)H_2(z),$$

where $H_1(z) = 1/(1 - 0.4z^{-1})$ and $H_2(z) = 1/(1 - 0.5z^{-1})$.

Assuming that this IIR filter is implemented in a 4-bit (a sign bit plus 3 data bits) DSP hardware, 0.9, 0.2, 0.4, and 0.5 are quantized to 0.875, 0.125, 0.375, and 0.5, respectively. Therefore the direct-form realization is described as

$$H'(z) = \frac{1}{1 - 0.875z^{-1} + 0.125z^{-2}}$$

and the cascade realization is expressed as

$$H''(z) = \frac{1}{1 - 0.375z^{-1}} \cdot \frac{1}{1 - 0.5z^{-1}}.$$

The pole locations of the direct-form $H'(z)$ are $z = 0.18$ and $z = 0.695$, and the pole locations of the cascade form $H''(z)$ are $z = 0.375$ and $z = 0.5$. Therefore the poles of cascade realization are closer to the desired $H(z)$.

In practice, one must always check the stability of the filter with the quantized coefficients. The problem of coefficient quantization may be studied by examining pole locations in the z-plane. For a second-order IIR filter given in (6.4.1), we can place the poles near $z = \pm 1$ with much less accuracy than elsewhere in the z-plane. Since the second-order IIR filters are the building blocks of the cascade and parallel forms, we can conclude that narrowband lowpass (or highpass) filters will be most sensitive to coefficient quantization because their poles close to $z = 1$ or $(z = -1)$. In summary, the cascade form is recommended for the implementation of high-order narrowband IIR filters that have closely clustered poles.

As discussed in Chapter 3, the effect of the input quantization noise on the output can be computed as

$$\sigma_{y,e}^2 = \frac{\sigma_e^2}{2\pi j} \oint z^{-1} H(z) H(z^{-1}) dz, \tag{6.6.13}$$

where $\sigma_e^2 = 2^{-2B}/3$ is defined by (3.5.6). The integration around the unit circle $|z| = 1$ in the counterclockwise direction can be evaluated using the residue method introduced in Chapter 4 for the inverse z-transform.

Example 6.16: Consider an IIR filter expressed as

$$H(z) = \frac{1}{1 - az^{-1}}, \quad |a| < 1,$$

and the input signal $x(n)$ is an 8-bit data. The noise power due to input quantization is $\sigma_e^2 = 2^{-16}/3$. Since

$$R_{z=a} = (z - a)\frac{1}{(z - a)(1 - az)}\bigg|_{z=a} = \frac{1}{1 - a^2},$$

the noise power at the output of the filter is calculated as

$$\sigma_{y,e}^2 = \frac{2^{-16}}{3(1 - a^2)}.$$

As shown in (3.5.11), the rounding of $2B$-bit product to B bits introduces the roundoff noise, which has zero-mean and its power is defined by (3.5.6). Roundoff errors can be trapped into the feedback loops of IIR filters and can be amplified. In the cascade realization, the output noise power due to the roundoff noise produced at the previous section may be evaluated using (6.6.13). Therefore the order in which individual sections are cascaded also influences the output noise power due to roundoff. Most modern DSP chips (such as the TMS320C55x) solve this problem by using double-precision accumulator(s) with additional guard bits that can perform many multiplication–accumulation operations without roundoff errors before the final result in the accumulator is rounded.

As discussed in Section 3.6, when digital filters are implemented using finite word-length, we try to optimize the ratio of signal power to the power of the quantization noise. This involves a trade-off with the probability of arithmetic overflow. The most effective technique in preventing overflow of intermediate results in filter computation is by introducing appropriate scaling factors at various nodes within the filter stages. The optimization is achieved by introducing scaling factors to keep the signal level as high as possible without getting overflow. For IIR filters, since the previous output is fed back, arithmetic overflow in computing an output value can be a serious problem. A detailed analysis related to scaling is available in a reference text [12].

Example 6.17: Consider the first-order IIR filter with scaling factor α described by

$$H(z) = \frac{\alpha}{1 - az^{-1}},$$

where stability requires that $|a| < 1$. The actual implementation of this filter is illustrated in Figure 6.21. The goal of including the scaling factor α is to ensure that the values of $y(n)$ will not exceed 1 in magnitude. Suppose that $x(n)$ is a sinusoidal signal of frequency ω_0, then the amplitude of the output is a factor of $|H(\omega_0)|$. For such signals, the gain of $H(z)$ is

$$\max_\omega |H(\omega)| = \frac{\alpha}{1 - |a|}.$$

Thus if the signals being considered are sinusoidal, a suitable scaling factor is given by

$$\alpha < 1 - |a|.$$

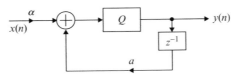

Figure 6.21   Implementation of a first-order IIR section with scaling factor

6.6.3   Software Implementations

As discussed in Chapter 1, the implementation of a digital filtering algorithm is often carried out on a general-purpose computer to verify that the designed filter indeed meets the goals of the application. In addition, such a software implementation may be adequate if the application does not require real-time processing.

For computer software implementation, we describe the filter in the form of a set of I/O difference equations. For example, a direct-form II realization of IIR filter is defined by (6.3.2). The MATLAB function `filter` in the *Signal Processing Toolbox* implements the IIR filter. The basic forms of this function are

```
y = filter(b, a, x);
y = filter(b, a, x, zi);
```

The numerator and denominator coefficients are contained in the vectors b and a respectively. The first element of vector a, $a(1)$, has been assumed to be equal to 1. The input vector is x and the output vector generated by the filter is y. At the beginning, the initial conditions (data in the signal buffer) are set to zero. However, they can be specified in the vector zi to reduce transients.

The direct-form realization of IIR filters can be implemented using following C function (`iir.c` in the software package):

```
/*****************************************************************
 *    IIR.C - This function performs IIR filtering               *
 *                                                               *
 *             na-1                    nb-1                       *
 *      yn = sum ai*x(n-i)   -    sum bj*y(n-j)                   *
 *             i=0                    j=1                         *
 *                                                               *
 *****************************************************************/
float iir(float *x, int na, float *a, float *y, int nb, float *b,
          int maxa, int maxb)
{
   float yn;                      /* Output of IIR filter */
   float yn1, yn2;                /* Temporary storage    */
   int i, j;                      /* Indexes              */

   yn1 = (float) 0.0;             /* y1(n) = 0. */
   yn2 = (float) 0.0;             /* y2(n) = 0. */

   for(i = 0; i <= na-1; ++i)
   {
      yn1 = yn1 + x[maxa-1-i] * a[i];
                                  /* FIR filtering of x(n) to get y1(n) */
   }
```

```
for(j = 1; j <= nb-1; ++j)
{
    yn2 = yn2 + y[maxb-j] * b[j];
                              /* FIR filtering of y(n) to get y2(n) */
}
yn = yn1-yn2;                 /* y(n) = y1(n)-y2(n)                  */
return(yn);                   /* Return y(n) to the main function   */
}
```

6.6.4 Practical Applications

Consider a simple second-order resonator filter whose frequency response is dominated by a single peak at frequency ω_0. To make a peak at $\omega = \omega_0$, we place a pair of complex-conjugate poles at

$$p_i = r_p e^{\pm j\omega_0}, \tag{6.6.14}$$

where $0 < r_p < 1$. The transfer function can be expressed as

$$H(z) = \frac{A}{(1 - r_p e^{j\omega_0} z^{-1})(1 - r_p e^{-j\omega_0} z^{-1})} = \frac{A}{1 - 2r_p\cos(\omega_0)z^{-1} + r_p^2 z^{-2}}$$

$$= \frac{A}{1 + a_1 z^{-1} + a_2 z^{-2}}, \tag{6.6.15}$$

where A is a fixed gain used to normalize the filter to unity at ω_0. That is, $|H(\omega_0)| = 1$. The direct-form realization is shown in Figure 6.22.

The magnitude response of this normalized filter is given by

$$|H(\omega_0)|_{z=e^{-j\omega_0}} = \frac{A}{\left|\left(1 - r_p e^{j\omega_0} e^{-j\omega_0}\right)\left(1 - r_p e^{-j\omega_0} e^{-j\omega_0}\right)\right|} = 1. \tag{6.6.16}$$

This condition can be solved to obtain the gain

$$A = \left|(1 - r_p)\left(1 - r_p e^{-2j\omega_0}\right)\right| = (1 - r_p)\sqrt{1 - 2r_p\cos(2\omega_0) + r_p^2}. \tag{6.6.17}$$

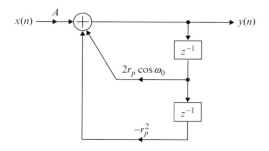

Figure 6.22 Signal flow graph of second-order resonator filter

The 3-dB bandwidth of the filter is defined as

$$20 \log_{10} \left| \frac{H(\omega)}{H(\omega_0)} \right| = 10 \log_{10} \left(\frac{1}{2} \right) = -3 \, \text{dB}. \tag{6.6.18}$$

This equation is equivalent to

$$|H(\omega)|^2 = \frac{1}{2} |H(\omega_0)|^2 = \frac{1}{2}. \tag{6.6.19}$$

There are two solutions on both sides of ω_0, and the bandwidth is the difference between these two frequencies. When the poles are close to the unit circle, the BW is approximated as

$$BW \cong 2(1 - r_p). \tag{6.6.20}$$

This design criterion determines the value of r_p for a given BW. The closer r_p is to one, the sharper the peak, and the longer it takes for the filter to reach its steady-state response. From (6.6.15), the I/O difference equation of resonator is given by

$$y(n) = Ax(n) - a_1 y(n-1) - a_2 y(n-2), \tag{6.6.21}$$

where

$$a_1 = -2r_p \cos \omega_0 \tag{6.6.22a}$$

and

$$a_2 = r_p^2. \tag{6.6.22b}$$

A recursive oscillator is a very useful tool for generating sinusoidal waveforms. The method is to use a marginally stable two-pole resonator where the complex-conjugate poles lie on the unit circle ($r_p = 1$). This recursive oscillator is the most efficient way for generating a sinusoidal waveform, particularly if the quadrature signals (sine and cosine signals) are required.

Consider two causal impulse responses

$$h_c(n) = \cos(\omega_0 n) u(n) \tag{6.6.23a}$$

and

$$h_s(n) = \sin(\omega_0 n) u(n), \tag{6.6.23b}$$

where $u(n)$ is the unit step function. The corresponding system transfer functions are

$$H_c(z) = \frac{1 - \cos(\omega_0) z^{-1}}{1 - 2 \cos(\omega_0) z^{-1} + z^{-2}} \tag{6.6.24a}$$

and

$$H_s(z) = \frac{\sin(\omega_0)z^{-1}}{1 - 2\cos(\omega_0)z^{-1} + z^{-2}}. \qquad (6.6.24b)$$

A two-output recursive structure with these system transfer functions is illustrated in Figure 6.23. The implementation requires just two data memory locations and two multiplications per sample. The output equations are

$$y_c(n) = w(n) - \cos(\omega_0)w(n-1) \qquad (6.6.25a)$$

and

$$y_s(n) = \sin(\omega_0)w(n-1), \qquad (6.6.25b)$$

where $w(n)$ is an internal state variable that is updated as

$$w(n) = 2\cos(\omega_0)w(n-1) - w(n-2). \qquad (6.6.26)$$

An impulse signal $A\delta(n)$ is applied to excite the oscillator, which is equivalent to presetting $w(n)$ and $w(n-1)$ to the following initial conditions:

$$w(0) = A \qquad (6.6.27a)$$

and

$$w(-1) = 0. \qquad (6.6.27b)$$

The waveform accuracy is limited primarily by the DSP processor wordlength. For example, quantization of the coefficient $\cos(\omega_0)$ causes the actual output frequency to differ slightly from the ideal frequency ω_0.

For some applications, only a sinewave is required. From equations (6.6.21), (6.2.22a) and (6.6.22b) using the conditions that $x(n) = A\delta(n)$ and $r_p = 1$, we can obtain the sinusoidal function

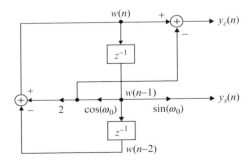

Figure 6.23 Recursive quadrature oscillator

$$y_s(n) = Ax(n) - a_1 y_s(n-1) - a_2 y_s(n-2)$$
$$= 2\cos(\omega_0) y_s(n-1) - y_s(n-2) \tag{6.6.28}$$

with the initial conditions

$$y_s(1) = A\sin(\omega_0) \tag{6.6.29a}$$

and

$$y_s(0) = 0. \tag{6.6.29b}$$

The oscillating frequency defined by Equation (6.6.28) is determined from its coefficient a_1 and its sampling frequency f_s, expressed as

$$f = \cos^{-1}\left(\frac{|a_1|}{2}\right)\frac{f_s}{2\pi} \text{ Hz}, \tag{6.6.30}$$

where the coefficient $|a_1| \leq 2$.

The sinewave generator using resonator can be realized from the recursive computation given in (6.6.28). The implementation using the TMS320C55x assembly language is listed as follows:

```
        mov     cos_w, T1
        mpym    * AR1+, T1, AC0      ; AC0 = cos(w) * y[n−1]
        sub     * AR1-≪#16, AC0, AC1 ; AC1 = cos(w) * y[n−1] − y[n−2]
        add     AC0, AC1             ; AC1 = 2* cos(w) * y[n−1] − y[n−2]
|| delay    * AR1                    ; y[n−2] = y[n−1]
        mov     rnd(hi(AC1)), *AR1   ; y[n−1] = y[n]
        mov     rnd(hi(AC1)), *AR0+  ; y[n] = 2* cos(w) * y[n−1] − y[n−2]
```

In the program, AR1 is the pointer for the signal buffer. The output sinewave samples are stored in the output buffer pointed by AR0. Due to the limited wordlength, the quantization error of fixed-point DSPs such as the TMSC320C55x could be severe for the recursive computation.

A simple parametric equalizer filter can be designed from a resonator given in (6.6.15) by adding a pair of zeros near the poles at the same angles as the poles. That is, placing the complex-conjugate poles at

$$z_i = r_z e^{\pm j\omega_0}, \tag{6.6.31}$$

where $0 < r_z < 1$. Thus the transfer function given in (6.6.15) becomes

$$H(z) = \frac{(1 - r_z e^{j\omega_0} z^{-1})(1 - r_z e^{-j\omega_0} z^{-1})}{(1 - r_p e^{j\omega_0} z^{-1})(1 - r_p e^{-j\omega_0} z^{-1})} = \frac{1 - 2r_z \cos(\omega_0) z^{-1} + r_z^2 z^{-2}}{1 - 2r_p \cos(\omega_0) z^{-1} + r_p^2 z^{-2}}$$
$$= \frac{1 + b_1 z^{-1} + b_2 z^{-2}}{1 + a_1 z^{-1} + a_2 z^{-2}}. \tag{6.6.32}$$

When $r_z < r_p$, the pole dominates over the zero because it is closer to the unit circle than the zero does. Thus it generates a peak in the frequency response at $\omega = \omega_0$. When $r_z > r_p$, the zero dominates over the pole, thus providing a dip in the frequency response. When the pole and zero are very close to each other, the effects of the poles and zeros are reduced, resulting in a flat response. Therefore Equation (6.6.32) provides a boost if $r_z < r_p$, or a reduction if $r_z > r_p$. The amount of gain and attenuation is controlled by the difference between r_p and r_z. The distance from r_p to the unit circle will determine the bandwidth of the equalizer.

6.7   Software Developments and Experiments Using the TMS320C55x

The digital IIR filters are widely used for practical DSP applications. In the previous sections, we discussed the characteristics, design, realization, and implementation of IIR filters. The experiments given in this section demonstrate DSP system design process using an IIR filter as example. We will also discuss some practical considerations for real-time applications.

As shown in Figure 1.8, a DSP system design usually consists of several steps, such as system requirements and specifications, algorithm development and simulation, software development and debugging, as well as system integration and testing. In this section, we will use an IIR filter as an example to show these steps with an emphasis on software development. First, we define the filter specifications such as the filter type, passband and stopband frequency ranges, passband ripple, and stopband attenuation. We then use MATLAB to design the filter and simulate its performance. After the simulation results meet the given specifications, we begin the software development process. We start with writing a C program with the floating-point data format in order to compare with MATLAB simulation results. We then measure the filter performance and improve its efficiency by using fixed-point C implementation and C55x assembly language. Finally, the design is integrated into the DSP system and is tested again.

Figure 6.24 shows a commonly used flow chart of DSP software development. In the past, software development was heavily concentrated in stage 3, while stage 2 was skipped. With the rapid improvement of DSP compiler technologies in recent years, C compilers have been widely used throughout stages 1 and 2 of the design process. Aided by compiler optimization features such as intrinsic as well as fast DSP processor speed, real-time DSP applications are widely implemented using the mixed C and assembly code. In the first experiment, we will use the floating-point C code to implement an IIR filter in the first stage as shown in Figure 6.24. Developing code in stage 1 does not require knowledge of the DSP processors and is suitable for algorithm development and analysis. The second and third experiments emphasize the use of C compiler optimization, data type management, and intrinsic for stage 2 of the design process. The third stage requires the longest development time because assembly language programming is much more difficult than C language programming. The last experiment uses the assembly code in order to compare it with previous experiments. In general, the assembly code is proven to be the most efficient in implementing DSP algorithms such as filtering that require intensive multiply/accumulate operations, while C code can do well in data manipulation such as data formatting and arrangement.

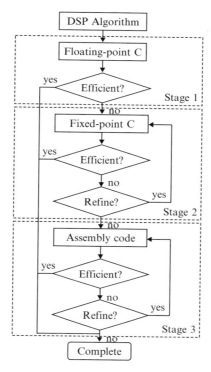

Figure 6.24 Flow chart for DSP software development

6.7.1 Design of IIR Filter

Digital filter coefficients can be determined by filter design software packages such as MATLAB for given specifications. As mentioned in Section 6.4, high-order IIR filters are often implemented in the form of cascade or parallel second-order sections for real-time applications. For instance, the fourth-order Butterworth filter given by Example 6.12 can be realized in the cascade direct-form II structure. The following MATLAB script (s671.m in the software package) shows a lowpass IIR Butterworth filter design process.

```
%
% Filter specifications
%
Fs = 8000;      % Sampling frequency 8 kHz
fc = 800;       % Passband cutoff frequency 800 Hz
fs = 1600;      % Stopband frequency 1.6 kHz
Rp = 1;         % Passband ripple in dB
Rs = 20;        % Stopband attenuation in dB
Wp = 2*fc/Fs;   % Normalized passband edge frequency
Ws = 2*fs/Fs;   % Normalized stopband edge frequency
%
```

```
% Filter design
%
[N,Wn] = buttord(Wp,Ws,Rp,Rs);   % Filter order selection
[b,a] = butter(N,Wn);            % Butterworth filter design
[Z,P,K] = tf2zp(b,a);            % Transfer function to zero-pole
[sos,G] = zp2sos(Z,P,K);         % Zero-pole to second-order section
```

This program generates a fourth-order IIR filter with the following coefficient vectors:

$$b = [0.0098, 0.0393, 0.0590, 0.0393, 0.0098]$$
$$a = [1.0000, -1.9908, 1.7650, -0.7403, 0.1235].$$

The fourth-order IIR filter is then converted into two second-order sections represented by the coefficients matrix sos and the overall system gain G. By decomposing the coefficient matrix sos defined in (6.4.19), we obtain two matrices

$$b = \begin{bmatrix} 0.0992 & 0.1984 & 0.0992 \\ 0.0992 & 0.1984 & 0.0992 \end{bmatrix} \quad \text{and} \quad a = \begin{bmatrix} 1.0 & -0.8659 & 0.2139 \\ 1.0 & -1.1249 & 0.5770 \end{bmatrix}, \quad (6.7.1)$$

where we equally distribute the overall gain factor into each second-order cascade configuration for simplicity. In the subsequent sections, we will use this Butterworth filter for the TMS320C55x experiments.

6.7.2   Experiment 6A – Floating-Point C Implementation

For an IIR filter consists of K second-order sections, the I/O equation of the cascade direct-form II realization is given by Equation (6.4.14). The C implementation of general cascade second-order sections can be written as follows:

```
temp = input[n];
for(k = 0; k < IIR_SECTION; k++)
{
    w[k][0] = temp - a[k][1] * w[k][1] - a[k][2] * w[k][2];
    temp = b[k][0] * w[k][0] + b[k][1] * w[k][1] + b[k][2] * w[k][2];
    w[k][2] = w[k][1];          /* w(n-2) <- w(n-1)   */
    w[k][1] = w[k][0] ;         /* w(n-1) <- w(n)     */
}
output[n] = temp;
```

where a[][] and b[][] are filter coefficient matrices defined in (6.7.1), and w[][] is the signal buffer for $w_k(n - m)$, $m = 0, 1, 2$. The row index k represents the kth second-order IIR filter section, and the column index points at the filter coefficient or signal sample in the buffer.

As mentioned in Chapter 5, the zero-overhead repeat loop, multiply–accumulate instructions, and circular buffer addressing modes are three important features of DSP processors. To better understand these features, we write the IIR filter function in C using data pointers to simulate the circular buffers instead of two-dimensional arrays. We also arrange the C statements to mimic the DSP multiply/accumulate operations. The following C program is an IIR filter that consists of Ns second-order sections in cascade form. The completed block IIR filter function (iir.c) written in floating-point C language is provided in the experimental software package.

```
m = Ns * 5;                        /* Setup for circular buffer C[m]   */
k = Ns * 2;                        /* Setup for circular buffer w[k]   */
j = 0;
w_0 = x[n];                        /* Get input signal                 */
for (i = 0; i < Ns; i++)
{
    w_0 -= *(w+1)* *(C+j);  j++; l = (1+Ns)%k;
    w_0 -= *(w+1)* *(C+j);  j++;
    temp = *(w+1);
    *(w+1) = w_0;
    w_0 = temp * *(C+j);    j++;
    w_0 += *(w+1) * *(C+j);  j++; l = (1+Ns)%k;
    w_0 += *(w+1) * *(C+j);  j = (j+1)%m; l = (1+1)%k;
}
y[n] = w_0                          /* Save output */
```

The coefficient and signal buffers are configured as circular buffers shown in Figure 6.25. The signal buffer contains two elements, $w_k(n-1)$ and $w_k(n-2)$, for each second-order section. The pointer address is initialized pointing at the first sample $w_1(n-1)$ in the buffer. The coefficient vector is arranged with five coefficients (a_{1k}, a_{2k}, b_{2k}, b_{0k},

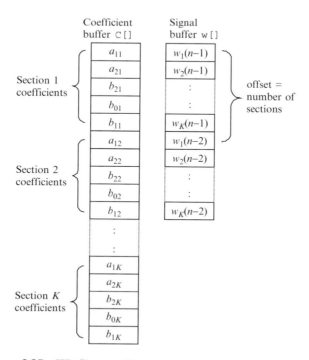

Figure 6.25   IIR filter coefficient and signal buffers configuration

and b_{1k}) per section with the coefficient pointer initialized pointing at the first coefficient, a_{11}. The circular pointers are updated by $j = (j+1)\%m$ and $l = (l+1)\%k$, where m
and k are the sizes of the coefficient and signal buffers, respectively.

The C function `exp6a.c` used for Experiment 6A is listed in Table 6.3. This program
calls the software signal generator `signal_gen2()` to create a block of signal samples
for testing. It then calls the IIR filter function `iir` to perform the lowpass filtering
process. The lowpass filter used for the experiment is the fourth-order Butterworth IIR

Table 6.3   List of floating-point C implementation of IIR filter

```
/*
    exp6a.c — Direct-form II IIR function implementation
               in floating-point C and using signal generator
*/
#define M  128        /* Number of samples per block    */
#define Ns   2        /* Number of second-order sections */

/* Low-pass IIR filter coefficients */
float C[Ns*5] = { /* i is section index */
                 /* A[i][1],A[i][2],B[i][2],B[i][0],B[i][1] */
                 −0.8659, 0.2139, 0.0992, 0.0992, 0.1984,
                 −1.1249, 0.5770, 0.0992, 0.0992, 0.1984};
/* IIR filter signal buffer:
w[] = w[i][n−1],w[i+1][n−1],...,w[i][n−2],w[i+1][n−2],... */
float w[Ns*2];

int out[M];
int in[M];

/* IIR filter function */
extern void iir(int *, int, int *, float *, int, float *);

/* Software signal generator */
extern void signal_gen2(int *, int);

void main(void)
{
   int i;

   /* Initialize IIR filter signal buffer */
   for(i = 0; i < Ns*2;i++)
     w[i] = 0;

   /* IIR filtering */
   for(;;)
   {
     signal_gen2(in, M);          /* Generate a block of samples */
     iir(in, M, out, C, Ns, w);   /* Filter a block of samples   */
   }
}
```

filter designed in Section 6.7.1. We rearranged the filter coefficients for using circular buffer. Two temporary variables, `temp` and `w_0`, are used for intermediate storage.

Go through the following steps for Experiment 6A:

1. Create the project `exp6a` and add the linker command file `exp6.cmd`, the C functions `iir.c`, `epx6a.c`, `signal_gen2.c` and `sine.asm` to the project. The lowpass filter `iir()` will attenuate high-frequency components from the input signal generated by the signal generator `signal_gen2()`, which uses the recursive sinewave generator `sine()` to generate three sinewaves at 800 Hz, 1.8 kHz, and 3.3 kHz.

2. Use `rts55.lib` for initializing the C function `main()` and build the project `exp6a`.

3. Set a breakpoint at the statement `for(;;)` of the `main()` function, and use the CCS graphic function to view the 16-bit integer output samples in the buffer `out[]`. Set data length to 128 for viewing one block of data at a time. Animate the filtering process, and observe the filter output as a clean 800 Hz sinewave.

4. Profile the IIR filter performance by measuring the average DSP clock cycles. Record the clock cycles and memory usage of the floating-point C implementation.

5. Overflow occurs when the results of arithmetic operations are larger than the fixed-point DSP can represent. Before we move on to fixed-point C implementation, let us examine the IIR filter for possible overflow. First, change the conditional compiling bit `CHECK_OVERFLOW` defined in `iir.c` from 0 to 1 to enable the sections that search for maximum and minimum intermediate values. Then, add `w_max` and `w_min` to the CCS watch window. Finally, run the experiment in the animation mode and examine the values of `w_max` and `w_min`. If $|w\_max| \geq 1$ or $|w\_min| \geq 1$, an overflow will happen when this IIR filter is implemented by a 16-bit fixed-point processor. If the overflow is detected, modify the IIR filter routine by scaling down the input signal until the values $|w\_max|$ and $|w\_min|$ are less than 1. Remember to scale up the filter output if the input is scaled down.

6.7.3   Experiment 6B – Fixed-Point C Implementation Using Intrinsics

Since the TMS320C55x is a fixed-point device, the floating-point implementation of the IIR filter given in Experiment 6A is very inefficient for real-time applications. This is because the floating-point implementation needs the floating-point math library functions that use fixed-point hardware to realize floating-point operations. In order to improve the performance, fixed-point C implementation should be considered. To ease the burden of fixed-point C programming, the TMS320C55x C compiler provides a set of intrinsics to handle specific signal processing operations.

The C55x intrinsics produce assembly language statements directly into the programs. The intrinsics are specified with a leading underscore and are used as C functions. The intrinsic function names are similar to their mnemonic assembly counterparts. For example, the following signed multiply–accumulate intrinsic

```
z = _smac(z,x,y);   /* Perform signed z = z+x* y */
```

will perform the equivalent assembly instruction

```
macm Xmem,Ymem,ACx
```

Table 6.4 lists the intrinsics supported by the TMS320C55x.

Table 6.4   Intrinsics provided by the TMS320C55x C compiler

C Compiler Intrinsic (a,b,c are 16-bit and d,e,f are 32-bit data)	Description
`c = _sadd(int a, int b);`	Adds 16-bit integers a and b, with SATA set, producing a saturated 16-bit result c.
`f = _lsadd(long d, long e);`	Adds 32-bit integers d and e, with SATD set, producing a saturated 32-bit result f.
`c = _ssub(int a, int b);`	Subtracts 16-bit integer b from a with SATA set, producing a saturated 16-bit result c.
`f = _lssub(long d, long e);`	Subtracts 32-bit integer e from d with SATD set, producing a saturated 32-bit result f.
`c = _smpy(int a, int b);`	Multiplies a and b, and shifts the result left by 1. Produces a saturated 16-bit result c. (upper 16-bit, SATD and FRCT set)
`f = _lsmpy(int a, int b);`	Multiplies a and b, and shifts the result left by 1. Produces a saturated 32-bit result f. (SATD and FRCT set)
`f = _smac(long d, int a, int b);`	Multiplies a and b, shifts the result left by 1, and adds it to d. Produces a saturated 32-bit result f. (SATD, SMUL and FRCT set)
`f = _smas(long d, int a, int b);`	Multiplies a and b, shifts the result left by 1, and subtracts it from d. Produces a 32-bit result f. (SATD, SMUL and FRCT set)
`c = _abss(int a);`	Creates a saturated 16-bit absolute value. c = |a|, _abss(0x8000) => 0x7FFF (SATA set)
`f = _labss(long d);`	Creates a saturated 32-bit absolute value. f = |d|, _labss(0x8000000) => 0x7FFFFFFF (SATD set)
`c = _sneg(int a);`	Negates the 16-bit value with saturation. c = − a, _sneg(0xffff8000) => 0x00007FFF

Table 6.4 *(continued)*

C Compiler Intrinsic (a, b, c are 16-bit and d, e, f are 32-bit data)	Description
`f = _lsneg(long d);`	Negates the 32-bit value with saturation. $f = -d$, _lsneg(0x80000000) => 0x7FFFFFFF
`c = _smpyr(int a, int b);`	Multiplies a and b, shifts the result left by 1, and rounds the result c. (SATD and FRCT set)
`c = _smacr(long d, int a, int b);`	Multiplies a and b, shifts the result left by 1, adds the result to d, and then rounds the result c. (SATD, SMUL and FRCT set)
`c = _smasr(long d, int a, int b);`	Multiplies a and b, shifts the result left by 1, subtracts the result from d, and then rounds the result c. (SATD, SMUL and FRCT set)
`c = _norm(int a);`	Produces the number of left shifts needed to normalize a and places the result in c.
`c = _lnorm(long d);`	Produces the number of left shifts needed to normalize d and places the result in c.
`c = _rnd(long d);`	Rounds d to produces the 16-bit saturated result c. (SATD set)
`c = _sshl(int a, int b);`	Shifts a left by b and produces a 16-bit result c. The result is saturated if b is greater than or equal to 8. (SATD set)
`f = _lsshl(long d, int a);`	Shifts a left by b and produces a 32-bit result f. The result is saturated if a is greater than or equal to 8. (SATD set)
`c = _shrs(int a, int b);`	Shifts a right by b and produces a 16-bit result c. (SATD set)
`f = _lshrs(long d, int a);`	Shifts d right by a and produces a 32-bit result f. (SATD set)
`c = _addc(int a, int b);`	Adds a, b, and Carry bit and produces a 16-bit result c.
`f = _laddc(long d, int a);`	Adds d, a, and Carry bit and produces a 32-bit result f.

The floating-point IIR filter function given in the previous experiment can be converted to the fixed-point C implementation using these intrinsics. To prevent intermediate overflow, we scale the input samples to Q14 format in the fixed-point implementation. Since the largest filter coefficient is between 1 and 2, we use Q14 representation for the fixed-point filter coefficients defined in (6.7.1). The implementation of the IIR filter in the fixed-point Q14 format is given as follows:

```
m = Ns* 5;                    /* Setup for circular buffer coef[m]  */
k = Ns* 2;                    /* Setup for circular buffer w[k]     */

j = 0;
w_0 = (long) x[n] ≪14;        /* Q14 input (scaled)*/
for (i = 0; i < Ns; i++)
{
    w_0 = _smas(w_0,*(w+1),*(coef+j)); j++; l = (1+Ns)%k;
    w_0 = _smas(w_0,*(w+1),*(coef+j)); j++;
    temp = *(w+1);
    *(w+1) = w_0≫15;
    w_0 = _lsmpy(temp,*(coef+j)); j++;
    w_0 = _smac(w_0,*(w+1), *(coef+j)); j++; l = (1+Ns)%k;
    w_0 = _smac(w_0,*(w+1), *(coef+j)); j = (j+1)%m; l = (1+1)%k;

}
y[n] = w_0≫14;                /* Q15 output */
```

Go through the following steps for Experiment 6B:

1. Create the project exp6b that include the linker command file exp6.cmd, the C functions exp6b.c, iir_i1.c, signal_gen2.c, and the assembly routine sine.asm.

2. Use rts55.lib for initializing the C function main(), and build the project exp6b.

3. Set a breakpoint at the statement for (;;) of the main() function, and use the CCS graphic function to view the 16-bit integer output samples in the buffer out[]. Set data length to 128 for viewing one block of data at a time. Animate the filtering process and observe the filter output as a clean 800 Hz sinewave.

4. Profile the IIR filter function iir_i1(), and compare the results with those obtained in Experiment 6A.

5. In file iir_i1(), set the scaling factor SCALE to 0 so the samples will not be scaled. Rebuild the project, and run the IIR filter experiment in the animation mode. We will see the output distortions caused by the intermediate overflow.

6.7.4   Experiment 6C – Fixed-Point C Programming Considerations

From the previous experiments, the fixed-point IIR filter implementation using intrinsics has greatly improved the efficiency of the fixed-point IIR filter using the floating-point implementation. We can further enhance the C function performance by taking advantage of the compiler optimization and restructuring the program to let the C compiler generate a more efficient assembly code.

Run-time support library functions

The TMS320C55x C compiler has many built-in C functions in its run-time support library `rts55.lib`. Although these functions are helpful, most of them may run at a slower speed due to the nested library function calls. We should try to avoid using them in real-time applications if possible. For example, the MOD (%) operation we use to simulate the circular addressing mode can be replaced by a simple AND (&) operation if the size of the buffer is a base 2 number, such as 2, 4, 8, 16, and so on. The example given in Table 6.5 shows the compiler will generate a more efficient assembly code (by avoiding calling the library function I$$MOD) when using the logic operator AND than the MOD operator, because the logic operation AND does not invoke any function calls.

Loop counters

The for-loop is the most commonly used loop control operations in C programs for DSP applications. The assembly code generated by the C55x C compiler varies depending on how the for-loop is written. Because the compiler must verify both the positive and negative conditions of the integer loop counter against the loop limit, it creates more lines of assembly code to check the entrance and termination of the loop. By using an unsigned integer as a counter, the C compiler only needs to generate a code that compares the positive loop condition. Another important loop-control method is to use a down counter instead of an up counter if possible. This is because most of the built-in conditional instructions act upon zero conditions. The example given in Table 6.6 shows the assembly code improvement when it uses an unsigned integer as a down counter.

Local repeat loop

Using local repeat-loop is another way to improve the DSP run-time efficiency. The local repeat-loop uses the C55x instruction-buffer-queue (see Figure 2.2) to store all the

Table 6.5 Example to avoid using library function when applying modulus operation

`;Ns = 2;`
`;k = 2*Ns;`
`;l = (1+Ns)%k;`
	`;Ns = 2;`
	`;k = 2*Ns-1;`
	`;l = (1+Ns)&k;`

`MOV *SP(#04h),AR1`
`MOV *SP(#0ah),T1`
`ADD *SP(#0bh),AR1,T0`
`CALL I$$MOD`
`MOV T0,*SP(#0bh)`
	`MOV T0,*SP(#07h)`
	`ADD *SP(#0bh),AR1`
	`AND *SP(#0ah),AR1`
	`MOV AR1,*SP(#0bh)`

Table 6.6   Example of using unsigned integer as down counter for loop control

``
; int i
; for (i = 0; i<Ns; i++)
; {
 	
   ... ...
 	
; }
``
	``
	; unsigned int i
	; for (i = Ns; i>0; i--)
	; {
	
	   ... ...
	
	; }
	``

``
    MOV *SP(#04h), AR1
    MOV #0, *SP(#06h)
    MOV *SP(#06h), AR2
    CMP AR2 >= AR1, TC1
    BCC L3, TC1
    ... ...
 	
    MOV *SP(#04h), AR2
    ADD #1, *SP(#06h)
    MOV *SP(#06h), AR1
    CMP AR1 < AR2, TC1
    BCC L2, TC1
``
	``
	    MOV *SP(#04h), AR1
	    MOV AR1, *SP(#06h)
	    BCC L3, AR1 == #0
	
	    ... ...
	
	    ADD #-1, *SP(#06h)
	    MOV *SP(#06h), AR1
	    BCC L2, AR1 != #0
	``

instructions within a loop. Local repeat-loop can execute the instructions repeatedly within the loop without additional instruction fetches. To allow the compiler to generate local-repeat loops, we should reduce the number of instructions within the loop because the size of instruction buffer queue is only 64 bytes.

Compiler optimization

The C55x C compiler has many options. The −on option ($n = 0, 1, 2,$ or 3) controls the compiler optimization level of the assembly code it generated. For example, the −o3 option will perform loop optimization, loop unrolling, local copy/constant propagation, simplify expression statement, allocate variables to registers, etc. The example given in Table 6.7 shows the code generated with and without the −o3 optimization.

Go through the following steps for Experiment 6C:

1. Create the project exp6c, add the linker command file exp6.cmd, the C functions exp6c.c, iir_i2.c, signal_gen2.c, and the assembly routine sine.asm into the project. The C function iir_i2.c uses unsigned integers for loop counters, and replaces the MOD operation with AND operation for the signal buffer.

2. Use rts55.lib for initializing the C function main(), and build the project exp6c.

3. Relocate the C program and data variables into SARAM and DARAM sections defined by the linker command files. Use pragma to allocate the program and data memory as follows:

Table 6.7   Example of compiler without and with -o3 optimization option

-o3 disabled	-o3 enabled

`; Ns = 2;`
`; for (i = Ns*2; i > 0; i--)`
`;  *ptr++ = 0;`
	`; Ns = 2;`
	`; for (i = Ns*2; i > 0; i--)`
	`;  *ptr++ = 0;`

` MOV #4,*SP(#00)`
` MOV *SP(#00),AR1`
` BCC L2,AR1 == #0`
 	
` MOV *SP(#01h),AR3`
` ADD #1,AR3,AR1`
` MOV AR1,*SP(#01h)`
` MOV #0,*AR3`
 	
` ADD #-1,*SP(#00h)`
` MOV *SP(#00h),AR1`
` BCC L1,AR1 != #0`
	` RPT #3`
	` MOV #0,*AR3+`

- Place the `main()` and `iir()` functions into the program SARAM, and name the section `iir_code`.
- Allocate the input and output buffers `in[]` and `out[]` to data SARAM, and name the sections input and output.
- Put the IIR filter coefficient buffer `C[]` in a separate data SARAM section, and name the section `iir_coef`.
- Place the temporary buffer `w[]` and temporary variables in a DARAM section, and name it `iir_data`.

4. Enable `-o3` and `-mp` options to rebuild the project.

5. Set a breakpoint at the statement `for (;;)` of the `main()` function, and use the CCS graphic function to view the 16-bit integer output samples in the buffer `out[]`. Set data length to 128 for viewing one block of data at a time. Animate the filtering process and observe the filter output as a clean 800 Hz sinewave.

6. Profile the IIR filter `iir_i2()`, and compare the result with those obtained in Experiment 6B.

6.7.5   Experiment 6D – Assembly Language Implementations

The advantages of C programming are quick prototyping, portable to other DSP processors, easy maintenance, and flexibility for algorithm evaluation and analysis. However, the IIR filter implementations given in previous experiments can be more efficient if the program is written using the C55x assembly instructions. By examining the IIR filter function used for Experiments 6B and 6C, we anticipate that the filter inner-loop can be implemented by the assembly language in seven DSP clock cycles. Obviously, from the previous experiments, the IIR filter implemented in C requires more cycles. To get the best performance, we can write the IIR filtering routine in assembly language. The trade-off between C and assembly programming is the time needed as well as the difficulties encountered for program development, maintenance, and system migration from one DSP system to the others.

The IIR filter realized by cascading the second-order sections can be implemented in assembly language as follows:

```
masm *AR3+,*AR7+,AC0        ; AC0 -= AC0-a1*w(n-1)
masm T3 = *AR3,*AR7+,AC0    ; AC0 -= AC0-a2*w(n-2)
mov  rnd(hi(AC0)),*AR3-     ; Update w(n)buffer
mpym *AR7+,T3,AC0           ; AC0 = bi2*w(n-2)
macm *(AR3+T1),*AR7+,AC0    ; AC0 += AC0+bi0*w(n)
macm *AR3+,*AR7+,AC0        ; AC0 += AC0+bi1*w(n-1)
mov  rnd(hi(AC0)),*AR1+     ; Store result
```

The assembly program uses three pointers. The IIR filter signal buffer for $w_k(n)$ is addressed by the auxiliary register AR3, while the filter coefficient buffer is pointed at by AR7. The filter output is rounded and placed in the output buffer pointed at by AR1.

The code segment can be easily modified for filtering either a single sample of data or a block of samples. The second-order IIR filter sections can be implemented using the inner repeat loop, while the outer loop can be used for controlling samples in blocks. The input sample is scaled down to Q14 format, and the IIR filter coefficients are also represented in Q14 format to prevent overflow. To compensate the Q14 format of coefficients and signal samples, the final result $y(n)$ is multiplied by 4 (implemented by shifting two bits to the left) to scale it back to Q15 format and store it with rounding. Temporary register T3 is used to hold the second element $w_k(n-2)$ when updating the signal buffer.

Go through the following steps for Experiment 6D:

1. Create the project exp6d, and include the linker command file exp6.cmd, the C functions exp6d.c, signal_gen2.c, the assembly routine sine.asm, and iirform2.asm into the project. The prototype of the IIR filter routine is written as

   ```
   void iirform2(int *x, unsigned int M, int *y, int *h,
                 unsigned int N, int *w);
   ```

   where
   x is the pointer to the input data buffer in[]
   h is the pointer to the filter coefficient buffer C[]
   y is the pointer to the filter output buffer out[]
   w is the pointer to the signal buffer w[]
   M is the number of samples in the input buffer
   N is the number of second-order IIR filter sections.

2. Use rts55.lib for initializing the C function main(), and build the project exp6d.

3. Set a breakpoint at the statement `for(;;)` of the `main()` function, and use the CCS graphic function to view the 16-bit integer output samples in the buffer `out[]`. Set data length to 128 for viewing one block of data at a time. Animate the filtering process, and observe the filter output as a clean 800 Hz sinewave.

4. Profile the IIR filter `iirform2()`, and compare the profile result with those obtained in Experiments 6B and 6C.

References

[1] N. Ahmed and T. Natarajan, *Discrete-Time Signals and Systems*, Englewood Cliffs, NJ: Prentice-Hall, 1983.

[2] V. K. Ingle and J. G. Proakis, *Digital Signal Processing Using MATLAB V.4*, Boston: PWS Publishing, 1997.

[3] *Signal Processing Toolbox for Use with MATLAB*, The Math Works Inc., 1994.

[4] A. V. Oppenheim and R. W. Schafer, *Discrete-Time Signal Processing*, Englewood Cliffs, NJ: Prentice-Hall, 1989.

[5] S. J. Orfanidis, *Introduction to Signal Processing*, Englewood Cliffs, NJ: Prentice-Hall, 1996.

[6] J. G. Proakis and D. G. Manolakis, *Digital Signal Processing – Principles, Algorithms, and Applications*, 3rd Ed., Englewood Cliffs, NJ: Prentice-Hall, 1996.

[7] S. K. Mitra, *Digital Signal Processing: A Computer-Based Approach*, New York, NY: McGraw-Hill, 1998.

[8] D. Grover and J. R. Deller, *Digital Signal Processing and the Microcontroller*, Englewood Cliffs, NJ: Prentice-Hall, 1999.

[9] F. Taylor and J. Mellott, *Hands-On Digital Signal Processing*, New York, NY: McGraw-Hill, 1998.

[10] S. D. Stearns and D. R. Hush, *Digital Signal Analysis*, 2nd Ed., Englewood Cliffs, NJ: Prentice-Hall, 1990.

[11] S. S. Soliman and M. D. Srinath, *Continuous and Discrete Signals and Systems*, 2nd Ed., Englewood Cliffs, NJ: Prentice-Hall, 1998.

[12] L. B. Jackson, *Digital Filters and Signal Processing*, 2nd Ed., Boston, MA: Kluwer Academic Publishers, 1989.

Exercises

Part A

1. Find the Laplace transform of

$$x(t) = e^{-at} \sin(\Omega_0 t) u(t).$$

2. Find the inverse Laplace transform of

(a) $X(s) = \dfrac{2s + 1}{s^3 + 3s^2 - 4s}$.

(b) $X(s) = \dfrac{2s^2 - 3s}{s^3 - 4s^2 + 5s - 2}$.

(c) $X(s) = \dfrac{s + 3}{s^2 + 4s + 13}$.

3. Given the transfer function of a continuous-time system as

$$H(s) = \frac{s(s-5)(s^2+s+1)}{(s+1)(s+2)(s+3)(s^2+cs+5)}, \quad c \geq 0.$$

(a) Show a plot of the poles and zeros of $H(s)$.

(b) Discuss the stability of this system for the cases $c = 0$ and $c > 0$.

4. Consider the circuit shown in Figure 6.3 with $R = 1\Omega$ and $C = 1F$. The input signal to the circuit is expressed as

$$x(t) = \begin{cases} 1 & 0 \leq t \leq 1 \\ 0 & \text{elsewhere.} \end{cases}$$

Show that the output is

$$y(t) = (1 - e^{-t})u(t) - [1 - e^{-(t-1)}]u(t-1).$$

5. Given the transfer function

$$H(s) = \frac{1}{(s+1)(s+2)},$$

find the $H(z)$ using the impulse-invariant method.

6. Given the transfer function of an analog IIR notch filter as

$$H(s) = \frac{s^2+1}{s^2+s+1},$$

design a digital filter using bilinear transform with notch frequency 100 Hz and sampling rate 1 kHz.

7. Given an analog IIR bandpass filter that has resonance at 1 radian/second with the transfer function

$$H(s) = \frac{5s+1}{s^2+0.4s+1},$$

design a digital resonant filter that resonates at 100 Hz with the sampling rate at 1 kHz.

8. Design a second-order digital Butterworth filter using bilinear transform. The cut-off frequency is 1 kHz at a sampling frequency of 10 kHz.

9. Repeat the previous problem for designing a highpass filter with the same specifications.

10. Design a second-order digital Butterworth bandpass filter with the lower cut-off frequency 200 Hz, upper cut-off frequency 400 Hz, and sampling frequency 2000 Hz.

11. Design a second-order digital Butterworth bandstop filter that has the lower cut-off frequency 200 Hz, upper cut-off frequency 400 Hz, and sampling frequency 2000 Hz.

12. Given the transfer function

$$H(z) = \frac{0.5(z^2 - 1.1z + 0.3)}{z^3 - 2.4z^2 + 1.91z - 0.504},$$

find the following realizations:

(a) Direct-form II.

(b) Cascade of first-order sections.

(c) Parallel form in terms of first-order sections.

13. Given an IIR filter transfer function

$$H(z) = \frac{(3 - 2.1z^{-1})(2.7 + 4.2z^{-1} - 5z^{-2})}{(1 + 0.52z^{-1})(1 + z^{-1} - 0.34z^{-2})},$$

realize the transfer function in

(a) cascade canonical form, and

(b) parallel form.

14. Draw the direct-form I and II realizations of the transfer function

$$H(z) = \frac{(z^2 + 2z + 2)(z + 0.6)}{(z - 0.8)(z + 0.8)(z^2 + 0.1z + 0.8)}.$$

15. Given a third-order IIR filter transfer function

$$H(z) = \frac{0.44z^2 + 0.36z + 0.02}{z^3 + 0.4z^2 + 0.18z - 0.2},$$

find and draw the following realizations:

(a) direct-form II,

(b) cascade realization based on direct-form II realization of each section, and

(c) parallel direct-form II realization.

16. The difference filter is defined as

$$y(n) = b_0 x(n) - b_1 x(n - 1),$$

derive the frequency response of the filter and show that this is a crude highpass filter.

17. Given an IIR filter with transfer function

$$H(z) = \frac{(1 + 1.414z^{-1} + z^{-2})(1 + 2z^{-1} + z^{-2})}{(1 - 0.8z^{-1} + 0.64z^{-2})(1 - 1.0833z^{-1} + 0.25z^{-2})},$$

(a) find the poles and zeros of the filter,

(b) using the stability triangle to check if $H(z)$ is a stable filter.

18. Consider the second-order IIR filter with the I/O equation

$$y(n) = x(n) + a_1 y(n - 1) + a_2 y(n - 2), \quad n \geq 0,$$

where a_1 and a_2 are constants.

(a) Find the transfer function $H(z)$.

(b) Discuss the stability conditions related to the cases:

(1) $\dfrac{a_1^2}{4} + a_2 < 0.$

(2) $\dfrac{a_1^2}{4} + a_2 > 0.$

(3) $\dfrac{a_1^2}{4} + a_2 = 0.$

19. An allpass filter has a magnitude response that is unity for all frequencies, that is, $|H(\omega)| = 1$ for all ω. Such filters are useful for phase equalization of IIR designs. Show that the transfer function of an allpass filter is of the form

$$H(z) = \frac{z^{-L} + b_1 z^{-L+1} + \cdots + b_L}{1 + b_1 z^{-1} + \cdots + b_L z^{-L}},$$

where all coefficients are real.

20. A first-order allpass filter has the transfer function

$$H(z) = \frac{z^{-1} - a}{1 - az^{-1}}.$$

(a) Draw the direct-form I and II realizations.

(b) Show that $|H(\omega)| = 1$ for all ω.

(c) Sketch the phase response of this filter.

21. Design a second-order resonator with peak at 500 Hz, bandwidth 32 Hz, and operating at the sampling rate 10 kHz.

Part B

22. Given the sixth-order IIR transfer function

$$H(z) = \frac{6 + 17z^{-1} + 33z^{-2} + 25z^{-3} + 20z^{-4} - 5z^{-5} + 8z^{-6}}{1 + 2z^{-1} + 3z^{-2} + z^{-3} + 0.2z^{-4} - 0.3z^{-5} - 0.2z^{-6}},$$

find the factored form of the IIR transfer function in terms of second-order section using MATLAB.

23. Given the fourth-order IIR transfer function

$$H(z) = \frac{12 - 2z^{-1} + 3z^{-2} + 20z^{-4}}{6 - 12z^{-1} + 11z^{-2} - 5z^{-3} + z^{-4}},$$

(a) using MATLAB to express $H(z)$ in factored form,

(b) develop two different cascade realizations, and

(c) develop two different parallel realizations.

24. Design and plot the magnitude response of an elliptic IIR lowpass filter with the following specifications using MATLAB:

Passband edge at 800 Hz

Stopband edge at 1000 Hz

Passband ripple of 0.5 dB

Minimum stopband attenuation of 40 dB

Sampling rate of 4 kHz.

25. Design an IIR Butterworth bandpass filter with the following specifications:

Passband edges at 450 Hz and 650 Hz

Stopband edges at 300 Hz and 750 Hz

Passband ripple of 1 dB

Minimum stopband attenuation of 40 dB

Sampling rate of 4 kHz.

26. Design a type I Chebyshev IIR highpass filter with passband edge at 700 Hz, stopband edge at 500 Hz, passband ripple of 1 dB, and minimum stopband attenuation of 32 dB. The sampling frequency is 2 kHz. Plot the magnitude response of the design filter.

27. Given an IIR lowpass filter with transfer function

$$H(z) = \frac{0.0662(1 + 3z^{-1} + 3z^{-2} + z^{-3})}{1 - 0.9356z^{-1} + 0.5671z^{-2} - 0.1016z^{-3}},$$

(a) plot the first 32 samples of the impulse response using MATLAB,

(b) filter the input signal that consists of two sinusoids of normalized frequencies 0.1 and 0.8 using MATLAB.

28. It is interesting to examine the frequency response of the second-order resonator filter given in (6.6.15) as the radius r_p and the pole angle ω_0 are varied. Using the MATLAB to compute and plot

(a) The magnitude response for $\omega_0 = \pi/2$ and various values of r_p.

(b) The magnitude response for $r_p = 0.95$ and various values of ω_0.

Part C

29. An IIR filter design and implementation using the direct-form II realization.

(a) Use MATLAB to design an elliptic bandpass IIR filter that meets the following specifications:

   – Sampling frequency is 8000 Hz

   – Lower stopband extends from 0 to 1200 Hz

   – Upper stopband extends from 2400 to 4000 Hz

   – Passband starts from 1400 Hz with bandwidth of 800 Hz

   – Passband ripple should be no more than 0.3 dB

   – Stopband attenuation should be at least 30 dB.

(b) For the elliptic bandpass IIR filter obtained above,

   – plot the amplitude and phase responses

    – realize the filter using the cascade of direct-form II second-order sections.

(c) Implement the filter using floating-point C language and verify the filter implementation.

(d) Modify the C program using the C55x intrincis.

(e) Implement the filter in C55x assembly language.

(f) Using the CCS to show the filter results in both time domain and frequency domain.

(g) Profile the different implementations (C, intrinsics, and assembly) of the elliptic IIR filter and compare the results.

30. The overflow we saw in Experiments 6A and 6B is called the intermediate overflow. It happens when the signal buffer of the direct-form II realization uses 16-bit wordlength. Realizing the IIR filter using the direct-form I structure can eliminate the intermediate overflow by keeping the intermediate results in the 40-bit accumulators. Write an assembly routine to realize the fourth-order lowpass Butterworth IIR filter in the direct-form I structure.

31. Implement the recursive quadrature oscillator shown in Figure 6.23 in TMS320C55x assembly language.

32. Verify the IIR filter design in real time using a C55x EVM/DSK. Use a signal generator and a spectrum analyzer to measure the amplitude response and plot it. Evaluate the IIR filter according to the following steps:

    – Set the TMS320C55x EVM/DSK to 8 kHz sampling rate

    – Connect the signal generator output to the audio input of the EVM/DSK

    – Write an interrupt service routine (ISR) to handle input samples

    – Process the random samples at 128 samples per input block

    – Verify the filter using a spectrum analyzer.

7

Fast Fourier Transform and Its Applications

Frequency analysis of digital signals and systems was discussed in Chapter 4. To perform frequency analysis on a discrete-time signal, we converted the time-domain sequence into the frequency-domain representation using the z-transform, the discrete-time Fourier transform (DTFT), or the discrete Fourier transform (DFT). The widespread application of the DFT to spectral analysis, fast convolution, and data transmission is due to the development of the fast Fourier transform (FFT) algorithm for its computation. The FFT algorithm allows a much more rapid computation of the DFT, was developed in the mid-1960s by Cooley and Tukey.

It is critical to understand the advantages and the limitations of the DFT and how to use it properly. We will discuss the important properties of the DFT in Section 7.1. The development of FFT algorithms will be covered in Section 7.2. In Section 7.3, we will introduce the applications of FFTs. Implementation considerations such as computational issues and finite-wordlength effects will be discussed in Section 7.4. Finally, implementation of the FFT algorithm using the TMS320C55x for experimental purposes will be given in Section 7.5.

7.1 Discrete Fourier Transform

As discussed in Chapter 4, we perform frequency analysis of a discrete-time signal $x(n)$ using the DTFT defined in (4.4.4). However, $X(\omega)$ is a continuous function of frequency ω and the computation requires an infinite-length sequence $x(n)$. Thus the DTFT cannot be implemented on digital hardware. We define the DFT in Section 4.4.3 for N samples of $x(n)$ at N discrete frequencies. The input to the N-point DFT is a digital signal containing N samples and the output is a discrete-frequency sequence containing N samples. Therefore the DFT is a numerically computable transform and is suitable for DSP implementations.

7.1.1 Definitions

Given the DTFT $X(\omega)$, we take N samples over the full Nyquist interval, $0 \le \omega < 2\pi$, at discrete frequencies $\omega_k = 2\pi k/N$, $k = 0, 1, \ldots, N - 1$. This is equivalent to evaluating $X(\omega)$ at N equally spaced frequencies ω_k, with a spacing of $2\pi/N$ radians (or f_s/N Hz) between successive samples. That is,

$$X(\omega_k) = \sum_{n=-\infty}^{\infty} x(n) e^{-j(2\pi/N)kn} = \sum_{n=0}^{N-1} \left[\sum_{l=-\infty}^{\infty} x(n - lN) \right] e^{-j(2\pi/N)kn}$$

$$= \sum_{n=0}^{N-1} x_p(n) e^{-j(2\pi/N)kn}, \quad k = 0, 1, \ldots, N - 1, \tag{7.1.1a}$$

where

$$x_p(n) = \sum_{l=-\infty}^{\infty} x(n - lN) \tag{7.1.1b}$$

is a periodic signal with period N.

In general, the equally spaced frequency samples do not represent the original spectrum $X(\omega)$ uniquely when $x(n)$ has infinite duration. Instead, these frequency samples correspond to a periodic sequence $x_p(n)$ as shown in (7.1.1). When the sequence $x(n)$ has a finite length N, $x_p(n)$ is simply a periodic repetition of $x(n)$. Therefore the frequency samples $X(\omega_k)$, $k = 0, 1, \ldots, N - 1$ uniquely represent the finite-duration sequence $x(n)$. Since $x(n) = x_p(n)$ over a single period, a finite-duration sequence $x(n)$ of length N has the DFT defined as

$$X(k) \equiv X(\omega_k) \equiv X\left(\frac{2\pi k}{N}\right)$$

$$= \sum_{n=0}^{N-1} x(n) e^{-j(2\pi/N)kn}, \quad k = 0, 1, \ldots, N - 1, \tag{7.1.2}$$

where $X(k)$ is the kth DFT coefficient and the upper and lower indices in the summation reflect the fact that $x(n) = 0$ outside the range $0 \le n \le N - 1$. Strictly speaking, the DFT is a mapping between an N-point sequence in the time domain and an N-point sequence in the frequency domain that is applicable in the computation of the DTFT of periodic and finite-length sequences.

Example 7.1: If the signals $\{x(n)\}$ are real valued and N is an even number, we can show that $X(0)$ and $X(N/2)$ are real values and can be computed as

$$X(0) = \sum_{n=0}^{N-1} x(n)$$

and

$$X(N/2) = \sum_{n=0}^{N-1} e^{-j\pi n} x(n) = \sum_{n=0}^{N-1} (-1)^n x(n).$$

The DFT defined in (7.1.2) can also be written as

$$X(k) = \sum_{n=0}^{N-1} x(n) W_N^{kn}, \quad k = 0, 1, \ldots N - 1, \tag{7.1.3}$$

where

$$W_N^{kn} = e^{-j\left(\frac{2\pi}{N}\right)kn} = \cos\left(\frac{2\pi kn}{N}\right) - j\sin\left(\frac{2\pi kn}{N}\right), \quad 0 \leq k, \ n \leq N - 1 \tag{7.1.4}$$

are the complex basis functions, or twiddle factors of the DFT. Each $X(k)$ can be viewed as a linear combination of the sample set $\{x(n)\}$ with the coefficient set $\{W_N^{kn}\}$. Thus we have to store the twiddle factors in terms of real and imaginary parts in the DSP memory. Note that W_N is the Nth root of unity since $W_N^N = e^{-j2\pi} = 1 = W_N^0$. All the successive powers W_N^k, $k = 0, 1, \ldots, N - 1$ are also Nth roots of unity, but in clockwise direction on the unit circle. It can be shown that $W_N^{N/2} = e^{-j\pi} = -1$, the symmetry property

$$W_N^{k+N/2} = -W_N^k, \quad 0 \leq k \leq N/2 - 1 \tag{7.1.5a}$$

and the periodicity property

$$W_N^{k+N} = W_N^k. \tag{7.1.5b}$$

Figure 7.1 illustrates the cyclic property of the twiddle factors for an eight-point DFT.

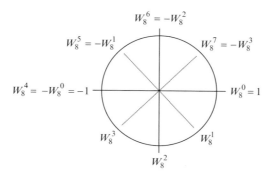

Figure 7.1 Twiddle factors for DFT, $N = 8$ case

The inverse discrete Fourier transform (IDFT) is used to transform the $X(k)$ back into the original sequence $x(n)$. Given the frequency samples $X(k)$, the IDFT is defined as

$$x(n) = \frac{1}{N}\sum_{k=0}^{N-1} X(k)e^{j(2\pi/N)kn} = \frac{1}{N}\sum_{k=0}^{N-1} X(k)W_N^{-kn}, \quad n = 0, 1, \ldots, N-1. \tag{7.1.6}$$

This is identical to the DFT with the exception of the normalizing factor $1/N$ and the sign of the exponent of the twiddle factors. The IDFT shows that there is no loss of information by transforming the spectrum $X(k)$ back into the original time sequence $x(n)$. The DFT given in (7.1.3) is called the analysis transform since it analyzes the signal $x(n)$ at N frequency components. The IDFT defined in (7.1.6) is called the synthesis transform because it reconstructs the signal $x(n)$ from the frequency components.

Example 7.2: Consider the finite-length sequence

$$x(n) = a^n, \quad n = 0, 1, \ldots, N-1,$$

where $0 < a < 1$. The DFT of the signal $x(n)$ is computed as

$$x(k) = \sum_{n=0}^{N-1} a^n e^{-j(2\pi k/N)n} = \sum_{n=0}^{N-1} \left(ae^{-j2\pi k/N}\right)^n$$

$$= \frac{1 - \left(ae^{-j2\pi k/N}\right)^N}{1 - ae^{-j2\pi k/N}} = \frac{1 - a^N}{1 - ae^{-j2\pi k/N}}, \quad k = 0, 1, \ldots, N-1.$$

The DFT and IDFT defined in (7.1.3) and (7.1.6), can be expressed in matrix–vector form as

$$\mathbf{X} = \mathbf{W}\mathbf{x} \tag{7.1.7a}$$

and

$$\mathbf{x} = \frac{1}{N}\mathbf{W}^*\mathbf{X}, \tag{7.1.7b}$$

where $\mathbf{x} = [x(0)\, x(1)\ldots x(N-1)]^T$ is the signal vector, the frequency-domain DFT coefficients are contained in the complex vector $\mathbf{X} = [X(0)\, X(1)\ldots X(N-1)]^T$, and the $N\times N$ twiddle-factor matrix (or DFT matrix) $\mathbf{W}$ is given by

$$\mathbf{W} = \left[W_N^{kn}\right]_{0\le k,\, n\le N-1}$$

$$= \begin{bmatrix} 1 & 1 & \cdots & 1 \\ 1 & W_N^1 & \cdots & W_N^{N-1} \\ \vdots & \vdots & \ddots & \vdots \\ 1 & W_N^{N-1} & \cdots & W_N^{(N-1)^2} \end{bmatrix}, \tag{7.1.8}$$

and $\mathbf{W}^*$ is the complex conjugate of the matrix $\mathbf{W}$. Since $\mathbf{W}$ is a symmetric matrix, the inverse matrix $\mathbf{W}^{-1} = \frac{1}{N}\mathbf{W}^*$ was used to derive (7.1.7b).

Example 7.3: Given $x(n) = \{1, 1, 0, 0\}$, the DFT of this four-point sequence can be computed using the matrix formulation as

$$
\mathbf{X} = \begin{bmatrix} 1 & 1 & 1 & 1 \\ 1 & W_4^1 & W_4^2 & W_4^3 \\ 1 & W_4^2 & W_4^4 & W_4^6 \\ 1 & W_4^3 & W_4^6 & W_4^9 \end{bmatrix} \mathbf{x}
$$

$$
= \begin{bmatrix} 1 & 1 & 1 & 1 \\ 1 & -j & -1 & j \\ 1 & -1 & 1 & -1 \\ 1 & j & -1 & -j \end{bmatrix} \begin{bmatrix} 1 \\ 1 \\ 0 \\ 0 \end{bmatrix} = \begin{bmatrix} 2 \\ 1-j \\ 0 \\ 1+j \end{bmatrix},
$$

where we used symmetry and periodicity properties given in (7.1.5) to obtain $W_4^0 = W_4^4 = 1$, $W_4^1 = W_4^9 = -j$, $W_4^2 = W_4^6 = -1$, and $W_4^3 = j$.
The IDFT can be computed with

$$
\mathbf{x} = \frac{1}{4} \begin{bmatrix} 1 & 1 & 1 & 1 \\ 1 & W_4^{-1} & W_4^{-2} & W_4^{-3} \\ 1 & W_4^{-2} & W_4^{-4} & W_4^{-6} \\ 1 & W_4^{-3} & W_4^{-6} & W_4^{-9} \end{bmatrix} \mathbf{X}.
$$

$$
= \frac{1}{4} \begin{bmatrix} 1 & 1 & 1 & 1 \\ 1 & j & -1 & -j \\ 1 & -1 & 1 & -1 \\ 1 & -j & -1 & j \end{bmatrix} \begin{bmatrix} 2 \\ 1-j \\ 0 \\ 1+j \end{bmatrix} = \begin{bmatrix} 1 \\ 1 \\ 0 \\ 0 \end{bmatrix}.
$$

As shown in Figure 7.1, the twiddle factors are equally spaced around the unit circle at frequency intervals of f_s/N (or $2\pi/N$). Therefore the frequency samples $X(k)$ represent discrete frequencies

$$
f_k = k\frac{f_s}{N}, \quad k = 0, 1, \ldots, N-1. \tag{7.1.9}
$$

The computational frequency resolution of the DFT is equal to the frequency increment f_s/N, and is sometimes referred to as the bin spacing of the DFT outputs. The spacing of the spectral lines depends on the number of data samples. This issue will be further discussed in Section 7.3.2.

Since the DFT coefficient $X(k)$ is a complex variable, it can be expressed in polar form as

$$X(k) = |X(k)|e^{j\phi(k)}, \tag{7.1.10}$$

where the DFT magnitude spectrum is defined as

$$|X(k)| = \sqrt{\{\text{Re}[X(k)]\}^2 + \{\text{Im}[X(k)]\}^2} \tag{7.1.11}$$

and the phase spectrum is defined as

$$\phi(k) = \tan^{-1}\left\{\frac{\text{Im}[X(k)]}{\text{Re}[X(k)]}\right\}. \tag{7.1.12}$$

These spectra provide a complementary way of representing the waveform, which clearly reveals information about the frequency content of the waveform itself.

7.1.2   Important Properties of DFT

The DFT is important for the analysis of digital signals and the design of DSP systems. Like the Fourier, Laplace, and z-transforms, the DFT has several important properties that enhance its utility for analyzing finite-length signals. Many DFT properties are similar to those of the Fourier transform and the z-transform. However, there are some differences. For example, the shifts and convolutions pertaining to the DFT are circular. Some important properties are summarized in this section. The circular convolution property will be discussed in Section 7.1.3.

Linearity

If $\{x(n)\}$ and $\{y(n)\}$ are time sequences of the same length, then

$$\text{DFT}[ax(n) + by(n)] = a\text{DFT}[x(n)] + b\text{DFT}[y(n)] = aX(k) + bY(k), \tag{7.1.13}$$

where a and b are arbitrary constants. Linearity is a key property that allows us to compute the DFTs of several different signals and determine the combined DFT via the summation of the individual DFTs. For example, the frequency response of a given system can be easily evaluated at each frequency component. The results can then be combined to determine the overall frequency response.

Complex-conjugate property

If the sequence $\{x(n),\ 0 \leq n \leq N - 1\}$ is real, then

$$X(-k) = X^*(k) = X(N - k), \quad 0 \leq k \leq N - 1, \tag{7.1.14}$$

where $X^*(k)$ is the complex conjugate of $X(k)$. Or equivalently,

$$X(M + k) = X^*(M - k), \quad 0 \leq k \leq M, \tag{7.1.15}$$

where $M = N/2$ if N is even, and $M = (N - 1)/2$ if N is odd. This property shows that only the first $(M + 1)$ DFT coefficients are independent. Only the frequency components from $k = 0$ to $k = M$ are needed in order to completely define the output. The rest can be obtained from the complex conjugate of corresponding coefficients, as illustrated in Figure 7.2.

The complex-conjugate (or symmetry) property shows that

$$\mathrm{Re}[X(k)] = \mathrm{Re}[X(N - k)], \quad k = 1, 2, \ldots, M - 1 \tag{7.1.16}$$

and

$$\mathrm{Im}[X(k)] = -\mathrm{Im}[X(N - k)], \quad k = 1, 2, \ldots, M - 1. \tag{7.1.17}$$

Thus the DFT of a real sequence produces symmetric real frequency components and anti-symmetric imaginary frequency components about $X(M)$. The real part of the DFT output is an even function, and the imaginary part of the DFT output is an odd function. From (7.1.16) and (7.1.17), we obtain

$$|X(k)| = |X(N - k)|, \quad k = 1, 2, \ldots, M - 1 \tag{7.1.18}$$

and

$$\phi(k) = -\phi(N - k), \quad k = 1, 2, \ldots, M - 1. \tag{7.1.19}$$

Because of the symmetry of the magnitude spectrum and the anti-symmetry of the phase spectrum, only the first $M + 1$ outputs represent unique information from the input signal. If the input to the DFT is a complex signal, however, all N complex outputs could carry information.

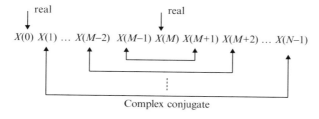

Figure 7.2   Complex-conjugate property

Periodicity

Because of the periodicity property shown in Figure 7.1, the DFT and IDFT produce periodic results with period N. Therefore the frequency and time samples produced by (7.1.3) and (7.1.6), respectively, are periodic with period N. That is,

$$X(k) = X(k + N) \quad \text{for all } k \qquad (7.1.20a)$$

and

$$x(n) = x(n + N) \quad \text{for all } n. \qquad (7.1.20b)$$

The finite-length sequence $x(n)$ can be considered as one period of a periodic function with period N. Also, the DFT $X(k)$ is periodic with period N.

As discussed in Section 4.4, the spectrum of a discrete-time signal is periodic. For a real-valued signal, the frequencies ranged from 0 to $f_s/2$ were reversed for the range of 0 to $-f_s/2$, and the entire range from $-f_s/2$ to $f_s/2$ was repeated infinitely in both directions in the frequency domain. The DFT outputs represent a single period (from 0 to f_s) of the spectrum.

Circular shifts

Let $\{X(k)\}$ be the DFT of a given N-periodic sequence $\{x(n)\}$, and let $y(n)$ be a circular shifted sequence defined by

$$y(n) = x(n - m)_{\text{mod } N}, \qquad (7.1.21)$$

where m is the number of samples by which $x(n)$ is shifted to the right (or delayed) and the modulo operation

$$(j)_{\text{mod } N} = j \pm iN \qquad (7.1.22a)$$

for some integer i such that

$$0 \leq (j)_{\text{mod } N} < N. \qquad (7.1.22b)$$

For example, if $m = 1$, $x(N - 1)$ replaces $x(0)$, $x(0)$ replaces $x(1)$, $x(1)$ replaces $x(2)$, etc. Thus a circular shift of an N-point sequence is equivalent to a linear shift of its periodic extension.

For a given $y(n)$ in (7.1.21), we have

$$Y(k) = W_N^{mk} X(k) = e^{-j(2\pi k/N)m} X(k). \qquad (7.1.23)$$

This equation states that the DFT coefficients of a circular-shifted N-periodic sequence by m samples are a linear shift of $X(k)$ by W_N^{mk}.

DFT and z-transform

Consider a sequence $x(n)$ having the z-transform $X(z)$ with an ROC that includes the unit circle. If $X(z)$ is sampled at N equally spaced points on the unit circle at $z_k = e^{j2\pi k/N}$, $k = 0, 1, \ldots, N-1$, we obtain

$$X(z)|_{z=e^{j2\pi k/N}} = \sum_{n=-\infty}^{\infty} x(n)z^{-n}\Big|_{z=e^{j2\pi k/N}} = \sum_{n=-\infty}^{\infty} x(n)e^{-j(2\pi k/N)n}. \tag{7.1.24}$$

This is identical to evaluating the discrete-time Fourier transform $X(\omega)$ at the N equally spaced frequencies $\omega_k = 2\pi k/N$, $k = 0, 1, \ldots, N-1$. If the sequence $x(n)$ has a finite duration of length N, the DFT of a sequence yields its z-transform on the unit circle at a set of points that are $2\pi/N$ radians apart, i.e.,

$$X(k) = X(z)|_{z = e^{j\left(\frac{2\pi}{N}\right)k}}, \quad k = 0, 1, \ldots, N-1. \tag{7.1.25}$$

Therefore the DFT is equal to the z-transform of a sequence $x(n)$ of length N, evaluated at N equally spaced points on the unit circle in the z-plane.

Example 7.4: Consider a finite-length DC signal

$$x(n) = c, \quad n = 0, 1, \ldots, N-1.$$

From (7.1.3), we obtain

$$X(k) = c\sum_{n=0}^{N-1} W_N^{kn} = c\frac{1 - W_N^{kN}}{1 - W_N^{k}}.$$

Since $W_N^{kN} = e^{-j\left(\frac{2\pi}{N}\right)kN} = 1$ for all k and for $W_N^{k} \neq 1$ for $k \neq iN$, we have $X(k) = 0$ for $k = 1, 2, \ldots, N-1$. For $k = 0$, $\sum_{n=0}^{N-1} W_N^{kn} = N$. Therefore we obtain

$$X(k) = cN\delta(k), \quad k = 0, 1, \ldots, N-1.$$

7.1.3 Circular Convolution

The Fourier transform, the Laplace transform, and the z-transform of the linear convolution of two time functions are simply the products of the transforms of the individual functions. A similar result holds for the DFT, but instead of a linear convolution of two sequences, we have a circular convolution. If $x(n)$ and $h(n)$ are real-valued N-periodic sequences, $y(n)$ is the circular convolution of $x(n)$ and $h(n)$ defined as

$$y(n) = x(n) \otimes h(n), \quad n = 0, 1, \ldots, N-1, \tag{7.1.26}$$

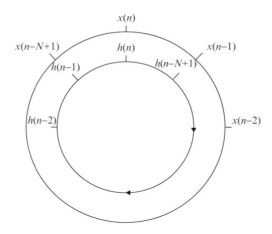

Figure 7.3   Circular convolution of two sequences using the concentric circle approach

where $\otimes$ denotes circular convolution. Then

$$Y(k) = X(k)H(k), \quad k = 0, 1, \ldots, N - 1. \tag{7.1.27}$$

Thus the circular convolution in time domain is equivalent to multiplication in the DFT domain. Note that to compute the product defined in (7.1.27), the DFTs must be of equal length. This means that the shorter of the two original sequences must be padded with zeros to the length of the other before its DFT is computed.

The circular convolution of two periodic signals with period N can be expressed as

$$y(n) = \sum_{m=0}^{N-1} x(m)h(n-m)_{\mathrm{mod}\,N} = \sum_{m=0}^{N-1} h(m)x(n-m)_{\mathrm{mod}\,N}, \tag{7.1.28}$$

where $y(n)$ is also periodic with period N. This cyclic property of circular convolution can be illustrated in Figure 7.3 by using two concentric rotating circles. To perform circular convolution, N samples of $x(n)$ [or $h(n)$] are equally spaced around the outer circle in the clockwise direction, and N samples of $h(n)$ [or $x(n)$] are displayed on the inner circle in the counterclockwise direction starting at the same point. Corresponding samples on the two circles are multiplied, and the resulting products are summed to produce an output. The successive value of the circular convolution is obtained by rotating the inner circle one sample in the clockwise direction; the result is computed by summing the corresponding products. The process is repeated to obtain the next result until the first sample of inner circle lines up with the first sample of the exterior circle again.

Example 7.5: Given two 8-point sequences $x(n) = \{1, 1, 1, 1, 1, 0, 0, 0\}$ and $h(n) = \{0, 0, 0, 1, 1, 1, 1, 1\}$. Using the circular convolution method illustrated in Figure 7.3, we can obtain

$n = 0, \ y(0) = 1 + 1 + 1 + 1 = 4$

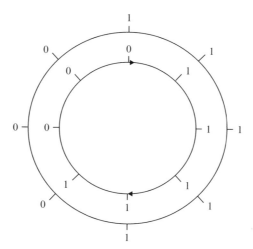

$n = 1, \ y(1) = 1 + 1 + 1 = 3$

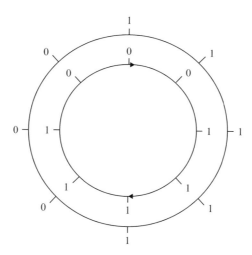

Repeating the process, we obtain

$$y(n) = x(n) \otimes h(n) = \{4, 3, 2, 2, 2, 3, 4, 5\}.$$

This circular convolution is due to the periodicity of the DFT. In circular convolution, the two sequences are always completely overlapping. As the end of one period is shifted out, the beginning of the next is shifted in as shown in Figure 7.3. To eliminate the circular effect and ensure that the DFT method results in a linear convolution, the signals must be zero-padded so that the product terms from the end of the period being shifted out are zero. Zero padding refers to the operation of extending a sequence of length N_1 to a length $N_2 \ (> N_1)$ by appending $(N_2 - N_1)$ zero samples to the tail of the given sequence. Note that the padding number of zeros at the end of signal has no effect on its DTFT.

Circular convolution can be used to implement linear convolution if both sequences contain sufficient zero samples. The linear convolution of two sequences of lengths L and M will result in a sequence of length $L + M - 1$. Thus the two sequences must be extended to the length of $L + M - 1$ or greater by zero padding. That is, for the circular convolution to yield the same result as the linear convolution, the sequence of length L must be appended with at least $M - 1$ zeros, while the sequence of length M must be appended with at least $L - 1$ zeros.

Example 7.6: Consider the previous example. If these 8-point sequences $h(n)$ and $x(n)$ are zero-padded to 16 points, the resulting circular convolution is

$$y(n) = x(n) \otimes h(n) = \{0, 0, 0, 1, 2, 3, 4, 5, 4, 3, 2, 1, 0, 0, 0, 0\}.$$

This result is identical to the linear convolution of the two sequences. Thus the linear convolution discussed in Chapter 5 can be realized by the circular convolution with proper zero padding.

In MATLAB, zero padding can be implemented using the function `zeros`. For example, the 8-point sequence $x(n)$ given in example 7.5 can be zero-padded to 16 points with the following command:

```
x = [1, 1, 1, 1, zeros(1, 11)];
```

where the MATLAB function `zeros(1, N)` generates a row vector of N zeros.

7.2   Fast Fourier Transforms

The DFT is a very effective method for determining the frequency spectrum of a time-domain signal. The only drawback with this technique is the amount of computation necessary to calculate the DFT coefficients $X(k)$. To compute each $X(k)$, we need approximately N complex multiplications and N complex additions based on the DFT defined in (7.1.3). Since we need to compute N samples of $X(k)$ for $k = 0, 1, \ldots, N - 1$, a total of approximately N^2 complex multiplications and $N^2 - N$ complex additions are required. When a complex multiplication is carried out using digital hardware, it requires four real multiplications and two real additions. Therefore the number of arithmetic operations required to compute the DFT is proportional to $4N^2$, which becomes very large for a large number of N. In addition, computing and storing the twiddle factors W_N^{kn} becomes a formidable task for large values of N.

The same values of the twiddle factors W_N^{kn} defined in (7.1.4) are calculated many times during the computation of DFT since W_N^{kn} is a periodic function with a limited number of distinct values. Because $W_N^N = 1$,

$$W_N^{kn} = W_N^{(kn)\bmod N} \quad \text{for } kn > N. \tag{7.2.1}$$

For example, different powers of W_N^{kn} have the same value as shown in Figure 7.1 for $N = 8$. In addition, some twiddle factors have real or imaginary parts equal to 1 or 0. By reducing these redundancies, a very efficient algorithm, called the FFT, exists. For example, if N is a power of 2, then the FFT makes it possible to calculate the DFT with $N \log_2 N$ operations instead of N^2 operations. For $N = 1024$, FFT requires about 10^4 operations instead of 10^6 of operations for DFT.

The generic term FFT covers many different algorithms with different features, advantages, and disadvantages. Each FFT has different strengths and makes different tradeoffs between code complexity, memory usage, and computation requirements. The FFT algorithm introduced by Cooley and Tukey requires approximately $N \log_2 N$ multiplications, where N is a power of 2. The FFT can also be applied to cases where N is a power of an integer other than 2. In this chapter, we introduce FFT algorithms for the case where N is a power of 2, the radix-2 FFT algorithm.

There are two classes of FFT algorithms: decimation-in-time and decimation-in-frequency. In the decimation-in-time algorithm, the input time sequence is successively divided up into smaller sequences, and the DFTs of these subsequences are combined in a certain pattern to yield the required DFT of the entire sequence with fewer operations. Since this algorithm was derived by separating the time-domain sequence into successively smaller sets, the resulting algorithm is referred to as a decimation-in-time algorithm. In the decimation-in-frequency algorithm, the frequency samples of the DFT are decomposed into smaller and smaller subsequences in a similar manner.

7.2.1   Decimation-in-Time

In the decimation-in-time algorithm, the N-sample sequence $\{x(n), \; n = 0, 1, \ldots, N - 1\}$ is first divided into two shorter interwoven sequences: the even numbered sequence

$$x_1(m) = x(2m), \quad m = 0, 1, \ldots, (N/2) - 1 \tag{7.2.2a}$$

and the odd numbered sequence

$$x_2(m) = x(2m + 1), \quad m = 0, 1, \ldots, (N/2) - 1. \tag{7.2.2b}$$

The DFT expressed in (7.1.3) can be divided into two DFTs of length $N/2$. That is,

$$
\begin{aligned}
X(k) &= \sum_{n=0}^{N-1} x(n) W_N^{kn} \\
&= \sum_{m=0}^{(N/2)-1} x(2m) W_N^{2mk} + \sum_{m=0}^{(N/2)-1} x(2m + 1) W_N^{(2m+1)k}.
\end{aligned}
\tag{7.2.3}
$$

Since

$$W_N^{2mk} = e^{-j \frac{2\pi}{N} 2mk} = e^{-j \frac{2\pi}{N/2} mk} = W_{N/2}^{mk}, \tag{7.2.4}$$

Equation (7.2.3) can be written as

$$X(k) = \sum_{m=0}^{(N/2)-1} x_1(m) W_{N/2}^{mk} + W_N^k \sum_{m=0}^{(N/2)-1} x_2(m) W_{N/2}^{mk}, \tag{7.2.5}$$

where each of the summation terms is reduced to an $N/2$ point DFT. Furthermore, from symmetry and periodicity properties given in (7.1.5), Equation (7.2.5) can be written as

$$X(k) = X_1(k) + W_N^k X_2(k), \quad k = 0, 1, \ldots, N - 1$$

$$= \begin{cases} X_1(k) + W_N^k X_2(k), & k = 0, 1, \ldots, (N/2) - 1 \\ X_1(k) - W_N^k X_2(k), & k = N/2, \ldots, N - 1, \end{cases} \quad (7.2.6)$$

where $X_1(k) = \text{DFT}[x_1(m)]$ and $X_2(k) = \text{DFT}[x_2(m)]$ using the $N/2$-point DFT.

The important point about this result is that the DFT of N samples becomes a linear combination of two smaller DFTs, each of $N/2$ samples. This procedure is illustrated in Figure 7.4 for the case $N = 8$. The computation of $X_1(k)$ and $X_2(k)$ requires $2(N/2)^2$ multiplications, the computation of $W_N^k X_2(k)$ requires $N/2$ multiplications. This gives a total of approximately $(N^2 + N)/2$ multiplications. Compared with N^2 operations for direct evaluation of the DFT, there is a saving in computation when N is large after only one stage of splitting signals into even and odd sequences. If we continue with this process, we can break up the single N-point DFT into $\log_2 N$ DFTs of length 2. The final algorithm requires computation proportional to $N \log_2 N$, a significant saving over the original N^2.

Equation (7.2.6) is commonly referred to as the butterfly computation because of its crisscross appearance, which can be generalized in Figure 7.5. The upper group generates the upper half of the DFT coefficient vector **X**, and the lower group generates the

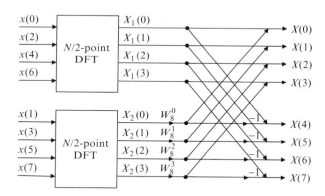

Figure 7.4   Decomposition of an N-point DFT into two $N/2$ DFTs, $N = 8$

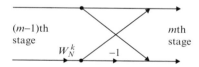

Figure 7.5   Flow graph for a butterfly computation

lower half. Each butterfly involves just a single complex multiplication by a twiddle factor W_N^k, one addition, and one subtraction. For this first decomposition, the twiddle factors are indexed consecutively, and the butterfly values are separated by $N/2$ samples. The order of the input samples has also been rearranged (split between even and odd numbers), which will be discussed in detail later.

Since N is a power of 2, $N/2$ is even. Each of these $N/2$-point DFTs in (7.2.6) can be computed via two smaller $N/4$-point DFTs, and so on. The second step process is illustrated in Figure 7.6. Note that the order of the input samples has been rearranged as $x(0)$, $x(4)$, $x(2)$, and $x(6)$ because $x(0)$, $x(2)$, $x(4)$, and $x(6)$ are considered to be the 0th, 1st, 2nd, and 3rd inputs in a 4-point DFT. Similarly, the order of $x(1)$, $x(5)$, $x(3)$, and $x(7)$ is used in the second 4-point DFT.

By repeating the process associated with (7.2.6), we will finally end up with a set of 2-point DFTs since N is a power of 2. For example, in Figure 7.6, the $N/4$-point DFT became a 2-point DFT since $N = 8$. Since the twiddle factor for the first stage, $W_N^0 = 1$, the 2-point DFT requires only one addition and one subtraction. The 2-point DFT illustrated in Figure 7.7 is identical to the butterfly network.

Example 7.7: Consider the two-point DFT algorithm which has two input time-domain samples $x(0)$ and $x(1)$. The output frequency-domain samples are $X(0)$ and $X(1)$. For this case, the DFT can be expressed as

$$X(k) = \sum_{n=0}^{1} x(n) W_2^{nk}, \quad k = 0, 1.$$

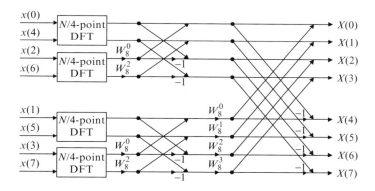

Figure 7.6   Flow graph illustrating second step of N-point DFT, $N = 8$

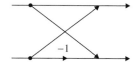

Figure 7.7   Flow graph of 2-point DFT

Since $W_2^0 = 1$ and $W_2^1 = e^{-\pi} = -1$, we have

$$X(0) = x(0) + x(1)$$

and

$$X(1) = x(0) - x(1).$$

The signal flow graph is shown in Figure 7.7. Note that the results are agreed with the results obtained in Example 7.1.

As shown in Figure 7.6, the output sequence is in natural order, while the input sequence has the unusual order. Actually the order of the input sequence is arranged as if each index was written in binary form and then the order of binary digits was reversed. The bit-reversal process is illustrated in Table 7.1 for the case $N = 8$. Each of the time sample indices in decimal is converted to its binary representation. The binary bit streams are then reversed. Converting the reversed binary numbers to decimal values gives the reordered time indices. If the input is in natural order the output will be in bit-reversed order. We can either shuffle the input sequence with a bit-reversal algorithm to get the output sequence in natural order, or let the input sequence be in natural order and shuffle the bit-reversed results to obtain the output in natural order. Note that most modern DSP chips such as the TMS320C55x provide the bit-reversal addressing mode to support this bit-reversal process. Therefore the input sequence can be stored in memory with the bit-reversed addresses computed by the hardware.

For the FFT algorithm shown in Figure 7.6, once all the values for a particular stage are computed, the old values that were used to obtain them are never required again. Thus the FFT needs to store only the N complex values. The memory locations used for the FFT outputs are the same as the memory locations used for storing the input data. This observation is used to produce in-place FFT algorithms that use the same memory locations for input samples, all intermediate calculations, and final output numbers.

Table 7.1 Example of bit-reversal process, $N = 8$ (3-bit)

Input sample index		Bit-reversed sample index	
Decimal	Binary	Binary	Decimal
0	000	000	0
1	001	100	4
2	010	010	2
3	011	110	6
4	100	001	1
5	101	101	5
6	110	011	3
7	111	111	7

7.2.2   Decimation-in-Frequency

The development of the decimation-in-frequency FFT algorithm is similar to the decimation-in-time algorithm presented in the previous section. The first step consists of dividing the data sequence into two halves of $N/2$ samples. Then $X(k)$ in (7.1.3) can be expressed as the sum of two components to obtain

$$X(k) = \sum_{n=0}^{(N/2)-1} x(n) W_N^{nk} + \sum_{n=N/2}^{N-1} x(n) W_N^{nk}$$

$$= \sum_{n=0}^{(N/2)-1} x(n) W_N^{nk} + W_N^{(N/2)k} \sum_{n=0}^{(N/2)-1} x\left(n + \frac{N}{2}\right) W_N^{nk}. \qquad (7.2.7)$$

Using the fact that

$$W_N^{N/2} = e^{-j\frac{2\pi}{N}(N/2)} = e^{-j\pi} = -1, \qquad (7.2.8)$$

Equation (7.2.7) can be simplified to

$$X(k) = \sum_{n=0}^{(N/2)-1} \left[x(n) + (-1)^k x\left(n + \frac{N}{2}\right) \right] W_N^{nk}. \qquad (7.2.9)$$

The next step is to separate the frequency terms $X(k)$ into even and odd samples of k. Since $W_N^{2kn} = W_{N/2}^{kn}$, Equation (7.2.9) can be written as

$$X(2k) = \sum_{n=0}^{(N/2)-1} \left[x(n) + x\left(n + \frac{N}{2}\right) \right] W_{N/2}^{kn} \qquad (7.2.10a)$$

and

$$X(2k+1) = \sum_{n=0}^{(N/2)-1} \left[x(n) - x\left(n + \frac{N}{2}\right) \right] W_N^k W_{N/2}^{kn} \qquad (7.2.10b)$$

for $0 \le k \le (N/2) - 1$. Let $x_1(n) = x(n) + x\left(n + \frac{N}{2}\right)$ and $x_2(n) = x(n) - x\left(n + \frac{N}{2}\right)$ for $0 \le n \le (N/2) - 1$, the first decomposition of an N-point DFT into two $N/2$-point DFTs is illustrated in Figure 7.8.

Again, the process of decomposition is continued until the last stage is made up of two-point DFTs. The decomposition proceeds from left to right for the decimation-in-frequency development and the symmetry relationships are reversed from the decimation-in-time algorithm. Note that the bit reversal occurs at the output instead of the input and the order of the output samples $X(k)$ will be re-arranged as bit-reversed samples index given in Table 7.1. The butterfly representation for the decimation-in-frequency FFT algorithm is illustrated in Figure 7.9.

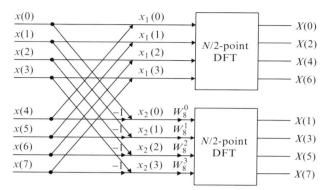

Figure 7.8 Decomposition of an N-point DFT into two $N/2$ DFTs using decimation-in-frequency algorithm, $N = 8$

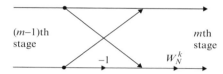

Figure 7.9 Butterfly network for decimation-in-frequency FFT algorithm

7.2.3 Inverse Fast Fourier Transform

The FFT algorithms introduced in the previous section can be easily modified to compute the IFFT efficiently. This is apparent from the similarities of the DFT and the IDFT definitions given in (7.1.3) and (7.1.6), respectively. Complex conjugating (7.1.6), we have

$$x^*(n) = \frac{1}{N} \sum_{k=0}^{N-1} X^*(k) W_N^{kn}, \quad n = 0, 1, \ldots, N-1. \tag{7.2.11}$$

Therefore an FFT algorithm can be used to compute the inverse DFT by first conjugating the DFT coefficients $X(k)$ to obtain $X^*(k)$, computing the DFT of $X^*(k)$ use an FFT algorithm, scaling the results by $1/N$ to obtain $x^*(n)$, and then complex conjugating $x^*(n)$ to obtain the output sequence $x(n)$. If the signal is real-valued, the final conjugation operation is not required.

All the FFT algorithms introduced in this chapter are based on two-input, two-output butterfly computations, and are classified as radix-2 complex FFT algorithms. It is possible to use other radix values to develop FFT algorithms. However, these algorithms do not work well when the length is a number with few factors. In addition, these algorithms are more complicated than the radix-2 FFT algorithms, and the routines are not as available for DSP processors. Radix-2 and radix-4 FFT algorithms are the most common, although other radix values can be employed. Different radix butterflies can be combined to form mixed-radix FFT algorithms.

7.2.4 MATLAB Implementations

As introduced in Section 4.4.4, MATLAB provides a function

```
y = fft(x);
```

to compute the DFT of time sequence $x(n)$ in the vector x. If x is a matrix, y is the DFT of each column of the matrix. If the length of the x is a power of 2, the fft function employs a high-speed radix-2 FFT algorithm. Otherwise a slower mixed-radix algorithm is employed.

An alternative way of using fft function is

```
y = fft(x, N);
```

to perform N-point FFT. If the length of x is less than N, then the vector x is padded with trailing zeros to length N. If the length of x is greater than N, fft function truncates the sequence x and only performs the FFT of the first N samples of data.

The execution time of the fft function depends on the input data type and the sequence length. If the input data is real-valued, it computes a real power-of-two FFT algorithm that is faster than a complex FFT of the same length. As mentioned earlier, the execution is fastest if the sequence length is exactly a power of 2. For this reason, it is usually better to use power-of-two FFT. For example, if the length of x is 511, the function y = fft(x, 512) will be computed faster than fft(x), which performs 511-point DFT.

It is important to note that the vectors in MATLAB are indexed from 1 to N instead of from 0 to $N-1$ given in the DFT and the IDFT definitions. Therefore the relationship between the actual frequency in Hz and the frequency index k given in (7.1.9) is modified as

$$f_k = (k-1)\frac{f_s}{N}, \quad k = 1, 2, \ldots, N \qquad (7.2.12)$$

for indexing into the y vector that contains $X(k)$.

The IFFT algorithm is implemented in the MATLAB function ifft, which can be used as

```
y = ifft(x);
```

or

```
y = ifft(x,N);
```

The characteristics and usage of ifft are the same as those for fft.

7.3 Applications

FFT has a wide variety of applications in DSP. Spectral analysis often requires the numerical computation of the frequency spectrum for a given signal with large sample sets. The DFT is also used in the coding of waveforms for efficient transmission or storage. In these cases the FFT may provide the only possible means for spectral computation within the limits of time and computing cost. In this section, we will also show how the FFT can be used to implement linear convolution for FIR filtering in a computationally efficient manner.

7.3.1 Spectrum Estimation and Analysis

The spectrum estimation techniques may be categorized as non-parametric and parametric. The non-parametric methods that include the periodogram have the advantage of using the FFT for efficient implementation, but have the disadvantage of having limited frequency resolution for short data lengths. Parametric methods can provide higher resolution. The most common parametric technique is to derive the spectrum from the parameters of an autoregressive model of the signal. In this section, we will introduce the principles and practice of the spectral estimation and analysis using non-parametric methods.

The inherent properties of the DFT directly relate to the performance of spectral analysis. If a discrete-time signal is periodic, it is possible to calculate its spectrum using the DFT. For an aperiodic signal, we can break up long sequence into smaller segments and analyze each individual segment using the FFT. This is reasonable because the very long signal probably consists of short segments where the spectral content does not change. The spectrum of a signal of length L can be computed using an N-point FFT. If $L < N$, we must increase the signal length from L to N by appending $N - L$ zero samples to the tail of the signal.

To compute the spectrum of an analog signal digitally, the signal is sampled first and then transformed to the frequency domain by a DFT or an FFT algorithm. As discussed in Chapter 3, the sampling rate f_s must be high enough to minimize aliasing effects. As discussed in Chapter 4, the spectrum of the sampled signal is the replication of the desired analog spectrum at multiples of the sampling frequency. The proper choice of sampling rate can guarantee that these two spectra are the same over the Nyquist interval. The DFT of an arbitrary set of sampled data may not be always the true DFT of the signal from which the data was obtained. This is because the signal is continuous, whereas the data set is truncated at its beginning and end. The spectrum estimated from a finite number of samples is correct only if the signal is periodic and the sample set has exactly one period. In practice, we may not have exactly one period of the periodic signal as the input.

As discussed in Section 7.1, the computational frequency resolution of the N-point DFT is f_s/N. The DFT coefficients $X(k)$ represent frequency components equally spaced at frequencies

$$f_k = \frac{kf_s}{N}, \quad k = 0, 1, \ldots, N - 1. \tag{7.3.1}$$

If there is a signal component that falls between two adjacent frequency components in the spectrum, it cannot be properly represented. Its energy will be shared between neighboring bins and the nearby spectral amplitude will be distorted.

Example 7.8: Consider a sinewave of frequency $f = 30\,\text{Hz}$ expressed as

$$x(n) = \sin(2\pi f n T), \quad n = 0, 1, \ldots, 127,$$

where the sampling period is $T = 1/128$ seconds. Because the computational frequency resolution (f_s/N) is 1 Hz using a 128-point FFT, the line component at 30 Hz can be represented by $X(k)$ at $k = 30$. The amplitude spectrum is shown in Figure 7.10a. The only non-zero term that occurs in $|X(k)|$ is at $k = 30$, which corresponds to 30 Hz. It is clear that a magnitude spectrum with a single non-zero spectral component is what one would expect for a pure sinusoid.

Example 7.9: Consider a different sinewave of frequency 30.5 Hz with sampling rate 128 Hz. The magnitude spectrum in Figure 7.10b shows several spectral components that are symmetric about 30.5 Hz. From evaluating the DFT coefficients, we cannot tell the exact frequency of the sinusoid. The reason why $|X(k)|$ does not have a single spectral component is that the line component at 30.5 Hz is in between $k = 30$ and $k = 31$. The frequency components in $x(n)$ that are not equal to f_k given in (7.3.1) tend to spread into portions of the spectrum that are adjacent to the correct spectral value. The MATLAB program (Fig7_10.m in the software package) for Examples 7.8 and 7.9 is listed as follows:

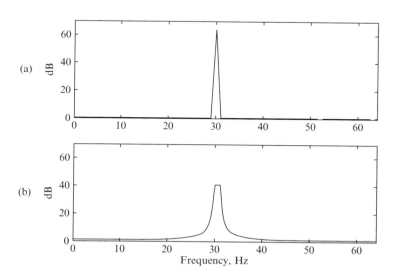

Figure 7.10 Effect of computational frequency resolution on sinewave spectra: (a) sinewave at 30 Hz, and (b) sinewave at 30.5 Hz

```
n = [0:127];
x1 = sin(2*pi*30*n/128); X1 = abs(fft(x1));
x2 = sin(2*pi*30.5*n/128); X2 = abs(fft(x2));
subplot(2,1,1),plot(n,X1), axis([0 64 0 70]),
title('(a) Sinewave at 30 Hz'),
subplot(2,1,2),plot(n,X2), axis([0 64 0 70]),
title('(b) Sinewave at 30.5 Hz'),
xlabel('Frequency, Hz');
```

A solution to this problem is to make the frequencies $f_k = kf_s/N$ more closely spaced, thus matching the signal frequencies. This may be achieved by using a larger DFT size N to increase the computational frequency resolution of the spectrum. If the number of data samples is not sufficiently large, the sequence may be expanded by adding additional zeros to the true data, thus increasing the length N. The added zeros serve to increase the computational frequency resolution of the estimated spectrum to the true spectrum without adding additional information. This process is simply the interpolation of the spectral curve between adjacent frequency components. A real improvement in frequency resolution can only be achieved if a longer data record is available.

A number of problems have to be avoided in performing non-parametric spectral analysis such as aliasing, finite data length, spectral leakage, and spectral smearing. The effects of spectral leakage and smearing may be minimized by windowing the data using a suitable window function. These issues will be discussed in the following section.

7.3.2 Spectral Leakage and Resolution

The data that represents the signal of length N is effectively obtained by multiplying all the sampled values in the interval by one, while all values outside this interval are multiplied by zero. This is equivalent to multiplying the signal by a rectangular window of width N and height 1, expressed as

$$w(n) = \begin{cases} 1, & 0 \le n \le N-1 \\ 0, & \text{otherwise.} \end{cases} \tag{7.3.2}$$

In this case, the sampled data $x_N(n)$ is obtained by multiplying the signal $x(n)$ with the window function $w(n)$. That is,

$$x_N(n) = w(n)x(n) = \begin{cases} x(n), & 0 \le n \le N-1 \\ 0, & \text{otherwise.} \end{cases} \tag{7.3.3}$$

The multiplication of $x(n)$ by $w(n)$ ensures that $x_N(n)$ vanishes outside the window. As the length of the window increases, the windowed signal $x_N(n)$ becomes a better approximation of $x(n)$, and thus $X(k)$ becomes a better approximation of the DTFT $X(\omega)$.

The time-domain multiplication given in (7.3.3) is equivalent to the convolution in the frequency domain. Thus the DFT of $x_N(n)$ can be expressed as

$$X_N(k) = W(k) * X(k) = \sum_{l=-N}^{N} W(k-l)X(k), \tag{7.3.4}$$

where $W(k)$ is the DFT of the window function $w(n)$, and $X(k)$ is the true DFT of the signal $x(n)$. Equation (7.3.4) shows that the computed spectrum consists of the true spectrum $X(k)$ convoluted with the window function's spectrum $W(k)$. This means that when we apply a window to a signal, the frequency components of the signal will be corrupted in the frequency domain by a shifted and scaled version of the window's spectrum.

As discussed in Section 5.3.3, the magnitude response of the rectangular window defined in (7.3.2) can be expressed as

$$|W(\omega)| = \left| \frac{\sin(\omega N/2)}{\sin(\omega/2)} \right|. \tag{7.3.5}$$

It consists of a mainlobe of height N at $\omega = 0$, and several smaller sidelobes. The frequency components that lie under the sidelobes represent the sharp transition of $w(n)$ at the endpoints. The sidelobes are between the zeros of $W(\omega)$, with center frequencies at $\omega = \frac{(2k+1)\pi}{N}$, $k = \pm 1, \pm 2, \ldots$, and the first sidelobe is down only 13 dB from the mainlobe level. As N increases, the height of the mainlobe increases and its width becomes narrower. However, the peak of the sidelobes also increases. Thus the ratio of the mainlobe to the first sidelobe remains the same about 13 dB.

The sidelobes introduce spurious peaks into the computed spectrum, or to cancel true peaks in the original spectrum. The phenomenon is known as spectral leakage. To avoid spectral leakage, it is necessary to use a shaped window to reduce the sidelobe effects. Suitable windows have a value of 1 at $n = M$, and are tapered to 0 at points $n = 0$ and $n = N - 1$ to smooth out both ends of the input samples to the DFT.

If the signal $x(n)$ consists of a single sinusoid, that is,

$$x(n) = \cos(\omega_0 n), \tag{7.3.6}$$

the spectrum of the infinite-length sampled signal over the Nyquist interval is given as

$$X(\omega) = 2\pi\delta(\omega \pm \omega_0), \quad -\pi \le \omega \le \pi, \tag{7.3.7}$$

which consists of two line components at frequencies $\pm\omega_0$. However, the spectrum of the windowed sinusoid defined in (7.3.3) can be obtained as

$$X_N(\omega) = \frac{1}{2}[W(\omega - \omega_0) + W(\omega + \omega_0)], \tag{7.3.8}$$

where $W(\omega)$ is the spectrum of the window function.

Equation (7.3.8) shows that the windowing process has the effect of smearing the original sharp spectral line $\delta(\omega - \omega_0)$ at frequency ω_0 and replacing it with $W(\omega - \omega_0)$.

Thus the power of the infinite-length signal that was concentrated at a single frequency has been spread into the entire frequency range by the windowing operation. This undesired effect is called spectral smearing. Thus windowing not only distorted the spectrum due to leakage effects, but also reduced spectral resolution. For example, a similar analysis can be made in the case when the signal consists of two sinusoidal components. That is,

$$x(n) = \cos(\omega_1 n) + \cos(\omega_2 n). \tag{7.3.9}$$

The spectrum of the windowed signal is

$$X_N(\omega) = \frac{1}{2}[W(\omega - \omega_1) + W(\omega + \omega_1) + W(\omega - \omega_2) + W(\omega + \omega_2)]. \tag{7.3.10}$$

Again, the sharp spectral lines are replaced with their smeared versions. From (7.3.5), the spectrum $W(\omega)$ has its first zero at frequency $\omega = 2\pi/N$. If the frequency separation, $\Delta\omega = |\omega_1 - \omega_2|$, of the two sinusoids is

$$\Delta\omega \leq \frac{2\pi}{N}, \tag{7.3.11}$$

the mainlobe of the two window functions $W(\omega - \omega_1)$ and $W(\omega - \omega_2)$ overlap. Thus the two spectral lines in $X_N(\omega)$ are not distinguishable. This undesired effect starts when $\Delta\omega$ is approximately equal to the mainlobe width $2\pi/N$. Therefore the frequency resolution of the windowed spectrum is limited by the window's mainlobe width.

To guarantee that two sinusoids appear as two distinct ones, their frequency separation must satisfy the condition

$$\Delta\omega > \frac{2\pi}{N}, \tag{7.3.12a}$$

in radians per sample, or

$$\Delta f > \frac{f_s}{N}, \tag{7.3.12b}$$

in Hz. Thus the minimum DFT length to achieve a desired frequency resolution is given as

$$N > \frac{f_s}{\Delta f} = \frac{2\pi}{\Delta\omega}. \tag{7.3.13}$$

In summary, the mainlobe width determines the frequency resolution of the windowed spectrum. The sidelobes determine the amount of undesired frequency leakage. The optimum window used for spectral analysis must have narrow mainlobe and small sidelobes. Although adding to the record length by zero padding increases the FFT size and thereby results in a smaller Δf, one must be cautious to have sufficient record length to support this resolution.

In Section 5.3.4, we used windows to smooth out the truncated impulse response of an ideal filter for designing an FIR filter. In this section, we showed that those window functions can also be used to modify the spectrum estimated by the DFT. If the window function, $w(n)$, is applied to the input signal, the DFT outputs are given by

$$X(k) = \sum_{n=0}^{N-1} w(n)x(n) W_N^{kn}, \quad k = 0, 1, \ldots, N-1. \tag{7.3.14}$$

The rectangular window has the narrowest mainlobe width, thus providing the best spectral resolution. However, its high-level sidelobes produce undesired spectral leakage. The amount of leakage can be substantially reduced at the cost of decreased spectral resolution by using appropriate non-rectangular window functions introduced in Section 5.3.4.

As discussed before, frequency resolution is directly related to the window's mainlobe width. A narrow mainlobe will allow closely spaced frequency components to be identified; while a wide mainlobe will cause nearby frequency components to blend. For a given window length N, windows such as rectangular, Hanning, and Hamming have relatively narrow mainlobe compared with Blackman or Kaiser windows. Unfortunately, the first three windows have relatively high sidelobes, thus having more spectral leakage. There is a trade-off between frequency resolution and spectral leakage in choosing windows for a given application.

Example 7.10: Consider the sinewave used in Example 7.9. Using the Kaiser window defined in (5.3.26) with $L = 128$ and $\beta = 8.96$, the magnitude spectrum is shown in Figure 7.11 using the MATLAB script Exam7_10.m included in the software package.

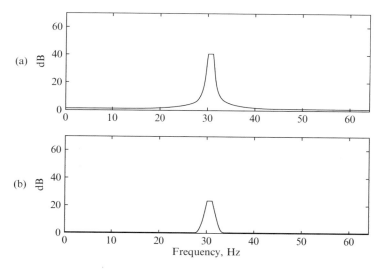

Figure 7.11 Effect of Kaiser window function for reducing spectral leakage: (a) rectangular window, and (b) Kaiser window

An effective method for decreasing the mainlobe width is by increasing the window length. For a given window, increasing the length of the window reduces the width of the mainlobe, which leads to better frequency resolution. However, if the signal changes frequency content over time, the window cannot be too long in order to provide a meaningful spectrum. In addition, a longer window implies using more data, so there is a trade-off between frequency resolution and the cost of implementation. If the number of available signal samples is less than the required length, we can use the zero padding technique. Note that the zeros are appended after windowing is performed.

7.3.3   Power Density Spectrum

The finite-energy signals possess a Fourier transform and are characterized in the frequency domain by their power density spectrum. Consider a sequence $x(n)$ of length N whose DFT is $X(k)$, Parseval's theorem can be expressed as

$$E = \sum_{n=0}^{N-1} |x(n)|^2 = \frac{1}{N} \sum_{k=0}^{N-1} |X(k)|^2. \tag{7.3.15}$$

The term $|X(k)|^2$ is called the power spectrum and is a measure of power in signal at frequency f_k defined in (7.3.1). The DFT magnitude spectrum $|X(k)|$ is defined in (7.1.11). Squaring the magnitude of the DFT coefficient produces a power spectrum, which is also called the periodogram.

As discussed in Section 3.3, stationary random processes do not have finite energy and thus do not possess Fourier transform. Such signals have a finite average power and are characterized by the power density spectrum (PDS) defined as

$$P(k) = \frac{1}{N} |X(k)|^2 = \frac{1}{N} X(k) X^*(k), \tag{7.3.16}$$

which is also commonly referred to as the power spectral density, or simply power spectrum.

The PDS is a very useful concept in the analysis of random signals since it provides a meaningful measure for the distribution of the average power in such signals. There are many different techniques developed for estimating the PDS. Since the periodogram is not a consistent estimate of the true PDS, the periodogram averaging method may be used to reduce statistical variation of the computed spectra. Given a signal vector xn which consists N samples of digital signal $x(n)$, a crude estimate of the PDS using MATLAB is

```
pxn = abs(fft(xn, 1024)).^2/N;
```

In practice, we only have a finite-length sequence whose PDS is desired. One way of computing the PDS is to decompose $x(n)$ into M segments, $x_m(n)$, of N samples each. These signal segments are spaced $N/2$ samples apart, i.e., there is 50 percent overlap between successive segments as illustrated in Figure 7.12.

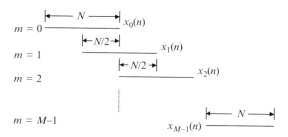

Figure 7.12 Segments used to estimate power density spectrum

In order to reduce spectral leakage, each $x_m(n)$ is multiplied by a non-rectangular window function $w(n)$ of length N. This results in 'windowed segments' $x'_m(n)$, which are given by

$$x'_m(n) = w(n)x_m(n), \quad n = 0, \ldots, N - 1, \tag{7.3.17}$$

for $0 \le m \le M - 1$. The windowing operation results in reduction of frequency resolution, which may be compensated by increasing the length N.

We then compute the DFT of $x'_m(n)$ given in (7.3.17) to get $X'_m(k)$. The mth periodogram is defined as

$$P_m(k) = \frac{1}{NP_w}\left|X'_m(k)\right|^2, \quad 0 \le k \le N - 1, \tag{7.3.18}$$

where

$$P_w = \frac{1}{N}\sum_{n=0}^{N-1} w^2(n) \tag{7.3.19}$$

is a normalization factor for the average power in the window sequence $w(n)$. Finally, the desired PDS is the average of these periodograms. That is,

$$P(k) = \frac{1}{M}\sum_{m=0}^{M-1} P_m(k) = \frac{1}{MNP_w}\sum_{m=0}^{M-1}\left|X'_m(k)\right|^2, \quad 0 \le k \le N - 1. \tag{7.3.20}$$

Therefore the PDS estimate given in (7.3.20) is a weighted sum of the periodograms of each of the individual overlapped segments. The 50 percent overlap between successive segments helps to improve certain statistical properties of this estimation.

The *Signal Processing Toolbox* provides the function psd to average the periodograms of windowed segments of a signal. This MATLAB function estimates the PDS of the signal given in the vector x using the following statement:

```
pxx = psd(x, nfft, Fs, window, noverlap);
```

where `nfft` specifies the FFT length, `Fs` is the sampling frequency, `window` specifies the selected window function, and `noverlap` is the number of samples by which the segments overlap.

7.3.4 Fast Convolution

As discussed in Chapter 5, FIR filtering is a linear convolution of the finite impulse response $h(n)$ with the input sequence $x(n)$. If the FIR filter has L coefficients, we need L real multiplications and $L - 1$ real additions to compute each output $y(n)$. To obtain L output samples, the number of operations (multiplication and addition) needed is proportional to L^2. To reduce the computational requirements, we consider the alternate approach defined in (7.1.26) and (7.1.27), which uses the property that the convolution of the sequences $x(n)$ and $h(n)$ is equivalent to multiplying their DFTs $X(k)$ and $H(k)$.

As described in Section 7.1.3, the linear convolution can be implemented using zero padding if the data sequence $x(n)$ also has finite duration. To take advantage of efficient FFT and IFFT algorithms, we use these computational efficient algorithms as illustrated in Figure 7.13. The procedure of using FFT to implement linear convolution in a computationally efficient manner is called the fast convolution. Compared to the direct implementation of FIR filtering, fast convolution will provide a significant reduction in computational requirements for higher order FIR filters, thus it is often used to implement FIR filtering in applications having long data samples.

It is important to note that the fast convolution shown in Figure 7.13 produces the circular convolution discussed in Section 7.1.3. In order to produce a filter result equal to a linear convolution, it is necessary to append zeros to the signals in order to overcome the circular effect. If the data sequence $x(n)$ has finite duration M, the first step is to pad both sequences with zeros to a length corresponding to an allowable FFT size N ($\geq L + M - 1$), where L is the length of the sequence $h(n)$. The FFT is computed for both sequences, the complex products defined in (7.1.27) are calculated, and the IFFT is used to obtain the results. The desired linear convolution is contained in the first $L + M - 1$ terms of these results.

The FFT of the zero-padded data requires about $N \log_2 N$ complex computations. Since the filter impulse response $h(n)$ is known as a priori, the FFT of the impulse response can be pre-calculated and stored. The computation of product $Y(k) = H(k)X(k)$ takes N complex multiplications, and the inverse FFT of the products $Y(k)$ requires another $N \log_2 N$ complex multiplications. Therefore the fast convolution shown in Figure 7.13 requires about $(8N \log_2 N + 4N)$ real multiplications.

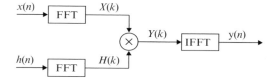

Figure 7.13   Fast convolution algorithm

Compared with LM required by direct FIR filtering, the computational saving is significant when both L and M are large.

For many applications, the input sequence is very long compared to the order of FIR filter L. This is especially true in real-time applications where the input sequence is of infinite duration. In order to use the efficient FFT and IFFT algorithms, the input sequence must be grouped into segments of N ($N > L$ and N is a size supported by the FFT algorithm) samples, process each segment using the FFT, and finally assemble the output sequence from the outputs of each segment. This procedure is called the block processing operation. The selection of the block size is a function of the filter order. There will be end effects from each convolution that must be accounted for as the segments are recombined to produce the output sequence. There are two techniques for the segmentation and recombination of the data: the overlap-save and overlap-add algorithms.

Overlap-save technique

The overlap-save process is carried out by overlapping L input samples on each segment, where L is the order of the FIR filter. The output segments are truncated to be non-overlapping and then concatenated. The process is illustrated in Figure 7.14 and is described by the following steps:

1. Perform N-point FFT of the expanded (zero padded) impulse response sequence

$$h'(n) = \begin{cases} h(n), & n = 0, 1, \ldots, L - 1 \\ 0, & n = L, L + 1, \ldots, N - 1, \end{cases} \qquad (7.3.21)$$

to obtain $H'(k)$ $k = 0, 1, \ldots, N - 1$, where $h(n)$ is the impulse response of the FIR filter. Note that for a time-invariant filter, this process can be pre-calculated off-line and stored in memory.

2. Select N signal samples $x_m(n)$ (where m is the segment index) from the input sequence $x(n)$ based on the overlap illustrated in Figure 7.14, and then use N-point

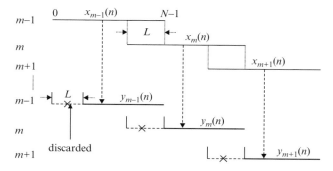

Figure 7.14 Overlap data segments for the overlap-save technique

FFT to obtain $X_m(k)$. Clearly, to avoid excessive overlap, we usually choose $N \gg L$ and N is the size supported by the FFT algorithms such as power of 2 for using radix-2 algorithms.

3. Multiply the stored $H'(k)$ (obtained in step 1) by the $X_m(k)$ of segment m (obtained in step 2) to get

$$Y_m(k) = H'(k)X_m(k), \quad k = 0, 1, \ldots, N - 1. \tag{7.3.22}$$

4. Perform N-point IFFT of $Y_m(k)$ to obtain $y_m(n)$, $n = 0, 1, \ldots, N - 1$.

5. Discard the first L samples from each successive IFFT output since they are circularly wrapped and superimposed as discussed in Section 7.1.3. The resulting segments of $(N - L)$ samples are concatenated to produce $y(n)$.

Overlap-add technique

In the overlap-add process, the input sequence $x(n)$ is divided into non-overlapping segments of length $(N - L)$. Each segment is zero-padded to produce $x_m(n)$ of length N. Following the steps 2, 3, and 4 of the overlap-save method to obtain N-point segments $y_m(n)$. Since the convolution is the linear operation, the output sequence $y(n)$ is simply the summation of all segments expressed as

$$y(n) = \sum_m y_m(n). \tag{7.3.23}$$

Because each output segment $y_m(n)$ overlaps the following segment $y_{m+1}(n)$ by L samples, (7.3.23) implies the actual addition of the last L samples in segment $y_m(n)$ with the first L samples in segment $y_{m+1}(n)$.

   This efficient FIR filtering using the overlap-add technique is implemented by the MATLAB function

```
y = fftfilt(h, x);
```

or

```
y = fftfilt(h, x, N);
```

The `fftfilt` function filters the input signal in the vector `x` with the FIR filter described by the coefficient vector `h`. The function `y = fftfilt(h, x)` chooses an FFT and a data block length that automatically guarantees efficient execution time. However, we can specify the FFT length N by using `y = fftfilt(h, x, N)`.

7.3.5 Spectrogram

The PDS introduced in Section 7.3.3 is a powerful technique to show how the power of the signal is distributed among the various frequency components. However, this method will result in a distorted (blurred) spectrum when the signal is non-stationary.

For a time-varying signal, it is more useful to compute a local spectrum that measures spectral contents over a short time interval.

In this section, we use a sliding window defined in (7.3.17) to break up a long sequence into several short finite-length blocks of N samples $x'_m(n)$, and then perform the FFT to obtain the time-dependent frequency spectrum at each short segment to obtain

$$X_m(k) = \sum_{n=0}^{N-1} x'_m(n) W_N^{kn}, \quad k = 0, 1, \ldots, N - 1. \tag{7.3.24}$$

This process is repeated for the next block of N samples as illustrated in Figure 7.12. This technique is also called the short-term Fourier transform, since $X_m(k)$ is just the DFT spectrum of the short segment of $x_m(n)$ that lies inside the sliding window $w(n)$. This form of time-dependent Fourier transform has several applications in speech, sonar, and radar signal processing.

Equation (7.3.24) shows that $X_m(k)$ is a two-dimensional sequence. The index k represents frequency as defined in (7.3.1), and the block index m represents time. Since the result is a function of both time and frequency, a three-dimensional graphical display is needed. This is done by plotting $|X_m(k)|$ as a function of both k and m using gray-scale (or color) images. The resulting three-dimensional graphic is called the spectrogram. It uses the x-axis to represent time and the y-axis to represent frequency. The gray level (or color) at point (m, k) is proportional to $|X_m(k)|$. The large values are black, and the small ones are white.

The *Signal Processing Toolbox* provides a function, specgram, to compute spectrogram. This MATLAB function has the form

```
B = specgram(a, nfft, Fs, window, noverlap);
```

where B is a matrix containing the complex spectrogram values $|X_m(k)|$, and other arguments are defined in the function psd. It is common to pick the overlap to be around 50 percent as shown in Figure 7.12. The specgram function with no output arguments displays the scaled logarithm of the spectrogram in the current graphic window. See the Signal Processing Toolbox for Use with MATLAB [7] for details.

7.4 Implementation Considerations

The FFT algorithm can be realized as a program in a general-purpose computer, a DSP processor, or implemented in special-purpose hardware. Many FFT routines are available in C and assembly programs. We probably would not even have to write an FFT routine for a given application. However, it is important to understand the implementation issues in order to use FFT properly. In this section, we only consider the radix-2 FFT algorithms.

7.4.1   Computational Issues

As illustrated in Figure 7.5, the radix-2 FFT algorithm takes two input samples at a time from memory, performs the butterfly computations, and returns the resulting numbers to the same input memory locations. This process is repeated $N \log_2 N$ times in the computation of an N-point FFT. The FFT routines accept complex-valued inputs, therefore the number of memory locations required is $2N$. Complex-valued signals are quite common in communications such as modems. However, most signals such as speech are real-valued. To use the available FFT routine, we have to set the imaginary part of each sample value to 0 for real input data. Note that each complex multiplication is of the form

$$(a + jb)(c + jd) = (ac + bd) + j(bc + ad)$$

and therefore requires four real multiplications and two real additions.

The number of multiplications and the storage requirements can be reduced if the signal has special properties. For example, if $x(n)$ is real, only $N/2$ samples from $X(0)$ to $X(N/2)$ need to be computed as shown by complex-conjugated property (7.1.15). In addition, if $x(n)$ is an even function of n, only the real part of $X(k)$ is non-zero. If $x(n)$ is odd, only the imaginary part is non-zero.

The computation of twiddle factors W_N^{kn} usually takes longer than the computation of complex multiplications. In most FFT programs on general-purpose computers, the sine and cosine calculations defined in (7.1.4) are embedded in the program for convenience. If N is fixed, it is preferable to tabulate the values of twiddle factors so that they can be looked up during the computation of FFT algorithms. When the FFT is performed repeatedly with N being constant, the computation of twiddle factors need not be repeated. In addition, in an efficient implementation of FFT algorithm on a DSP processor, the twiddle factors are computed once and then stored in a table during the programming stage.

There are other implementation issues such as indexing, bit reversal, and parallelism in computations. The complexity of FFT algorithms is usually measured by the required number of arithmetic operations (multiplications and additions). However, in practical implementations on DSP chips, the architecture, instruction set, data structures, and memory organizations of the processors are critical factors. Modern DSP chips such as the TMS320C55x usually provide single-cycle multiplication-and-accumulation operation, bit-reversal addressing, and a high degree of instruction parallelism to efficiently implement FFT algorithms. These issues will be discussed further in Section 7.5.

7.4.2   Finite-Precision Effects

Since FFT is often employed in DSP hardware for real-time applications, it is important to analyze the finite-precision effects in FFT computations. We assume that the FFT computations are being carried out using fixed-point arithmetic. With clever scaling and checking for overflow, the most critical error in the computation is due to roundoff errors. Without loss of generality, we analyze the decimation-in-time radix-2 FFT algorithm introduced in Section 7.2.1.

From the flow-graph of the FFT algorithm shown in Figure 7.6, $X(k)$ are computed by a series of butterfly computations with a single complex multiplication per butterfly network. Note that some of the butterfly computations require multiplications by ± 1 (such as 2-point FFT in the first stage) that do not require multiplication in practical implementation, thus avoiding roundoff errors.

Figure 7.6 shows that the computation of N-point FFT requires $M = \log_2 N$ stages. There are $N/2$ butterflies in the first stage, $N/4$ in the second stage, and so on. Thus the total number of butterflies required to produce an output sample is

$$\frac{N}{2} + \frac{N}{4} + \cdots + 2 + 1 = 2^{M-1} + 2^{M-2} + \cdots + 2 + 1$$

$$= 2^{M-1}\left[1 + \left(\frac{1}{2}\right) + \cdots + \left(\frac{1}{2}\right)^{M-1}\right] = 2^{M-1}\sum_{m=0}^{M-1}\left(\frac{1}{2}\right)^m$$

$$= 2^M\left[1 - \left(\frac{1}{2}\right)^M\right] = N - 1. \tag{7.4.1}$$

The quantization errors introduced at the mth stage appear at the output after propagation through $(m-1)$ stages, while getting multiplied by the twiddle factors at each subsequent stage. Since the magnitude of the twiddle factor is always unity, the variances of the quantization errors do not change while propagating to the output. If we assume that the quantization errors in each butterfly are uncorrelated with the errors in other butterflies, the total number of roundoff error sources contributing to the output is $4(N-1)$. Therefore the variance of the output roundoff error is

$$\sigma_e^2 = 4(N-1)\frac{2^{-2B}}{12} \approx \frac{N2^{-2B}}{3}. \tag{7.4.2}$$

As mentioned earlier, some of the butterflies do not require multiplications in practical implementation, thus the total roundoff error is less than the one given in (7.4.2).

The definition of DFT given in (7.1.3) shows that we can scale the input sequence with the condition

$$|x(n)| < \frac{1}{N} \tag{7.4.3}$$

to prevent the overflow at the output because $|e^{-j(2\pi/N)kn}| = 1$. For example, in a 1024-point FFT, the input data must be shifted right by 10 bits. If the original data is 16-bit, the effective wordlength after scaling is reduced to only 6 bits. This worst-case scaling substantially reduces the resolution of the FFT results.

Instead of scaling the input samples by $1/N$ at the beginning, we can scale the signals at each stage since the FFT algorithm consists of a sequence of stages. Figure 7.5 shows that we can avoid overflow within the FFT by scaling the input at each stage by $1/2$ (right shift one bit in a fixed-point hardware) because the outputs of each butterfly involve the addition of two numbers. That is, we shift right the input by 1 bit, perform the first stage of FFT, shift right that result by 1 bit, perform the second stage of FFT, and so on. This unconditional scaling process does not affect the signal level at the output of the FFT, but it significantly reduces the variance of the quantization errors at the output. Thus it provides a better accuracy than unconditional scaling the input by $1/N$.

An alternative conditional scaling method examines the results of each FFT stage to determine whether all the results of that stage should be scaled. If all the results in a particular stage have magnitude less than 1, no scaling is necessary at that stage. Otherwise, all the inputs of that stage have to be scaled by 1/2. The conditional scaling technique achieves much better accuracy since we may scale less often than the unconditional scaling method. However, this conditional scaling method increases software complexity and may require longer execution time.

7.5   Experiments Using the TMS320C55x

In this section, we implement the decimation-in-time FFT algorithm using the floating-point C, fixed-point C, and assembly language. We then implement the IFFT algorithm using the same FFT routine. Finally, we apply both the FFT and IFFT for fast convolution.

7.5.1   Experiment 7A – Radix-2 Complex FFT

The decimation-in-time FFT algorithm based on (7.2.6) shows how to compute an N-point DFT by combining $N/2$-point DFT sections. The computation described by (7.2.6) is called the butterfly computation and is shown graphically in Figure 7.5. The floating-point C function for computing a radix-2 complex decimation-in-time FFT is listed in Table 7.2. This program will compute an N-point FFT using a sinusoidal signal as input. The output of this program should be all zeros except at the FFT bins of $X(k)$ and $X(N - k)$. By changing the constants N and EXP, we are able to perform different length of radix-2 complex FFT using routine shown in Table 7.3. Since this is a complex radix-2 FFT routine, the imaginary portion of the complex data buffer must be set to 0 if the data is real.

Table 7.2   Floating-point C program for testing the FFT algorithm

```
/*
    Example to test floating-point complex FFT
*/
#include <math.h>
#include "fcomplex.h"      /* Floating-point complex.h header file */
#include "input7_f.dat"    /* Floating-point testing data          */

extern void fft(complex *, unsigned int, complex *, unsigned int);
extern void bit_rev(complex *X, int M);

#define N    128            /* FFT size      */
#define EXP 7               /* EXP = log2(N) */
#define pi   3.1415926535897
```

Table 7.2 (*continued*)

```
complex X[N];                   /* Declare input buffer          */
complex W[EXP];                 /* Twiddle factors table         */
complex temp;
float spectrum[N];
float re1[N],im1[N];

void main()
{
  unsigned int i, j, L, LE, LE1;

  for(L = 1; L <= EXP; L++)   /* Create twiddle-factor table     */
  {
    LE = 1≪L;                   /* LE = 2^L = points of sub DFT    */
    LE1 = LE≫1;                 /* Number of butterflies in sub DFT */
    W[L−1].re = cos(pi/LE1);
    W[L−1].im = −sin(pi/LE1);
  }

  j = 0;
  for(;;)
  {
    for(i = 0; i < N; i++)
    {
      /* Generate input samples */
      X[i].re = input7_f[j++];
      X[i].im = 0.0;
      /* Copy to reference buffer */
      re1[i] = X[i].re;
      im1[i] = X[i].im;

      if(j == 1664)
        j = 0;
    }
    /* Start FFT */
    bit_rev(X, EXP);            /* Arrange X[] in bit-reversal order */
    fft(X, EXP, W, 1);          /* Perform FFT                     */

    /* Verify FFT results */
    for(i = 0; i < N; i++)
    {
      /* Compute power spectrum */
      temp.re = X[i].re* X[i].re;
      temp.im = X[i].im* X[i].im;
      spectrum[i] = (temp.re + temp.im)* 4;
    }
  }
}
```

The complex radix-2 FFT program listed in Table 7.3 computes the complex decimation-in-time FFT algorithm as shown in Figure 7.5. To prevent the results from overflowing, the intermediate results are scaled down in each stage as described in Section 7.4.2. The radix-2 FFT function contains two complex arguments and two unsigned integer arguments. They are the complex input sample, X[N], the power of the radix-2 FFT, EXP, the initial complex twiddle-factor table W[EXP], and the scaling flag SCALE. The FFT is performed in place, that is, the complex input array is overwritten by the output array. The initial twiddle factors are created by the C function listed in Table 7.2.

As discussed in Section 7.2.1, the data used for FFT need to be placed in the bit-reversal order. An N-point FFT bit-reversal example is given in Table 7.1. Table 7.4 illustrated the C function that performs the bit-reversal addressing task.

Table 7.3   Floating-point complex radix-2 FFT routine in C

```
/*
    fft_float.c – Floating-point complex radix-2 decimation-in-time FFT
    Perform in-place FFT, the output overwrite the input buffer
*/

#include "fcomplex.h"          /* Floating-point complex.h header file */

void fft(complex *X, unsigned int M, complex *W, unsigned int SCALE)
{
    complex temp;              /* Temporary storage of complex variable */
    complex U;                 /* Twiddle factor W^k                    */
    unsigned int i,j;
    unsigned int id;           /* Index for lower point in butterfly  */
    unsigned int N = 1≪EXP;    /* Number of points for FFT             */
    unsigned int L;            /* FFT stage                            */
    unsigned int LE;           /* Number of points in sub FFT at stage
                                  L and offset to next FFT in stage    */
    unsigned int LE1;          /* Number of butterflies in one FFT at
                                  stage L. Also is offset to lower
                                  point in butterfly at stage L        */
    float scale;

    scale = 0.5;
    if (SCALE == 0)
      scale = 1.0;

    for(L = 1; L <= EXP; L++)   /* FFT butterfly                        */
    {
      LE = 1≪L;                 /* LE = 2^L = points of sub DFT         */
      LE1 = LE≫1;               /* Number of butterflies in sub-DFT     */
      U.re = 1.0;
      U.im = 0.;
```

Table 7.3 *(continued)*

```
   for (j = 0; j < LE1; j++)
   {
     for(i = j; i < N; i += LE) /* Do the butterflies */
     {
        id = i+LE1;
        temp.re = (X[id].re*U.re - X[id].im*U.im)*scale;
        temp.im = (X[id].im*U.re + X[id].re*U.im)*scale;

        X[id].re = X[i].re*scale - temp.re;
        X[id].im = X[i].im*scale - temp.im;

        X[i].re = X[i].re*scale + temp.re;
        X[i].im = X[i].im*scale + temp.im;
     }

     /* Recursive compute W^k as U*W^(k-1) */
     temp.re = U.re*W[L-1].re - U.im*W[L-1].im;
     U.im = U.re*W[L-1].im + U.im*W[L-1].re;
     U.re = temp.re;
   }
 }
}
```

Table 7.4 Floating-point bit-reversal function

```
/*
   fbit_rev.c - Arrange input samples in bit-reversal order
               The index j is the bit-reversal of i
*/

#include "fcomplex.h"        /* Floating-point complex.h header file */

void bit_rev(complex *X, unsigned int EXP)
{
  unsigned int i, j, k;
  unsigned int N = 1<<EXP;   /* Number of points for FFT              */
  unsigned int N2 = N>>1;
  complex temp;              /* Temp storage of the complex variable */

  for (j = 0, i = 1; i < N-1; i++)
  {
    k = N2;
    while(k <= j)
    {
```

continues overleaf

Table 7.4 (*continued*)

```
      j -= k;
      k >>= 1;
   }
   j += k;

   if (i < j)
   {
      temp = X[j];
      X[j] = X[i];
      X[i] = temp;
   }

 }
}
```

Based on the floating-point C function, the fixed-point conversion is done by employing the C intrinsic functions. The implementation of the decimation-in-time FFT butterfly computation using the C55x intrinsics is listed below:

```
ltemp.re = _lsmpy(X[id].re, U.re);
temp.re = (_smas(ltemp.re, X[id].im, U.im) >> 16);
temp.re = _sadd(temp.re, 1) >> scale;            /* Rounding & scale */
ltemp.im = _lsmpy(X[id].im, U.re);
temp.im = (_smac(ltemp.im, X[id].re, U.im) >> 16);
temp.im = _sadd(temp.im, 1) >> scale;            /* Rounding & scale */
X[id].re = _ssub(X[i].re >> scale, temp.re);
X[id].im = _ssub(X[i].im >> scale, temp.im);
X[i].re = _sadd(X[i].re >> scale, temp.re);
X[i].im = _sadd(X[i].im >> scale, temp.im);
```

In the program, X[] is the complex sample buffer and U is the complex twiddle factor. The scale is done by right-shifting 1-bit instead of multiply by 0.5.

Go through the following steps for Experiment 7A:

1. Verify the floating-point C programs test_fft.c, fft_float.c, fbit_rev.c, and fcomplex.h using a PC or C55x simulator. The output should be all zeros except $X(k)$ and $X(N-k)$. These squared values are equal to 1.0. The floating-point program will be used as reference for the code development using the fixed-point C and assembly language. The floating-point program uses the floating-point data file input7_f.dat, while the fixed-point data file input7_i.dat will be used for the rest of experiments.

2. Create the project exp7a using CCS. Add the command file exp7.cmd, the functions epx7a.c, fft_a.c, and ibit_rev.c, and the header file icomplex.h from the software package into the project.

3. Build the fixed-point FFT project and verify the results. Comparing the results with the floating-point complex radix-2 FFT results obtained by running it on the PC.

4. The FFT output samples are squared and placed in a data buffer named `spectrum[]`. Use CCS to plot the results by displaying the `spectrum[]` and `re1[]` buffer.

5. Profile the DSP run-time clock cycles for 128-point and 1024-point FFTs. Record the memory usage of the fixed-point functions `bit_rev()` and `fft()`.

7.5.2   Experiment 7B – Radix-2 Complex FFT Using Assembly Language

Although using intrinsics can improve the DSP performance, the assembly language implementation has been proven to have the fastest execution speed and memory efficiency for most applications, especially for computational intensive algorithms such as FFT. The development time for assembly code, however, will be much longer than that of C code. In addition, the maintenance and upgrade of assembly code are usually more difficult. In this experiment, we will use C55x assembly routines for computing the same radix-2 FFT algorithm as the fixed-point C function used in Experiment 7A. The assembly FFT routine is listed in Table 7.5. This routine is written based on the C function used for Experiment 7A, and it follows the C55x C calling convention. For readability, the assembly code has been written to mimic the C function of Experiment 7A closely. It optimizes the memory usage but not the run-time efficiency. By unrolling the loop and taking advantage of the FFT butterfly characteristics, the FFT execution speed can be further improved with the expense of the memory space, see the exercise problems at the end of this chapter.

In `fft.asm`, the local variables are defined as structure using the stack relative addressing mode when the assembly routine is called. The last memory location contains the return address of the caller function. Since the status registers ST1 and ST3 will be modified, we use two stack locations to store the contents of these status registers at entry. The status registers will be restored upon returning to the caller function. The complex temporary variable is stored in two consecutive memory locations by using a bracket with the numerical number to indicate the number of memory locations for the integer data type.

The FFT implementation is carried out in three nested loops. The butterfly computation is implemented in the inner loop and the group loop is in the middle, while the stages are managed by the outer loop. Among these three loops, the butterfly loop is repeated most often. We use the local block repeat instruction, `rptblocal`, for the butterfly loop and the middle loop to minimize the loop overhead. We also use parallel instructions, modulo addressing, and dual memory access instructions to further improve the efficiency of butterfly computation. By limiting the size of the loop, we can place the middle loop inside the DSP instruction buffer queue (IBQ) as well. The FFT computation is improved since the two inner loops are only fetched once from the program memory each time we compute the groups and butterflies. The twiddle-factor

Table 7.5 FFT routine using the C55x assembly language

```
outer_loop                        ; for(L = 1; L <= M; L++)
    mov  fft.d_L,T0               ; Note: Since the buffer is
||  mov  #2,AC0                   ;         arranged in re,im pairs
    sfts AC0, T0                  ;         the index to the buffer
    neg  T0                       ;         is doubled
||  mov  fft.d_N, AC1             ;         But the repeat counters
    sftl AC1,T0                   ;         are not doubled
    mov  AC0,T0                   ; LE = 2 ≪ L
||  sfts AC0,#−1
    mov  AC0,AR0                  ; LE1 = LE≫1
||  sfts AC0,#−1
    sub  #1,AC0                   ; Initialize mid_loop counter
    mov  mmap(AC0L),BRC0          ;   BRC0 = LE1−1
    sub  #1,AC1                   ; Initialize inner loop counter
    mov  mmap(AC1L),BRC1          ;   BRC1 = (N≫L)−1
    add  AR1,AR0
    mov  #0,T2                    ; j = 0
||  rptblocal mid_loop-1          ; for(j = 0; j < LE1; j++)
    mov  T2,AR5                   ; AR5 = id = i+LE1
    mov  T2,AR3
    add  AR0,AR5                  ; AR5 = pointer to X[id].re
    add  #1,AR5,AR2               ; AR2 = pointer to X[id].im
    add  AR1,AR3                  ; AR3 = pointer to X[i].re
||  rptblocal inner_loop-1        ; for(i = j; i < N; i += LE)
    mpy  *AR5+, *CDP+, AC0        ; AC0 = (X[id].re*U.re
::  mpy  *AR2−, *CDP+, AC1        ;       −X[id].im*U.im)/SCALE
    masr *AR5−, *CDP−, AC0        ; AC1 = (X[id].im*U.re
::  macr *AR2+, *CDP−, AC1        ;       +X[id].re*U.im)/SCALE
    mov  pair(hi(AC0)),dbl(*AR4); AC0H = temp.re AC1H = temp.im
||  mov  dbl(*AR3), AC0
    xcc  scale,TC1
||  mov  AC0 ≫ #1,dual(*AR3)      ; Scale X[i] by 1/SCALE
    mov  dbl(*AR3), AC0
scale
    add  T0,AR2
||  sub  dual(*AR4),AC0,AC1       ; X[id].re = X[i].re/SCALE-temp.re
    mov  AC1,dbl(*(AR5+T0))       ; X[id].im = X[i].im/SCALE-temp.im
||  add  dual(*AR4),AC0           ; X[i].re = X[i].re/SCALE+temp.re
    mov  AC0,dbl(*(AR3+T0))       ; X[i].im = X[i].im/SCALE+temp.im
inner_loop                        ; End of inner loop
    amar *CDP+
    amar *CDP+                    ; Update k for pointer to U[k]
||  add  #2,T2                    ; Update j
mid_loop                          ; End of mid-loop
    sub  #1,T1
    add  #1,fft.d_L               ; Update L
    bcc  outer_loop,T1 > 0        ; End of outer-loop
```

table is pre-calculated during the initialization phase. The calculation of the twiddle factors can be implemented as follows:

```
for (i = 0, l = 1; l <= EXP; l++)
{
    SL = 1<<l;                      /* LE = 2^L = points of sub FFT */
    SL1 = SL>>1;                    /* # of twiddle factors in sub-FFT */
    for (j = 0; j < SL1; j++)
    {
        W.re = (int)((0x7fff*cos(j*pi/SL1)) + 0.5);
        W.im = -(int)((0x7fff*sin(j*pi/SL1)) + 0.5);
        U[i++] = W;
    }
}
```

where U[] and W are defined as complex data type.

The bit-reversal addressing assembly routine is shown in Table 7.6. Since we use the special C55x bit-reversal addressing mode, the assembly routine is no longer the same as the C function used in the previous experiment. The bit-reversal addressing mode uses the syntax of *(AR1+T0B), where the temporary register T0 contains the size of the buffer, and the letter B indicates that the memory addressing is in a bit-reversal order. For Experiment 7B, we also use the data and program pragma directives to manage the memory space.

Complete the following steps for Experiment 7B:

1. Create the project exp7b, add files exp7.cmd, exp7b.c, w_table.c, fft.asm, and bit_rev.asm from the software package into the project.

2. Build and verify the FFT function, and compare the results with the results obtained from Experiment 7A. Make sure that the scale flag for the FFT routine is set to 1.

3. Profile the FFT run-time clock cycles and its memory usage again and compare these results with those obtained in Experiment 7A.

Table 7.6   Bit-reversal routine in the C55x assembly language

```
      rptblocal loop_end-1              ; Start bit-reversal loop
      mov  dbl(*AR0),AC0                ; Get a pair of samples
||    amov AR1,T1
      mov  dbl(*AR1),AC1                ; Get another pair
||    asub AR0,T1
      xccpart swap1,T1 >= #0
||    mov  AC1,dbl(*AR0+)               ; Swap samples if j >= i
swap1
      xccpart loop_end,T1 >= #0
||    mov  AC0,dbl(*(AR1+T0B))
loop_end                               ; End bit-reversal loop
```

7.5.3   Experiment 7C – FFT and IFFT

As discussed in Section 7.2.3, the inverse DFT defined by (7.2.11) is similar to the DFT defined in (7.1.6). Thus the FFT routine developed in Experiment 7B can be modified for computing the inverse FFT. Two simple changes are needed in order to use the same FFT routine for the IFFT calculation. First, the conjugating twiddle factors imply the sign change of the imaginary portion of the complex samples. That is, X[i].im = −X[I].im. Second, the normalization of $1/N$ is handled in the FFT routine by setting the scale flag to 0. Table 7.7 shows the example of computing both the FFT and IFFT.

Go over the following steps for Experiment 7C:

1. Create the project epx7c and include the files exp7.cmd, exp7c.c, w_table.c, fft.asm, and bit_rev.asm from the software package into the project.

2. Build and view the IFFT results by plotting and comparing the input array re1[] and IFFT output array re2[]. Make sure that the scale flag for the FFT calculation is set to 1 (one), and the IFFT calculation is set to 0 (zero).

7.5.4   Experiment 7D – Fast Convolution

As discussed in Section 7.3.4, the application of fast convolution using FFT/IFFT is the most efficient technique of FIR filtering for long time-domain sequence such as for high-fidelity digital audio systems, or FIR filtering in frequency-domain such as in the xDSL modems. The fast convolution algorithm is shown in Figure 7.13. There are two basic methods for FFT convolution as mentioned in Section 7.3.4. This experiment will use the overlap-add technique. This method involves the following steps:

– Pad $M = N − L$ zeros to the FIR filter impulse response of length L where $N > L$, and process the sequence using an N-point FFT. Store the results in the complex buffer H[N].

Table 7.7   Perform FFT and IFFT using the same routine

```
/* Start FFT */
bit_rev(X,EXP);          /* Arrange X[] in bit-reversal order    */
fft(X,EXP,U,1);          /* Perform FFT                          */

/* Inverse FFT */
for(i = 0; i < N; i++)   /* Change the sign of imaginary part    */
{
   X[i].im = −X[i].im;
}
bit_rev(X,EXP);          /* Arrange sample in bit-reversal order */
fft(X,EXP,U,0);          /* Perform IFFT                         */
```

– Segment the input sequence of length M with $L - 1$ zeros padded at the end.

– Process each segment of data samples with an N-point FFT to obtain the complex array X[N].

– Multiply H and X in frequency domain to obtain Y.

– Perform N-point IFFT to find the time-domain filtered sequence.

– Add the first L samples that are overlapped with the previous segment to form the output. All resulting segments are combined to obtain $y(n)$.

The C program implementation of fast convolution using FFT and IFFT is listed in Table 7.8, where we use the same data file and FIR coefficients as the experiments given in Chapter 5. In general, for low- to median-order FIR filters, the direct FIR routines introduced in Chapter 5 are more efficient. Experiment 5A shows that an FIR filter can be implemented as one clock cycle per filter tap, while Experiments 5B and 5C complete two taps per cycle. However, the computational complexity of those routines is linearly increased with the number of coefficients. When the application requires high-order FIR filters, the computation requirements can be reduced by using fast convolution as shown in this experiment.

Table 7.8 Fast convolution using FFT and IFFT

```
/* Initialization */
for(i = 0; i < L-1; i++) /* Initialize overlap buffer          */
  OVRLAP[i] = 0;

for(i = 0; i < L; i++)    /* Copy filter coefficients to buffer  */
{
  X[i].re = LP_h[i];
  X[i].im = 0;
}

for(i = i; i < N; i++)    /* Pad zeros to the buffer             */
{
  X[i].re = 0;
  X[i].im = 0;
}

w_table(U, EXP);          /* Create twiddle-factor table         */
bit_rev(X, EXP);          /* Bit-reversal arrangement of coefficients */
fft(X, EXP, U, 1);        /* FFT of filter coefficients          */

for(i = 0; i < N; i++)    /* Save frequency-domain coefficients  */
{
  H[i].re = X[i].re << EXP;
  H[i].im = X[i].im << EXP;
}
```

continues overleaf

Table 7.8   (*continued*)

```
/* Start fast convolution test                                         */
j = 0;
for(;;)
{
    for(i = 0; i < M; i++)
    {                       .
        X[i].re = input[j++];       /* Generate input samples          */
        X[i].im = 0;
        if(j==160)
            j = 0;
    }

    for(i = i; i < N; i++)          /* Fill zeros to data buffer       */
    {
        X[i].re = 0;
        X[i].im = 0;
    }

    /* Start FFT convolution */
    bit_rev(X, EXP);                /* Samples in bit-reversal order */
    fft(X, EXP, U, 1);              /* Perform FFT                     */

    freqflt(X, H, N);              /* Frequency domain filtering      */

    bit_rev(X, EXP);                /* Samples in bit-reversal order */
    fft(X, EXP, U, 0);              /* Perform IFFT                    */
    olap_add(X, OVRLAP, L, M, N);  /* Overlap-add algorithm           */
}
```

Go through the following steps for Experiment 7D:

1. Create the project `exp7d`, add the files `exp7.cmd`, `exp7d.c`, `w_table.c`, `fft.asm`, `bit_rev.asm`, `freqflt.asm`, and `olap_add.asm` from the software package to the project.

2. Build and verify the fast convolution results, and compare the results with the results obtained in Experiment 5A.

3. Profile the run-time clock cycles of the fast convolution using FFT/IFFT for various FIR filter lengths by using different filter coefficient files `firlp8.dat`, `firlp16.dat`, `firlp32.dat`, `firlp64.dat`, `firlp128.dat`, `firlp256.dat`, and `firlp512.dat`. These files are included in the experiment software package.

References

[1] D. J. DeFatta, J. G. Lucas, and W. S. Hodgkiss, *Digital Signal Processing: A System Design Approach*, New York: Wiley, 1988.

[2] N. Ahmed and T. Natarajan, *Discrete-Time Signals and Systems*, Englewood Cliffs, NJ: Prentice-Hall, 1983.

[3] V. K. Ingle and J. G. Proakis, *Digital Signal Processing Using MATLAB V.4*, Boston: PWS Publishing, 1997.

[4] L. B. Jackson, *Digital Filters and Signal Processing*, 2nd Ed., Boston: Kluwer Academic, 1989.

[5] *MATLAB User's Guide*, Math Works, 1992.

[6] *MATLAB Reference Guide*, Math Works, 1992.

[7] *Signal Processing Toolbox for Use with MATLAB*, Math Works, 1994.

[8] A. V. Oppenheim and R. W. Schafer, *Discrete-Time Signal Processing*, Englewood Cliffs, NJ: Prentice-Hall, 1989.

[9] S. J. Orfanidis, *Introduction to Signal Processing*, Englewood Cliffs, NJ: Prentice-Hall, 1996.

[10] J. G. Proakis and D. G. Manolakis, *Digital Signal Processing – Principles, Algorithms, and Applications*, 3rd Ed., Englewood Cliffs, NJ: Prentice-Hall, 1996.

[11] A Bateman and W. Yates, *Digital Signal Processing Design*, New York: Computer Science Press, 1989.

[12] S. D. Stearns and D. R. Hush, *Digital Signal Analysis*, 2nd Ed., Englewood Cliffs, NJ: Prentice-Hall, 1990.

Exercises

Part A

1. Compute the four-point DFT of the sequence $\{1, 1, 1, 1\}$ using the matrix equations given in (7.1.7) and (7.1.8).

2. Repeat Problem 1 with eight-point DFT of sequence $\{1, 1, 1, 1, 0, 0, 0, 0\}$. Compare the results with the results of Problem 1.

3. Calculate the DFTs of the following signals:

    (a) $x(n) = \delta(n)$.

    (b) $x(n) = \delta(n - n_0), \quad 0 < n_0 < N$.

    (c) $x(n) = c^n, \quad 0 \leq n \leq N - 1$.

    (d) $x(n) = \cos(\omega_0 n), \quad 0 \leq n \leq N$.

    (e) $x(n) = \sin(\omega_0 n), \quad 0 \leq n \leq N$.

4. Prove the symmetry and periodicity properties of the twiddle factors defined as

    (a) $W_N^{k+N/2} = -W_N^k$.

    (b) $W_N^{k+N} = W_N^k$.

5. Generalize the derivation of Example 7.7 to a four-point DFT and show a detailed signal-flow graph of four-point DFT.

6. Consider the following two sequences:

$$x_1(n) = x_2(n) = 1, \quad 0 \leq n \leq N - 1.$$

(a) Compute the circular convolution of the two sequences using DFT and IDFT.

(b) Show that the linear convolution of these two sequences is the triangular sequence given by

$$x_3(n) = \begin{cases} n+1, & 0 \leq n < N \\ 2N - n - 1, & N \leq n < 2N \\ 0, & \text{otherwise.} \end{cases}$$

(c) How to make the circular convolution of the two sequences becomes a triangular sequence defined in (b)?

7. Construct the signal-flow diagram of FFT for $N = 16$ using the decimation-in-time method with bit-reversal input.

8. Construct the signal-flow diagram of FFT for $N = 8$ using the decimation-in-frequency method without bit-reversal input.

9. Complete the development of decimation-in-frequency FFT algorithm for $N = 8$. Show the detailed flow graph.

10. Consider a digitized signal of one second with the sampling rate 20 kHz. The spectrum is desired with a computational frequency resolution of 100 Hz or less. Is this possible? If possible, what FFT size N should be used?

11. A 1 kHz sinusoid is sampled at 8 kHz. The 128-point FFT is performed to compute $X(k)$. At what frequency indices k we expect to observe any peaks in $|X(k)|$?

12. A touch-tone phone with a dual-tone multi-frequency (DTMF) transmitter encodes each keypress as a sum of two sinusoids, with one frequency taken from each of the following groups:

Vertical group: 697, 770, 852, 941 Hz

Horizontal group: 1209, 1336, 1477, 1633 Hz

What is the smallest DFT size N that we can distinguish these two sinusoids from the computed spectrum? The sampling rate used in telecommunications is 8 kHz.

13. Compute the linear convolution $y(n) = x(n)^*h(n)$ using 512-point FFT, where $x(n)$ is of length 4096 and $h(n)$ is of length 256.

(a) How many FFTs and how many adds are required using the overlap-add method?

(b) How many FFTs are required using the overlap-save method?

(c) What is the length of output $y(n)$?

Part B

14. Write a C or MATLAB program to compute the fast convolution of a long sequence with a short sequence employing the overlap-save method introduced in Section 7.3.4. Compare the results with the MATLAB function `fftfilt` that use overlap-add method.

15. Experiment with the capability of the `psd` function in the MATLAB. Use a sinusoid embedded in white noise for testing signal.

16. Using the MATLAB function `specgram` to display the spectrogram of the speech file `timit1.asc` included in the software package.

Part C

17. The radix-2 FFT code used in the experiments is written in consideration of minimizing the code size. An alternative FFT implementation can be more efficient in terms of the execution speed with the expense of using more program memory locations. For example, the twiddle factors used by the first stage and the first group of other stages are constants, $W_N^0 = 1$. Therefore the multiplication operations in these stages can be simplified. Modify the assembly FFT routine given in Table 7.5 to incorporate this observation. Profile the run-time clock cycles and record the memory usage. Compare the results with those obtained by Experiment 7C.

18. The radix-2 FFT is the most widely used algorithm for FFT computation. When the number of data samples are a power of $2n$ (i.e., $N = 2^{2n} = 4^n$), we can further improve the run-time efficiency by employing the radix-4 FFT algorithm. Modify the assembly FFT routine give in Table 7.5 for the radix-4 FFT algorithm. Profile the run-time clock cycles, and record the memory space usage for a 1024-point radix-4 FFT ($2^{10} = 4^5 = 1024$). Compare the radix-4 FFT results with the results of 1024-point radix-2 FFT computed by the assembly routine.

19. Take advantage of twiddle factor, $W_N^0 = 1$, to further improve the radix-4 FFT algorithm run-time efficiency. Compare the results of 1024-point FFT implementation using different approaches.

20. Most of DSP applications have real input samples, our complex FFT implementation zeros out the imaginary components of the complex buffer (see `exp7c.c`). This approach is simple and easy, but it is not efficient in terms of the execution speed. For real input, we can split the even and odd samples into two sequences, and compute both even and odd sequences in parallel. This approach will reduce the execution time by approximately 50 percent. Given a real value input $x(n)$ of $2N$ samples, we can define $c(n) = a(n) + jb(n)$, where two inputs $a(n) = x(n)$ and $b(n) = x(n+1)$ are real sequences. We can represent these sequences as $a(n) = [c(n) + c^*(n)]/2$ and $b(n) = -j[c(n) - c^*(n)]/2$, then they can be written in terms of DFTs as and $A(k) = [C(k) + C^*(N - k)]/2$ and $B(k) = -j[C(k) - C^*(N - k)]/2$. Finally, the real input FFT can be obtained by $X(k) = A(k) + W_{2N}^k B(k)$ and $X(k + N) = A(k) - W_{2N}^k B(k)$, where $k = 0, 1, \ldots, N - 1$. Modify the complex radix-2 FFT assembly routine to efficiently compute $2N$ real input samples.

8

Adaptive Filtering

As discussed in previous chapters, filtering refers to the linear process designed to alter the spectral content of an input signal in a specified manner. In Chapters 5 and 6, we introduced techniques for designing and implementing FIR and IIR filters for given specifications. Conventional FIR and IIR filters are time-invariant. They perform linear operations on an input signal to generate an output signal based on the fixed coefficients. Adaptive filters are time varying, filter characteristics such as bandwidth and frequency response change with time. Thus the filter coefficients cannot be determined when the filter is implemented. The coefficients of the adaptive filter are adjusted automatically by an adaptive algorithm based on incoming signals. This has the important effect of enabling adaptive filters to be applied in areas where the exact filtering operation required is unknown or is non-stationary.

In Section 8.1, we will review the concepts of random processes that are useful in the development and analysis of various adaptive algorithms. The most popular least-mean-square (LMS) algorithm will be introduced in Section 8.2. Its important properties will be analyzed in Section 8.3. Two widely used modified adaptive algorithms, the normalized and leaky LMS algorithms, will be introduced in Section 8.4. In this chapter, we introduce and analyze the LMS algorithm following the derivation and analysis given in [8]. In Section 8.5, we will briefly introduce some important applications of adaptive filtering. The implementation considerations will be discussed in Section 8.6, and the DSP implementations using the TMS320C55x will be presented in Section 8.7.

8.1 Introduction to Random Processes

A signal is called a deterministic signal if it can be described precisely and be reproduced exactly and repeatedly. However, the signals encountered in practice are not necessarily of this type. A signal that is generated in a random fashion and cannot be described by mathematical expressions or rules is called a random (or stochastic) signal. The signals in the real world are often random in nature. Some common examples of random signals are speech, music, and noises. These signals cannot be reproduced and need to be modeled and analyzed using statistical techniques. We have briefly introduced probability and random variables in Section 3.3. In this section, we will review the important properties of the random processes and introduce fundamental techniques for processing and analyzing them.

A random process may be defined as a set of random variables. We associate a time function $x(n) = x(n, A)$ with every possible outcome A of an experiment. Each time function is called a realization of the random process or a random signal. The ensemble of all these time functions (called sample functions) constitutes the random process $x(n)$. If we sample this process at some particular time n_0, we obtain a random variable. Thus a random process is a family of random variables.

We may consider the statistics of a random process in two ways. If we fix the time n at n_0 and consider the random variable $x(n_0)$, we obtain statistics over the ensemble. For example, $E[x(n_0)]$ is the ensemble average, where $E[\cdot]$ is the expectation operation introduced in Chapter 3. If we fix A and consider a particular sample function, we have a time function and the statistics we obtain are temporal. For example, $E[x(n, A_i)]$ is the time average. If the time average is equal to the ensemble average, we say that the process is ergodic. The property of ergodicity is important because in practice we often have access to only one sample function. Since we generally work only with temporal statistics, it is important to be sure that the temporal statistics we obtain are the true representation of the process as a whole.

8.1.1 Correlation Functions

For many applications, one signal is often used to compare with another in order to determine the similarity between the pair, and to determine additional information based on the similarity. Autocorrelation is used to quantify the similarity between two segments of the same signal. The autocorrelation function of the random process $x(n)$ is defined as

$$r_{xx}(n, k) = E[x(n)x(k)]. \tag{8.1.1}$$

This function specifies the statistical relation of two samples at different time index n and k, and gives the degree of dependence between two random variables of $(n - k)$ units apart. For example, consider a digital white noise $x(n)$ as uncorrelated random variables with zero-mean and variance σ_x^2. The autocorrelation function is

$$r_{xx}(n, k) = E[x(n)x(k)] = E[x(n)]E[x(k)] = \begin{cases} 0, & n \neq k \\ \sigma_x^2, & n = k. \end{cases} \tag{8.1.2}$$

If we subtract the means in (8.1.1) before taking the expected value, we have the autocovariance function

$$\gamma_{xx}(n, k) = E\{[x(n) - m_x(n)][x(k) - m_x(k)]\} = r_{xx}(n, k) - m_x(n)m_x(k). \tag{8.1.3}$$

The objective in computing the correlation between two different random signals is to measure the degree in which the two signals are similar. The crosscorrelation and crosscovariance functions between two random processes $x(n)$ and $y(n)$ are defined as

$$r_{xy}(n, k) = E[x(n)y(k)] \tag{8.1.4}$$

and

$$\gamma_{xy}(n,k) = E\{[x(n) - m_x(n)][y(k) - m_y(k)]\} = r_{xy}(n,k) - m_x(n)m_y(k). \qquad (8.1.5)$$

Correlation is a very useful DSP tool for detecting signals that are corrupted by additive random noise, measuring the time delay between two signals, determining the impulse response of a system (such as obtain the room impulse response used in Section 4.5.2), and many others. Signal correlation is often used in radar, sonar, digital communications, and other engineering areas. For example, in CDMA digital communications, data symbols are represented with a set of unique key sequences. If one of these sequences is transmitted, the receiver compares the received signal with every possible sequence from the set to determine which sequence has been received. In radar and sonar applications, the received signal reflected from the target is the delayed version of the transmitted signal. By measuring the round-trip delay, one can determine the location of the target.

Both correlation functions and covariance functions are extensively used in analyzing random processes. In general, the statistical properties of a random signal such as the mean, variance, and autocorrelation and autocovariance functions are time-varying functions. A random process is said to be stationary if its statistics do not change with time. The most useful and relaxed form of stationary is the wide-sense stationary (WSS) process. A random process is called WSS if the following two conditions are satisfied:

1. The mean of the process is independent of time. That is,

$$E[x(n)] = m_x, \qquad (8.1.6)$$

where m_x is a constant.

2. The autocorrelation function depends only on the time difference. That is,

$$r_{xx}(k) = E[x(n+k)x(n)]. \qquad (8.1.7)$$

Equation (8.1.7) indicates that the autocorrelation function of a WSS process is independent of the time shift and $r_{xx}(k)$ denotes the autocorrelation function of a time lag of k samples.

The autocorrelation function $r_{xx}(k)$ of a WSS process has the following important properties:

1. The autocorrelation function is an even function of the time lag k. That is,

$$r_{xx}(-k) = r_{xx}(k). \qquad (8.1.8)$$

2. The autocorrelation function is bounded by the mean squared value of the process expressed as

$$|r_{xx}(k)| \le r_{xx}(0), \qquad (8.1.9)$$

where $r_{xx}(0) = E[x^2(n)]$ is equal to the mean-squared value, or the power in the random process.

In addition, if $x(n)$ is a zero-mean random process, we have

$$r_{xx}(0) = E[x^2(n)] = \sigma_x^2. \tag{8.1.10}$$

Thus the autocorrelation function of a signal has its maximum value at zero lag. If $x(n)$ has a periodic component, then $r_{xx}(k)$ will contain the same periodic component.

Example 8.1: Given the sequence

$$x(n) = a^n u(n), \quad 0 < a < 1,$$

the autocorrelation function can be computed as

$$r_{xx}(k) = \sum_{n=-\infty}^{\infty} x(n+k)x(n) = \sum_{n=0}^{\infty} a^{n+k}a^n = a^k \sum_{n=0}^{\infty} (a^2)^n.$$

Since $a < 0$, we obtain

$$r_{xx}(k) = \frac{a^k}{1-a^2}.$$

Example 8.2: Consider the sinusoidal signal expressed as

$$x(n) = \cos(\omega n),$$

find the mean and the autocorrelation function of $x(n)$.

(a) $m_x = E[\cos(\omega n)] = 0.$

(b) $r_{xx}(k) = E[x(n+k)x(n)] = E[\cos(\omega n + \omega k)\cos(\omega n)]$

$$= \frac{1}{2}E[\cos(2\omega n + \omega k)] + \frac{1}{2}\cos(\omega k) = \frac{1}{2}\cos(\omega k).$$

The crosscorrelation function of two WSS processes $x(n)$ and $y(n)$ is defined as

$$r_{xy}(k) = E[x(n+k)y(n)]. \tag{8.1.11}$$

This crosscorrelation function has the property

$$r_{xy}(k) = r_{yx}(-k). \tag{8.1.12}$$

Therefore $r_{yx}(k)$ is simply the folded version of $r_{xy}(k)$. Hence, $r_{yx}(k)$ provides exactly the same information as $r_{xy}(k)$, with respect to the similarity of $x(n)$ to $y(n)$.

In practice, we only have one sample sequence $\{x(n)\}$ available for analysis. As discussed earlier, a stationary random process $x(n)$ is ergodic if all its statistics can be determined from a single realization of the process, provided that the realization is long enough. Therefore time averages are equal to ensemble averages when the record length is infinite. Since we do not have data of infinite length, the averages we compute differ from the true values. In dealing with finite-duration sequence, the sample mean of $x(n)$ is defined as

$$\bar{m}_x = \frac{1}{N} \sum_{n=0}^{N-1} x(n), \tag{8.1.13}$$

where N is the number of samples in the short-time analysis interval. The sample variance is defined as

$$\bar{\sigma}_x^2 = \frac{1}{N} \sum_{n=0}^{N-1} [x(n) - m_x]^2. \tag{8.1.14}$$

The sample autocorrelation function is defined as

$$\bar{r}_{xx}(k) = \frac{1}{N-k} \sum_{n=0}^{N-k-1} x(n+k)x(n), \quad k = 0, 1, \ldots, N-1, \tag{8.1.15}$$

where N is the length of the sequence $x(n)$. Note that for a given sequence of length N, Equation (8.1.15) generates values for up to N different lags. In practice, we can only expect good results for lags of no more than 5–10 percent of the length of the signals.

The autocorrelation and crosscorrelation functions introduced in this section can be computed using the MATLAB function xcorr in the *Signal Processing Toolbox*. The crosscorrelation function $r_{xy}(k)$ of the two sequences $x(n)$ and $y(n)$ can be computed using the statement

```
c = xcorr(x, y);
```

where x and y are length N vectors and the crosscorrelation vector c has length $2N - 1$. The autocorrelation function $r_{xx}(k)$ of the sequence $x(n)$ can be computed using the statement

```
c = xcorr(x);
```

In addition, the crosscovariance function can be estimated using

```
v = xcov(x, y);
```

and the autocovariance function can be computed with

```
v = xcov(x);
```

See Signal Processing Toolbox User's Guide for details.

8.1.2 Frequency-Domain Representations

In the study of deterministic digital signals, we use the discrete-time Fourier transform (DTFT) or the z-transform to find the frequency contents of the signals. In this section, we will use the same transform for random signals. Consider an ergodic random process $x(n)$. This sequence cannot be really representative of the random process because the sequence $x(n)$ is only one of infinitely possible sequences. However, if we consider the autocorrelation function $r_{xx}(k)$, the result is always the same no matter which sample sequence is used to compute $r_{xx}(k)$. Therefore we should apply the transform to $r_{xx}(k)$ rather than $x(n)$.

The correlation functions represent the time-domain description of the statistics of a random process. The frequency-domain statistics are represented by the power density spectrum (PDS) or the autopower spectrum. The PDS is the DTFT (or the z-transform) of the autocorrelation function $r_{xx}(k)$ of a WSS signal $x(n)$ defined as

$$P_{xx}(\omega) = \sum_{k=-\infty}^{\infty} r_{xx}(k)e^{-j\omega k}, \tag{8.1.16}$$

or

$$P_{xx}(z) = \sum_{k=-\infty}^{\infty} r_{xx}(k)z^{-k}. \tag{8.1.17}$$

A sufficient condition for the existence of the PDS is that $r_{xx}(k)$ is summable. The PDS defined in (7.3.16) is equal to the DFT of the autocorrelation function. The windowing technique introduced in Section 7.3.3 can be used to improve the convergence properties of (7.3.16) and (7.3.17) if the DFT is used in computing the PDS of random signals.

Equation (8.1.16) implies that the autocorrelation function is the inverse DTFT of the PDS, which is expressed as

$$r_{xx}(k) = \frac{1}{2\pi} \int_{-\pi}^{\pi} P_{xx}(\omega)e^{j\omega k}d\omega. \tag{8.1.18}$$

From (8.1.10), we have the mean-square value

$$E[x^2(n)] = r_{xx}(0) = \frac{1}{2\pi} \int_{-\pi}^{\pi} P_{xx}(\omega)d\omega. \tag{8.1.19}$$

Thus $r_{xx}(0)$ represents the average power in the random signal $x(n)$. The PDS is a periodic function of the frequency ω, with the period equal to 2π. We can show (in the exercise problems) that $P_{xx}(\omega)$ of a WSS signal is a real-valued function of ω. If $x(n)$ is a real-valued signal, $P_{xx}(\omega)$ is an even function of ω. That is,

$$P_{xx}(\omega) = P_{xx}(-\omega) \tag{8.1.20}$$

or

$$P_{xx}(z) = P_{xx}(z^{-1}). \tag{8.1.21}$$

The DTFT of the crosscorrelation function $P_{xy}(\omega)$ of two WSS signals $x(n)$ and $y(n)$ is given by

$$P_{xy}(\omega) = \sum_{k=-\infty}^{\infty} r_{xy}(k)e^{-j\omega k}, \qquad (8.1.22)$$

or

$$P_{xy}(z) = \sum_{k=-\infty}^{\infty} r_{xy}(k)z^{-k}. \qquad (8.1.23)$$

This function is called the cross-power spectrum.

Example 8.3: The autocorrelation function of a WSS white random process can be defined as

$$r_{xx}(k) = \sigma_x^2 \delta(k) + m_x^2. \qquad (8.1.24)$$

The corresponding PDS is given by

$$P_{xx}(\omega) = \sigma_x^2 + 2\pi m_x^2 \delta(\omega), \quad |\omega| \leq \pi. \qquad (8.1.25)$$

An important white random signal is called white noise, which has zero mean. Thus its autocorrelation function is expressed as

$$r_{xx}(k) = \sigma_x^2 \delta(k), \qquad (8.1.26)$$

and the power spectrum is given by

$$P_{xx}(\omega) = \sigma_x^2, \quad |\omega| < \pi, \qquad (8.1.27)$$

which is of constant value for all frequencies ω.

Consider a linear and time-invariant digital filter defined by the impulse response $h(n)$, or the transfer function $H(z)$. The input of the filter is a WSS random signal $x(n)$ with the PDS $P_{xx}(\omega)$. As illustrated in Figure 8.1, the PDS of the filter output $y(n)$ can be expressed as

$$P_{yy}(\omega) = |H(\omega)|^2 P_{xx}(\omega) \qquad (8.1.28)$$

or

$$P_{yy}(z) = |H(z)|^2 P_{xx}(z), \qquad (8.1.29)$$

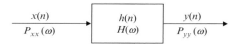

Figure 8.1 Linear filtering of random processes

where $H(\omega)$ is the frequency response of the filter. Therefore the value of the output PDS at frequency ω depends on the squared magnitude response of the filter and the input PDS at the same frequency.

Another important relationships between $x(n)$ and $y(n)$ are

$$m_y = E\left[\sum_{l=-\infty}^{\infty} h(l)x(n-l)\right] = \sum_{l=-\infty}^{\infty} h(l)E[x(n-l)] = m_x \sum_{l=-\infty}^{\infty} h(l), \qquad (8.1.30)$$

and

$$r_{yx}(k) = E[y(n+k)x(n)] = E\left[\sum_{l=-\infty}^{\infty} h(l)x(n+k-l)x(n)\right]$$

$$= \sum_{l=-\infty}^{\infty} h(l)r_{xx}(k-l) = h(k) * r_{xx}(k). \qquad (8.1.31)$$

Taking the z-transform of both sides, we obtain

$$P_{yx}(z) = H(z)P_{xx}(z). \qquad (8.1.32)$$

Similarly, the relationships between the input and the output signals are

$$r_{xy}(k) = \sum_{l=-\infty}^{\infty} h(l)r_{xx}(k+l) = h(k) * r_{xx}(-k) \qquad (8.1.33)$$

and

$$P_{yx}(z) = H^*(z)P_{xx}(z). \qquad (8.1.34)$$

If the input signal $x(n)$ is a zero-mean white noise with the autocorrelation function defined in (8.1.26), Equation (8.1.31) becomes

$$r_{yx}(k) = \sum_{l=-\infty}^{\infty} h(l)\sigma_x^2\delta(k-l) = \sigma_x^2 h(k). \qquad (8.1.35)$$

This equation shows that by computing the crosscorrelation function $r_{yx}(k)$, the impulse response $h(n)$ of a filter (or system) can be obtained. This fact can be used to estimate an unknown system such as the room impulse response used in Chapter 4.

Example 8.4: Let the system shown in Figure 8.1 be a second-order FIR filter. The input $x(n)$ is a zero-mean white noise given by Example 8.3, and the I/O equation is expressed as

$$y(n) = x(n) + 3x(n-1) + 2x(n-2).$$

Find the mean m_y and the autocorrelation function $r_{yy}(k)$ of the output $y(n)$.

(a) $m_y = E[y(n)] = E[x(n)] + 3E[x(n-1)] + 2E[x(n-2)] = 0.$

(b) $r_{yy}(k) = E[y(n+k)y(n)]$

$$= 14r_{xx}(k) + 9r_{xx}(k-1) + 9r_{xx}(k+1) + 2r_{xx}(k-2) + 2r_{xx}(k+2)$$

$$= \begin{cases} 14\sigma_x^2 & \text{if } k = 0 \\ 9\sigma_x^2 & \text{if } k = \pm 1 \\ 2\sigma_x^2 & \text{if } k = \pm 2 \\ 0 & \text{otherwise.} \end{cases}$$

8.2 Adaptive Filters

Many practical applications involve the reduction of noise and distortion for extraction of information from the received signal. The signal degradation in some physical systems is time varying, unknown, or possibly both. Adaptive filters provide a useful approach for these applications. Adaptive filters modify their characteristics to achieve certain objectives and usually accomplish the modification (adaptation) automatically. For example, consider a high-speed modem for transmitting and receiving data over telephone channels. It employs a filter called a channel equalizer to compensate for the channel distortion. Since the dial-up communication channels have different characteristics on each connection and are time varying, the channel equalizers must be adaptive.

Adaptive filters have received considerable attention from many researchers over the past 30 years. Many adaptive filter structures and adaptation algorithms have been developed for different applications. This chapter presents the most widely used adaptive filter based on the FIR filter with the LMS algorithm. Adaptive filters in this class are relatively simple to design and implement. They are well understood with regard to convergence speed, steady-state performance, and finite-precision effects.

8.2.1 Introduction to Adaptive Filtering

An adaptive filter consists of two distinct parts – a digital filter to perform the desired signal processing, and an adaptive algorithm to adjust the coefficients (or weights) of that filter. A general form of adaptive filter is illustrated in Figure 8.2, where $d(n)$ is a desired signal (or primary input signal), $y(n)$ is the output of a digital filter driven by a reference input signal $x(n)$, and an error signal $e(n)$ is the difference between $d(n)$ and $y(n)$. The function of the adaptive algorithm is to adjust the digital filter coefficients to

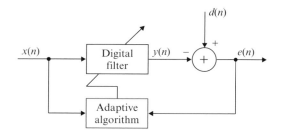

Figure 8.2   Block diagram of adaptive filter

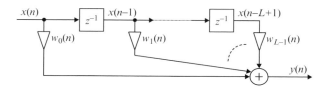

Figure 8.3   Block diagram of FIR filter for adaptive filtering

minimize the mean-square value of $e(n)$. Therefore the filter weights are updated so that the error is progressively minimized on a sample-by-sample basis.

In general, there are two types of digital filters that can be used for adaptive filtering: FIR and IIR filters. The choice of an FIR or an IIR filter is determined by practical considerations. The FIR filter is always stable and can provide a linear phase response. On the other hand, the IIR filter involves both zeros and poles. Unless they are properly controlled, the poles in the filter may move outside the unit circle and make the filter unstable. Because the filter is required to be adaptive, the stability problems are much difficult to handle. Thus the FIR adaptive filter is widely used for real-time applications. The discussions in the following sections will be restricted to the class of adaptive FIR filters.

The most widely used adaptive FIR filter is depicted in Figure 8.3. Given a set of L coefficients, $w_l(n)$, $l = 0, 1, \ldots, L - 1$, and a data sequence, $\{x(n)\ x(n - 1) \ldots x(n - L + 1)\}$, the filter output signal is computed as

$$y(n) = \sum_{l=0}^{L-1} w_l(n)x(n - l), \tag{8.2.1}$$

where the filter coefficients $w_l(n)$ are time varying and updated by the adaptive algorithms that will be discussed next.

We define the input vector at time n as

$$\mathbf{x}(n) \equiv [x(n)\ x(n - 1) \ldots x(n - L + 1)]^T \tag{8.2.2}$$

and the weight vector at time n as

$$\mathbf{w}(n) \equiv [w_0(n) \; w_1(n) \ldots w_{L-1}(n)]^T. \tag{8.2.3}$$

Then the output signal $y(n)$ in (8.2.1) can be expressed using the vector operation

$$y(n) = \mathbf{w}^T(n)\mathbf{x}(n) = \mathbf{x}^T(n)\mathbf{w}(n). \tag{8.2.4}$$

The filter output $y(n)$ is compared with the desired response $d(n)$, which results in the error signal

$$e(n) = d(n) - y(n) = d(n) - \mathbf{w}^T(n)\mathbf{x}(n). \tag{8.2.5}$$

In the following sections, we assume that $d(n)$ and $x(n)$ are stationary, and our objective is to determine the weight vector so that the performance (or cost) function is minimized.

8.2.2 Performance Function

The general block diagram of the adaptive filter shown in Figure 8.2 updates the coefficients of the digital filter to optimize some predetermined performance criterion. The most commonly used performance measurement is based on the mean-square error (MSE) defined as

$$\xi(n) \equiv E[e^2(n)]. \tag{8.2.6}$$

For an adaptive FIR filter, $\xi(n)$ will depend on the L filter weights $w_0(n)$, $w_1(n)$, $\ldots, w_{L-1}(n)$. The MSE function can be determined by substituting (8.2.5) into (8.2.6), expressed as

$$\xi(n) = E[d^2(n)] - 2\mathbf{p}^T\mathbf{w}(n) + \mathbf{w}^T(n)\mathbf{R}\mathbf{w}(n), \tag{8.2.7}$$

where $\mathbf{p}$ is the crosscorrelation vector defined as

$$\mathbf{p} \equiv E[d(n)\mathbf{x}(n)] = [r_{dx}(0) \; r_{dx}(1) \ldots r_{dx}(L-1)]^T, \tag{8.2.8}$$

and

$$r_{dx}(k) \equiv E[d(n)x(n-k)] \tag{8.2.9}$$

is the crosscorrelation function between $d(n)$ and $x(n)$. In (8.2.7), $\mathbf{R}$ is the input auto-correlation matrix defined as

$$\mathbf{R} \equiv E[\mathbf{x}(n)\mathbf{x}^T(n)] = \begin{bmatrix} r_{xx}(0) & r_{xx}(1) & \cdots & r_{xx}(L-1) \\ r_{xx}(1) & r_{xx}(0) & \cdots & r_{xx}(L-2) \\ \vdots & \cdots & \ddots & \vdots \\ r_{xx}(L-1) & r_{xx}(L-2) & \cdots & r_{xx}(0) \end{bmatrix}, \tag{8.2.10}$$

where

$$r_{xx}(k) \equiv E[x(n)x(n-k)] \tag{8.2.11}$$

is the autocorrelation function of $x(n)$.

Example 8.5: Given an optimum filter illustrated in the following figure:

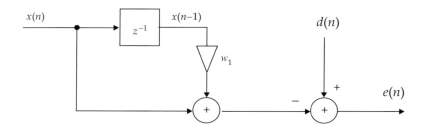

If $E[x^2(n)] = 1$, $E[x(n)x(n-1)] = 0.5$, $E[d^2(n)] = 4$, $E[d(n)x(n)] = -1$, and $E[d(n)x(n-1)] = 1$. Find ξ.

From (8.2.10), $\mathbf{R} = \begin{bmatrix} 1 & 0.5 \\ 0.5 & 1 \end{bmatrix}$, and from (8.2.8), we have $\mathbf{P} = \begin{bmatrix} -1 \\ 1 \end{bmatrix}$.

Therefore from (8.2.7), we obtain

$$\xi = E[d^2(n)] - 2\mathbf{p}^T\mathbf{w} + \mathbf{w}^T\mathbf{R}\mathbf{w}$$

$$= 4 - 2[-1 \quad 1]\begin{bmatrix} 1 \\ w_1 \end{bmatrix} + [1 \quad w_1]\begin{bmatrix} 1 & 0.5 \\ 0.5 & 1 \end{bmatrix}\begin{bmatrix} 1 \\ w_1 \end{bmatrix} = w_1^2 - w_1 + 7.$$

The optimum filter $\mathbf{w}^o$ minimizes the MSE cost function $\xi(n)$. Vector differentiation of (8.2.7) gives $\mathbf{w}^o$ as the solution to

$$\mathbf{R}\mathbf{w}^o = \mathbf{p}. \tag{8.2.12}$$

This system equation defines the optimum filter coefficients in terms of two correlation functions – the autocorrelation function of the filter input and the crosscorrelation function between the filter input and the desired response. Equation (8.2.12) provides a solution to the adaptive filtering problem in principle. However, in many applications, the signal may be non-stationary. This linear algebraic solution, $\mathbf{w}^o = \mathbf{R}^{-1}\mathbf{p}$, requires continuous estimation of $\mathbf{R}$ and $\mathbf{p}$, a considerable amount of computations. In addition, when the dimension of the autocorrelation matrix is large, the calculation of $\mathbf{R}^{-1}$ may present a significant computational burden. Therefore a more useful algorithm is obtained by developing a recursive method for computing $\mathbf{w}^o$, which will be discussed in the next section.

To obtain the minimum MSE, we substitute the optimum weight vector $\mathbf{w}^o = \mathbf{R}^{-1}\mathbf{p}$ for $\mathbf{w}(n)$ in (8.2.7), resulting in

$$\xi_{min} = E[d^2(n)] - \mathbf{p}^T\mathbf{w}^o. \tag{8.2.13}$$

Since **R** is positive semidefinite, the quadratic form on the right-hand side of (8.2.7) indicates that any departure of the weight vector $\mathbf{w}(n)$ from the optimum $\mathbf{w}^{\circ}$ would increase the error above its minimum value. In other words, the error surface is concave and possesses a unique minimum. This feature is very useful when we utilize search techniques in seeking the optimum weight vector. In such cases, our objective is to develop an algorithm that can automatically search the error surface to find the optimum weights that minimize $\xi(n)$ using the input signal $x(n)$ and the error signal $e(n)$.

Example 8.6: Consider a second-order FIR filter with two coefficients w_0 and w_1, the desired signal $d(n) = \sqrt{2}\sin(n\omega_0)$, $n \geq 0$, and the reference signal $x(n) = d(n-1)$. Find $\mathbf{w}^{\circ}$ and $\xi_{\min}$.

Similar to Example 8.2, we can obtain $r_{xx}(0) = E[x^2(n)] = E[d^2(n)] = 1$, $r_{xx}(1) = \cos(\omega_0)$, $r_{xx}(2) = \cos(2\omega_0)$, $r_{dx}(0) = r_{xx}(1)$, and $r_{dx}(1) = r_{xx}(2)$. From (8.2.12), we have

$$\mathbf{w}^{\circ} = \mathbf{R}^{-1}\mathbf{p} = \begin{bmatrix} 1 & \cos(\omega_0) \\ \cos(\omega_0) & 1 \end{bmatrix}^{-1} \begin{bmatrix} \cos(\omega_0) \\ \cos(2\omega_0) \end{bmatrix} = \begin{bmatrix} 2\cos(\omega_0) \\ -1 \end{bmatrix}.$$

From (8.2.13), we obtain

$$\xi_{\min} = 1 - [\cos(\omega_0) \quad \cos(2\omega_0)] \begin{bmatrix} 2\cos(\omega_0) \\ -1 \end{bmatrix} = 0.$$

Equation (8.2.7) is the general expression for the performance function of an adaptive FIR filter with given weights. That is, the MSE is a function of the filter coefficient vector $\mathbf{w}(n)$. It is important to note that the MSE is a quadratic function because the weights appear only to the first and second degrees in (8.2.7). For each coefficient vector $\mathbf{w}(n)$, there is a corresponding (scalar) value of MSE. Therefore the MSE values associated with $\mathbf{w}(n)$ form an $(L+1)$-dimensional space, which is commonly called the MSE surface, or the performance surface.

For $L = 2$, this corresponds to an error surface in a three-dimensional space. The height of $\xi(n)$ corresponds to the power of the error signal $e(n)$ that results from filtering the signal $x(n)$ with the coefficients $\mathbf{w}(n)$. If the filter coefficients change, the power in the error signal will also change. This is indicated by the changing height on the surface above w_0-w_1 the plane as the component values of $\mathbf{w}(n)$ are varied. Since the error surface is quadratic, a unique filter setting $\mathbf{w}(n) = \mathbf{w}^{\circ}$ will produce the minimum MSE, $\xi_{\min}$. In this two-weight case, the error surface is an elliptic paraboloid. If we cut the paraboloid with planes parallel to the w_0-w_1 plane, we obtain concentric ellipses of constant mean-square error. These ellipses are called the error contours of the error surface.

Example 8.7: Consider a second-order FIR filter with two coefficients w_0 and w_1. The reference signal $x(n)$ is a zero-mean white noise with unit variance. The desired signal is given as

$$d(n) = b_0 x(n) + b_1 x(n-1).$$

Plot the error surface and error contours.

From Equation (8.2.10), we obtain $\mathbf{R} = \begin{bmatrix} r_{xx}(0) & r_{xx}(1) \\ r_{xx}(1) & r_{xx}(0) \end{bmatrix} = \begin{bmatrix} 1 & 0 \\ 0 & 1 \end{bmatrix}$. From (8.2.8), we have $\mathbf{p} = \begin{bmatrix} r_{dx}(0) \\ r_{dx}(1) \end{bmatrix} = \begin{bmatrix} b_0 \\ b_1 \end{bmatrix}$. From (8.2.7), we get

$$\xi = E[d^2(n)] - 2\mathbf{p}^T\mathbf{w} + \mathbf{w}^T\mathbf{R}\mathbf{w} = (b_0^2 + b_1^2) - 2b_0 w_0 - 2b_1 w_1 + w_0^2 + w_1^2$$

Let $b_0 = 0.3$ and $b_1 = 0.5$, we have

$$\xi = 0.34 - 0.6w_0 - w_1 + w_0^2 + w_1^2.$$

The MATLAB script (exam8_7a.m in the software package) is used to plot the error surface shown in Figure 8.4(a) and the script exam8_7b.m is used to plot the error contours shown in Figure 8.4(b).

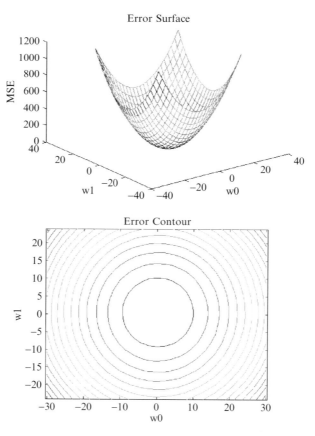

Figure 8.4 Performance surface and error contours, $L = 2$

One of the most important properties of the MSE surface is that it has only one global minimum point. At that minimum point, the tangents to the surface must be 0. Minimizing the MSE is the objective of many current adaptive methods such as the LMS algorithm.

8.2.3 Method of Steepest Descent

As shown in Figure 8.4, the MSE of (8.2.7) is a quadratic function of the weights that can be pictured as a positive-concave hyperparabolic surface. Adjusting the weights to minimize the error involves descending along this surface until reaching the 'bottom of the bowl.' Various gradient-based algorithms are available. These algorithms are based on making local estimates of the gradient and moving downward toward the bottom of the bowl. The selection of an algorithm is usually decided by the speed of convergence, steady-state performance, and the computational complexity.

The steepest-descent method reaches the minimum by following the direction in which the performance surface has the greatest rate of decrease. Specifically, an algorithm whose path follows the negative gradient of the performance surface. The steepest-descent method is an iterative (recursive) technique that starts from some initial (arbitrary) weight vector. It improves with the increased number of iterations. Geometrically, it is easy to see that with successive corrections of the weight vector in the direction of the steepest descent on the concave performance surface, we should arrive at its minimum, ξ_{min}, at which point the weight vector components take on their optimum values. Let $\xi(0)$ represent the value of the MSE at time $n = 0$ with an arbitrary choice of the weight vector $\mathbf{w}(0)$. The steepest-descent technique enables us to descend to the bottom of the bowl, $\mathbf{w}^o$, in a systematic way. The idea is to move on the error surface in the direction of the tangent at that point. The weights of the filter are updated at each iteration in the direction of the negative gradient of the error surface.

The mathematical development of the method of steepest descent is easily seen from the viewpoint of a geometric approach using the MSE surface. Each selection of a filter weight vector $\mathbf{w}(n)$ corresponds to only one point on the MSE surface, $[\mathbf{w}(n), \xi(n)]$. Suppose that an initial filter setting $\mathbf{w}(0)$ on the MSE surface, $[\mathbf{w}(0), \xi(0)]$ is arbitrarily chosen. A specific orientation to the surface is then described using the directional derivatives of the surface at that point. These directional derivatives quantify the rate of change of the MSE surface with respect to the $\mathbf{w}(n)$ coordinate axes. The gradient of the error surface $\nabla\xi(n)$ is defined as the vector of these directional derivatives.

The concept of steepest descent can be implemented in the following algorithm:

$$\mathbf{w}(n+1) = \mathbf{w}(n) - \frac{\mu}{2}\nabla\xi(n) \tag{8.2.14}$$

where μ is a convergence factor (or step size) that controls stability and the rate of descent to the bottom of the bowl. The larger the value of μ, the faster the speed of descent. The vector $\nabla\xi(n)$ denotes the gradient of the error function with respect to $\mathbf{w}(n)$, and the negative sign increments the adaptive weight vector in the negative gradient direction. The successive corrections to the weight vector in the direction of the steepest descent of the performance surface should eventually lead to the minimum mean-square error ξ_{min}, at which point the weight vector reaches its optimum value $\mathbf{w}^o$.

When $\mathbf{w}(n)$ has converged to $\mathbf{w}^o$, that is, when it reaches the minimum point of the performance surface, the gradient $\nabla\xi(n) = \mathbf{0}$. At this time, the adaptation in (8.2.14) is stopped and the weight vector stays at its optimum solution. The convergence can be viewed as a ball placed on the 'bowl-shaped' MSE surface at the point $[\mathbf{w}(0), \xi(0)]$. If the ball was released, it would roll toward the minimum of the surface, and would initially roll in a direction opposite to the direction of the gradient, which can be interpreted as rolling towards the bottom of the bowl.

8.2.4   The LMS Algorithm

From (8.2.14), we see that the increment from $\mathbf{w}(n)$ to $\mathbf{w}(n+1)$ is in the negative gradient direction, so the weight tracking will closely follow the steepest descent path on the performance surface. However, in many practical applications the statistics of $d(n)$ and $x(n)$ are unknown. Therefore the method of steepest descent cannot be used directly, since it assumes exact knowledge of the gradient vector at each iteration. Widrow [13] used the instantaneous squared error, $e^2(n)$, to estimate the MSE. That is,

$$\hat{\xi}(n) = e^2(n). \tag{8.2.15}$$

Therefore the gradient estimate used by the LMS algorithm is

$$\nabla\hat{\xi}(n) = 2[\nabla e(n)]e(n). \tag{8.2.16}$$

Since $e(n) = d(n) - \mathbf{w}^T(n)\mathbf{x}(x)$, $\nabla e(n) = -\mathbf{x}(n)$, the gradient estimate becomes

$$\nabla\hat{\xi}(n) = -2\mathbf{x}(n)e(n). \tag{8.2.17}$$

Substituting this gradient estimate into the steepest-descent algorithm of (8.2.14), we have

$$\mathbf{w}(n+1) = \mathbf{w}(n) + \mu\mathbf{x}(n)e(n). \tag{8.2.18}$$

This is the well-known LMS algorithm, or stochastic gradient algorithm. This algorithm is simple and does not require squaring, averaging, or differentiating. The LMS algorithm provides an alternative method for determining the optimum filter coefficients without explicitly computing the matrix inversion suggested in (8.2.12).

Widrow's LMS algorithm is illustrated in Figure 8.5 and is summarized as follows:

1.   Determine L, μ, and $\mathbf{w}(0)$, where L is the order of the filter, μ is the step size, and $\mathbf{w}(0)$ is the initial weight vector at time $n = 0$.

2.   Compute the adaptive filter output

$$y(n) = \sum_{l=0}^{L-1} w_l(n)x(n-l). \tag{8.2.19}$$

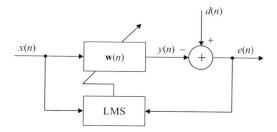

Figure 8.5 Block diagram of an adaptive filter with the LMS algorithm

3. Compute the error signal

$$e(n) = d(n) - y(n). \tag{8.2.20}$$

4. Update the adaptive weight vector from $\mathbf{w}(n)$ to $\mathbf{w}(n + 1)$ by using the LMS algorithm

$$w_l(n + 1) = w_l(n) + \mu x(n - l)e(n), \quad l = 0, 1, \ldots, L - 1. \tag{8.2.21}$$

8.3 Performance Analysis

A detailed discussion of the performance of the LMS algorithm is available in many textbooks. In this section, we present some important properties of the LMS algorithm such as stability, convergence rate, and the excess mean-square error due to gradient estimation error.

8.3.1 Stability Constraint

As shown in Figure 8.5, the LMS algorithm involves the presence of feedback. Thus the algorithm is subject to the possibility of becoming unstable. From (8.2.18), we observe that the parameter μ controls the size of the incremental correction applied to the weight vector as we adapt from one iteration to the next. The mean weight convergence of the LMS algorithm from initial condition $\mathbf{w}(0)$ to the optimum filter $\mathbf{w}^o$ must satisfy

$$0 < \mu < \frac{2}{\lambda_{\max}}, \tag{8.3.1}$$

where $\lambda_{\max}$ is the largest eigenvalue of the autocorrelation matrix $\mathbf{R}$ defined in (8.2.10). Applying the stability constraint on μ given in (8.3.1) is difficult because of the computation of $\lambda_{\max}$ when L is large.

In practical applications, it is desirable to estimate $\lambda_{\max}$ using a simple method. From (8.2.10), we have

$$\text{tr}[\mathbf{R}] = Lr_{xx}(0) = \sum_{l=0}^{L-1} \lambda_l, \tag{8.3.2}$$

where $\text{tr}[\mathbf{R}]$ denotes the trace of matrix $\mathbf{R}$. It follows that

$$\lambda_{\max} \leq \sum_{l=0}^{L-1} \lambda_l = Lr_{xx}(0) = LP_x, \tag{8.3.3}$$

where

$$P_x \equiv r_{xx}(0) = E\left[x^2(n)\right] \tag{8.3.4}$$

denotes the power of $x(n)$. Therefore setting

$$0 < \mu < \frac{2}{LP_x} \tag{8.3.5}$$

assures that (8.3.1) is satisfied.

Equation (8.3.5) provides some important information on how to select μ, and they are summarized as follows:

1. Since the upper bound on μ is inversely proportional to L, a small μ is used for large-order filters.

2. Since μ is made inversely proportional to the input signal power, weaker signals use a larger μ and stronger signals use a smaller μ. One useful approach is to normalize μ with respect to the input signal power P_x. The resulting algorithm is called the normalized LMS algorithm, which will be discussed in Section 8.4.

8.3.2 Convergence Speed

In the previous section, we saw that $\mathbf{w}(n)$ converges to $\mathbf{w}^\circ$ if the selection of μ satisfies (8.3.1). Convergence of the weight vector $\mathbf{w}(n)$ from $\mathbf{w}(0)$ to $\mathbf{w}^\circ$ corresponds to the convergence of the MSE from $\xi(0)$ to $\xi_{\min}$. Therefore convergence of the MSE toward its minimum value is a commonly used performance measurement in adaptive systems because of its simplicity. During adaptation, the squared error $e^2(n)$ is non-stationary as the weight vector $\mathbf{w}(n)$ adapts toward $\mathbf{w}^\circ$. The corresponding MSE can thus be defined only based on ensemble averages. A plot of the MSE versus time n is referred to as the learning curve for a given adaptive algorithm. Since the MSE is the performance criterion of LMS algorithms, the learning curve is a natural way to describe the transient behavior.

Each adaptive mode has its own time constant, which is determined by the overall adaptation constant μ and the eigenvalue λ_l associated with that mode. Overall convergence is clearly limited by the slowest mode. Thus the overall MSE time constant can be approximated as

$$\tau_{mse} \cong \frac{1}{\mu \lambda_{min}}, \tag{8.3.6}$$

where λ_{min} is the minimum eigenvalue of the **R** matrix. Because τ_{mse} is inversely proportional to μ, we have a large τ_{mse} when μ is small (i.e., the speed of convergence is slow). If we use a large value of μ, the time constant is small, which implies faster convergence.

The maximum time constant $\tau_{mse} = 1/\mu \lambda_{min}$ is a conservative estimate of filter performance, since only large eigenvalues will exert significant influence on the convergence time. Since some of the projections may be negligibly small, the adaptive filter error convergence may be controlled by fewer modes than the number of adaptive filter weights. Consequently, the MSE often converges more rapidly than the upper bound of (8.3.6) would suggest.

Because the upper bound of τ_{mse} is inversely proportional to λ_{min}, a small λ_{min} can result in a large time constant (i.e., a slow convergence rate). Unfortunately, if λ_{max} is also very large, the selection of μ will be limited by (8.3.1) such that only a small μ can satisfy the stability constraint. Therefore if λ_{max} is very large and λ_{min} is very small, from (8.3.6), the time constant can be very large, resulting in very slow convergence. As previously noted, the fastest convergence of the dominant mode occurs for $\mu = 1/\lambda_{max}$. Substituting this smallest step size into (8.3.6) results in

$$\tau_{mse} \leq \frac{\lambda_{max}}{\lambda_{min}}. \tag{8.3.7}$$

For stationary input and sufficiently small μ, the speed of convergence of the algorithm is dependent on the eigenvalue spread (the ratio of the maximum to minimum eigenvalues) of the matrix **R**.

As mentioned in the previous section, the eigenvalues λ_{max} and λ_{min} are very difficult to compute. However, there is an efficient way to estimate the eigenvalue spread from the spectral dynamic range. That is,

$$\frac{\lambda_{max}}{\lambda_{min}} \leq \frac{\max |X(\omega)|^2}{\min |X(\omega)|^2}, \tag{8.3.8}$$

where $X(\omega)$ is DTFT of $x(n)$ and the maximum and minimum are calculated over the frequency range $0 \leq \omega \leq \pi$. From (8.3.7) and (8.3.8), input signals with a flat (white) spectrum have the fastest convergence speed.

8.3.3 Excess Mean-Square Error

The steepest-descent algorithm in (8.2.14) requires knowledge of the gradient $\nabla \xi(n)$, which must be estimated at each iteration. The estimated gradient $\nabla \hat{\xi}(n)$ produces the gradient estimation noise. After the algorithm converges, i.e., $\mathbf{w}(n)$ is close to $\mathbf{w}^o$, the true gradient $\nabla \xi(n) \approx \mathbf{0}$. However, the gradient estimator $\nabla \hat{\xi}(n) \neq \mathbf{0}$. As indicated by the update of Equation (8.2.14), perturbing the gradient will cause the weight vector $\mathbf{w}(n+1)$ to move away from the optimum solution $\mathbf{w}^o$. Thus the gradient estimation noise prevents $\mathbf{w}(n+1)$ from staying at $\mathbf{w}^o$ in steady state. The result is that $\mathbf{w}(n)$ varies randomly about $\mathbf{w}^o$. Because $\mathbf{w}^o$ corresponds to the minimum MSE, when $\mathbf{w}(n)$ moves away from $\mathbf{w}^o$, it causes $\xi(n)$ to be larger than its minimum value, ξ_{min}, thus producing excess noise at the filter output.

The excess MSE, which is caused by random noise in the weight vector after convergence, is defined as the average increase of the MSE. For the LMS algorithm, it can be approximated as

$$\xi_{excess} \approx \frac{\mu}{2} L P_x \xi_{min}. \tag{8.3.9}$$

This approximation shows that the excess MSE is directly proportional to μ. The larger the value of μ, the worse the steady-state performance after convergence. However, Equation (8.3.6) shows that a larger μ results in faster convergence. There is a design trade-off between the excess MSE and the speed of convergence.

The optimal step size μ is difficult to determine. Improper selection of μ might make the convergence speed unnecessarily slow or introduce excess MSE. If the signal is non-stationary and real-time tracking capability is crucial for a given application, then use a larger μ. If the signal is stationary and convergence speed is not important, use a smaller μ to achieve better performance in a steady state. In some practical applications, we can use a larger μ at the beginning of the operation for faster convergence, then use a smaller μ to achieve better steady-state performance.

The excess MSE, ξ_{excess}, in (8.3.9) is also proportional to the filter order L, which means that a larger L results in larger algorithm noise. From (8.3.5), a larger L implies a smaller μ, resulting in slower convergence. On the other hand, a large L also implies better filter characteristics such as sharp cutoff. There exists an optimum order L for any given application. The selection of L and μ also will affect the finite-precision error, which will be discussed in Section 8.6.

In a stationary environment, the signal statistics are unknown but fixed. The LMS algorithm gradually learns the required input statistics. After convergence to a steady state, the filter weights jitter around the desired fixed values. The algorithm performance is determined by both the speed of convergence and the weight fluctuations in steady state. In the non-stationary case, the algorithm must continuously track the time-varying statistics of the input. Performance is more difficult to assess.

8.4 Modified LMS Algorithms

The LMS algorithm described in the previous section is the most widely used adaptive algorithm for practical applications. In this section, we present two modified algorithms that are the direct variants of the basic LMS algorithm.

8.4.1 Normalized LMS Algorithm

The stability, convergence speed, and fluctuation of the LMS algorithm are governed by the step size μ and the reference signal power. As shown in (8.3.5), the maximum stable step-size μ is inversely proportional to the filter order L and the power of the reference signal $x(n)$. One important technique to optimize the speed of convergence while maintaining the desired steady-state performance, independent of the reference signal power, is known as the normalized LMS algorithm (NLMS). The NLMS algorithm is expressed as

$$\mathbf{w}(n+1) = \mathbf{w}(n) + \mu(n)\mathbf{x}(n)e(n), \tag{8.4.1}$$

where $\mu(n)$ is an adaptive step size that is computed as

$$\mu(n) = \frac{\alpha}{L\hat{P}_x(n),} \tag{8.4.2}$$

where $\hat{P}_x(n)$ is an estimate of the power of $x(n)$ at time n, and α is a normalized step size that satisfies the criterion

$$0 < \alpha < 2. \tag{8.4.3}$$

The commonly used method to estimate $\hat{P}_x(n)$ sample-by-sample is introduced in Section 3.2.1. Some useful implementation considerations are given as follows:

1. Choose $\hat{P}_x(0)$ as the best a priori estimate of the reference signal power.

2. Since it is not desirable that the power estimate $\hat{P}_x(n)$ be zero or very small, a software constraint is required to ensure that $\mu(n)$ is bounded even if $\hat{P}_x(n)$ is very small when the signal is absent for a long time. This can be achieved by modifying (8.4.2) as

$$\mu(n) = \frac{\alpha}{L\hat{P}_x(n) + c,} \tag{8.4.4}$$

where c is a small constant.

8.4.2 Leaky LMS Algorithm

Insufficient spectral excitation of the LMS algorithm may result in divergence of the adaptive weights. In that case, the solution is not unique and finite-precision effects can cause the unconstrained weights to grow without bound, resulting in overflow during the weight update process. This long-term instability is undesirable for real-time applications.

Divergence can often be avoided by means of a 'leaking' mechanism used during the weight update calculation. This is called the leaky LMS algorithm and is expressed as

$$\mathbf{w}(n+1) = v\mathbf{w}(n) + \mu\mathbf{x}(n)e(n), \tag{8.4.5}$$

where v is the leakage factor with $0 < v \leq 1$. It can be shown that leakage is the deterministic equivalent of adding low-level white noise. Therefore this approach results in some degradation in adaptive filter performance. The value of the leakage factor is determined by the designer on an experimental basis as a compromise between robustness and loss of performance of the adaptive filter. The leakage factor introduces a bias on the long-term coefficient estimation. The excess error power due to the leakage is proportional to $[(1 - v)/\mu]^2$. Therefore $(1 - v)$ should be kept smaller than μ in order to maintain an acceptable level of performance. For fixed-point hardware realization, multiplication of each coefficient by v, as shown in (8.4.5), can lead to the introduction of roundoff noise, which adds to the excess MSE. Therefore the leakage effects must be incorporated into the design procedure for determining the required coefficient and internal data wordlength. The leaky LMS algorithm not only prevents unconstrained weight overflow, but also limits the output power in order to avoid nonlinear distortion.

8.5 Applications

The desirable features of an adaptive filter are the ability to operate in an unknown environment and to track time variations of the input signals, making it a powerful algorithm for DSP applications. The essential difference between various applications of adaptive filtering is where the signals $x(n)$, $d(n)$, $y(n)$, and $e(n)$ are connected. There are four basic classes of adaptive filtering applications: identification, inverse modeling, prediction, and interference canceling.

8.5.1 Adaptive System Identification

System identification is an experimental approach to the modeling of a process or a plant. The basic idea is to measure the signals produced by the system and to use them to construct a model. The paradigm of system identification is illustrated in Figure 8.6, where $P(z)$ is an unknown system to be identified and $W(z)$ is a digital filter used to model $P(z)$. By exciting both the unknown system $P(z)$ and the digital model $W(z)$ with the same excitation signal $x(n)$ and measuring the output signals $y(n)$ and $d(n)$, we can determine the characteristics of $P(z)$ by adjusting the digital model $W(z)$ to minimize the difference between these two outputs. The digital model $W(z)$ can be an FIR filter or an IIR filter.

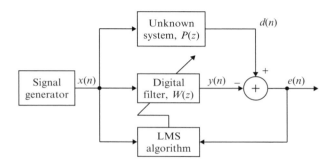

Figure 8.6 Block diagram of adaptive system identification using the LMS algorithm

Adaptive system identification is a technique that uses an adaptive filter for the model $W(z)$. This section presents the application of adaptive estimation techniques for direct system modeling. This technique has been widely applied in echo cancellation, which will be introduced in Sections 9.4 and 9.5. A further application for system modeling is to estimate various transfer functions in active noise control systems [8].

Adaptive system identification is a very important procedure that is used frequently in the fields of control systems, communications, and signal processing. The modeling of a single-input/single-output dynamic system (or plant) is shown in Figure 8.6, where $x(n)$, which is usually white noise, is applied simultaneously to the adaptive filter and the unknown system. The output of the unknown system then becomes the desired signal, $d(n)$, for the adaptive filter. If the input signal $x(n)$ provides sufficient spectral excitation, the adaptive filter output $y(n)$ will approximate $d(n)$ in an optimum sense after convergence.

Identification could mean that a set of data is collected from the system, and that a separate procedure is used to construct a model. Such a procedure is usually called off-line (or batch) identification. In many practical applications, however, the model is sometimes needed on-line during the operation of the system. That is, it is necessary to identify the model at the same time that the data set is collected. The model is updated at each time instant that a new data set becomes available. The updating is performed with a recursive adaptive algorithm such as the LMS algorithm.

As shown in Figure 8.6, it is desired to learn the structure of the unknown system from knowledge of its input $x(n)$ and output $d(n)$. If the unknown time-invariant system $P(z)$ can be modeled using an FIR filter of order L, the estimation error is given as

$$e(n) = d(n) - y(n) = \sum_{l=0}^{L-1} [p(l) - w_l(n)]x(n-l), \tag{8.5.1}$$

where $p(l)$ is the impulse response of the unknown plant.

By choosing each $w_l(n)$ close to each $p(l)$, the error will be made small. For white-noise input, the converse also holds: minimizing $e(n)$ will force the $w_l(n)$ to approach $p(l)$, thus identifying the system

$$w_l(n) \approx p(l), \quad l = 0, 1, \ldots, L - 1. \tag{8.5.2}$$

The basic concept is that the adaptive filter adjusts itself, intending to cause its output to match that of the unknown system. When the difference between the physical system response $d(n)$ and adaptive model response $y(n)$ has been minimized, the adaptive model approximates $P(z)$. In actual applications, there will be additive noise present at the adaptive filter input and so the filter structure will not exactly match that of the unknown system. When the plant is time varying, the adaptive algorithm has the task of keeping the modeling error small by continually tracking time variations of the plant dynamics.

8.5.2 Adaptive Linear Prediction

Linear prediction is a classic signal processing technique that provides an estimate of the value of an input process at a future time where no measured data is yet available. The techniques have been successfully applied to a wide range of applications such as speech coding and separating signals from noise. As illustrated in Figure 8.7, the time-domain predictor consists of a linear prediction filter in which the coefficients $w_l(n)$ are updated with the LMS algorithm. The predictor output $y(n)$ is expressed as

$$y(n) = \sum_{l=0}^{L-1} w_l(n)x(n - \Delta - l), \qquad (8.5.3)$$

where the delay Δ is the number of samples involved in the prediction distance of the filter. The coefficients are updated as

$$\mathbf{w}(n + 1) = \mathbf{w}(n) + \mu\mathbf{x}(n - \Delta)e(n), \qquad (8.5.4)$$

where $\mathbf{x}(n - \Delta) = [x(n - \Delta) \ x(n - \Delta - 1) \ \cdots \ x(n - \Delta - L + 1)]^T$ is the delayed reference signal vector, and $e(n) = x(n) - y(n)$ is the prediction error. Proper selection of the prediction delay Δ allows improved frequency estimation performance for multiple sinusoids in white noise.

Now consider the adaptive predictor for enhancing an input of M sinusoids embedded in white noise, which is of the form

$$x(n) = s(n) + v(n) = \sum_{m=0}^{M-1} A_m \sin(\omega_m n + \phi_m) + v(n), \qquad (8.5.5)$$

where $v(n)$ is white noise with uniform noise power σ_v^2. In this application, the structure shown in Figure 8.7 is called the adaptive line enhancer, which provides an efficient means for the adaptive tracking of the sinusoidal components of a received signal $x(n)$ and separates these narrowband signals $s(n)$ from broadband noise $v(n)$. This technique has been shown effective in practical applications when there is insufficient a priori knowledge of the signal and noise parameters.

As shown in Figure 8.7, we want the highly correlated components of $x(n)$ to appear in $y(n)$. This is accomplished by adjusting the weights to minimize the expected mean-square value of the error signal $e(n)$. This causes an adaptive filter $W(z)$ to form

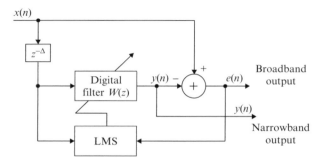

Figure 8.7 Block diagram of an adaptive predictor

narrowband bandpass filters centered at the frequency of the sinusoidal components. The noise component of the input is rejected, while the phase difference (caused by Δ) of the narrowband signals is readjusted so that they can cancel correlated components in $d(n)$ to minimize the error signal $e(n)$. In this case, the output $y(n)$ is the enhanced signal, which contains multiple sinusoidals as expressed in (8.5.5).

In many digital communications and signal detection applications, the desired broadband (spread-spectrum) signal $v(n)$ is corrupted by an additive narrowband interference $s(n)$. From a filtering viewpoint, the objective of an adaptive filter is to form a notch in the frequency band occupied by the interference, thus suppressing the narrowband noise. The narrowband characteristics of the interference allow $W(z)$ to estimate $s(n)$ from past samples of $x(n)$ and to subtract the estimate from $x(n)$. The error signal $e(n)$ in Figure 8.7 consists of desired broadband signals. In this application, the desired output from the overall interference suppression filter is $e(n)$.

8.5.3 Adaptive Noise Cancellation

The wide spread use of cellular phones has significantly increased the use of communication systems in high noise environments. Intense background noise, however, often corrupts speech and degrading the performance of many communication systems. Existing signal processing techniques such as speech coding, automatic speech recognition, speaker identification, channel transmission, and echo cancellation are developed under noise-free assumptions. These techniques could be employed in noisy environments if a front-end noise suppression algorithm sufficiently reduces additive noise. Noise reduction is becoming increasingly important with the development and application of hands-free and voice-activated cellular phones.

Single-channel noise reduction methods involve Wiener filtering, Kalman filtering, and spectral subtraction. In the dual-channel systems, a second sensor provides a reference noise to better characterize changing noise statistics, which is necessary for dealing with non-stationary noise. The most widely used dual-channel adaptive noise canceler (ANC) employs an adaptive filter with the LMS algorithm to cancel the noise component in the primary signal picked up by the primary sensor.

As illustrated in Figure 8.8, the basic concept of adaptive noise cancellation is to process signals from two sensors and to reduce the level of the undesired noise with adaptive filtering techniques. The primary sensor is placed close to the signal source in order to pick up the desired signal. However, the primary sensor output also contains noise from the noise source. The reference sensor is placed close to the noise source to sense only the noise. This structure takes advantage of the correlation between the noise signals picked up by the primary sensor and those picked up by the reference sensor.

A block diagram of the adaptive noise cancellation system is illustrated in Figure 8.9, where $P(z)$ represents the transfer function between the noise source and the primary sensor. The canceler has two inputs: the primary input $d(n)$ and the reference input $x(n)$. The primary input $d(n)$ consists of signal $s(n)$ plus noise $x'(n)$, i.e., $d(n) = s(n) + x'(n)$, which is highly correlated with $x(n)$ since they are derived from the same noise source. The reference input simply consists of noise $x(n)$. The objective of the adaptive filter is to use the reference input $x(n)$ to estimate the noise $x'(n)$. The filter output $y(n)$, which is an

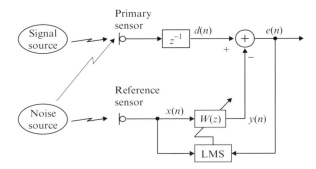

Figure 8.8 Basic concept of adaptive noise canceling

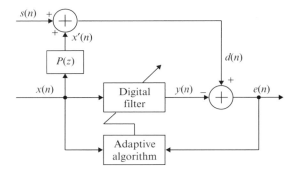

Figure 8.9 Block diagram of adaptive noise canceler

estimate of noise $x'(n)$, is then subtracted from the primary channel signal $d(n)$, producing $e(n)$ as the desired signal plus reduced noise.

To minimize the residual error $e(n)$, the adaptive filter $W(z)$ will generate an output
$y(n)$ that is an approximation of $x'(n)$. Therefore the adaptive filter $W(z)$ will converge
to the unknown plant $P(z)$. This is the adaptive system identification scheme discussed
in Section 8.5.1. To apply the ANC effectively, the reference noise picked up by the
reference sensor must be highly correlated with the noise components in the primary
signal. This condition requires a close spacing between the primary and reference
sensors. Unfortunately, it is also critical to avoid the signal components from the signal
source being picked up by the reference sensor. This 'crosstalk' effect will degrade the
performance of ANC because the presence of the signal components in reference signal
will cause the ANC to cancel the desired signal along with the undesired noise. The
performance degradation of ANC with crosstalk includes less noise reduction, slower
convergence, and reverberant distortion in the desired signal.

Crosstalk problems may be eliminated by placing the primary sensor far away from
the reference sensor. Unfortunately, this arrangement requires a large-order filter in
order to obtain adequate noise reduction. For example, a separation of a few meters
between the two sensors requires a filter with 1500 taps to achieve 20 dB noise reduction.
The long filter increases excess mean-square error and decreases the tracking ability of

ANC because the step size must be reduced to ensure stability. Furthermore, it is not always feasible to place the reference sensor far away from the signal source. The second method for reducing crosstalk is to place an acoustic barrier (an oxygen mask in an aircraft cockpit, for example) between the primary and reference sensors. However, many applications do not allow an acoustic barrier between sensors, and a barrier may reduce the correlation of the noise component in the primary and reference signals. The third technique involves allowing the adaptive algorithm to update filter coefficients during silent intervals in the speech. Unfortunately, this method depends on a reliable speech detector that is very application dependent. This technique also fails to track the environment changes during the speech periods.

8.5.4 Adaptive Notch Filters

In certain situations, the primary input is a broadband signal with an undesired narrowband (sinusoidal) interference. The conventional method of eliminating such sinusoidal interference is by using a notch filter tuned to the frequency of the interference. To design the filter, we need to estimate the precise frequency of the interference. A very narrow notch is usually desired in order to filter out the interference without seriously distorting the signal of interest. The advantages of the adaptive notch filter are that it offers an infinite null, and the capability to adaptively track the frequency of the interference. The adaptive notch filter is especially useful when the interfering sinusoid drifts slowly in frequency. In this section, we will present two adaptive notch filters.

The adaptive structure shown in Figure 8.7 can be applied to enhance the broadband signal, which is corrupted by multiple narrowband components. For example, the input signal expressed in (8.5.5) consists of a broadband signal $v(n)$, which is music. In this application, $e(n)$ is the desired output that consists of enhanced broadband music signals since the narrowband components are readjusted by $W(z)$ so that they can cancel correlated components in $d(n)$. The adaptive system between the input $x(n)$ and the output $e(n)$ is an adaptive notch filter, which forms several notch filters centered at the frequency of the sinusoidal components.

A sinusoid can be used as a reference signal for canceling each component of narrowband noise. When a sinewave is employed as the reference input, the LMS algorithm becomes an adaptive notch filter, which removes the primary spectral components within a narrowband centered about the reference frequency. Furthermore, multiple harmonic disturbances can be handled if the reference signal is composed of a number of sinusoids.

A single-frequency adaptive notch filter with two adaptive weights is illustrated in Figure 8.10. The reference input is a cosine signal

$$x(n) = x_0(n) = A \cos(\omega_0 n). \tag{8.5.6}$$

A 90° phase shifter is used to produce the quadrature reference signal

$$x_1(n) = A \sin(\omega_0 n). \tag{8.5.7}$$

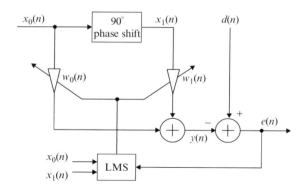

Figure 8.10   Single-frequency adaptive notch filter

Digital Hilbert transform filters can be employed for this purpose. Instead of using cosine generator and a phase shifter, the recursive quadrature oscillator given in Figure 6.23 can be used to generate both sine and cosine signals simultaneously. For a reference sinusoidal signal, two filter coefficients are needed.

The LMS algorithm employed in Figure 8.10 is summarized as follows:

$$y(n) = w_0(n)x_0(n) + w_1(n)x_1(n), \tag{8.5.8}$$

$$e(n) = d(n) - y(n), \tag{8.5.9}$$

$$w_l(n+1) = w_l(n) + \mu x_l(n)e(n), \quad l = 0, 1. \tag{8.5.10}$$

Note that the two-weight adaptive filter $W(z)$ shown in Figure 8.10 can be replaced with a general L-weight adaptive FIR filter for a multiple sinusoid reference input $x(n)$. The reference input supplies a correlated version of the sinusoidal interference that is used to estimate the composite sinusoidal interfering signal contained in the primary input $d(n)$.

The single-frequency adaptive notch filter has the property of a tunable notch filter. The center frequency of the notch filter depends on the sinusoidal reference signal, whose frequency is equal to the frequency of the primary sinusoidal noise. Therefore the noise at that frequency is attenuated. This adaptive notch filter provides a simple method for tracking and eliminating sinusoidal interference.

Example 8.8: For a stationary input and sufficiently small μ, the convergence speed of the LMS algorithm is dependent on the eigenvalue spread of the input autocorrelation matrix. For $L = 2$ and the reference input is given in (8.5.6), the autocorrelation matrix can be expressed as

$$\mathbf{R} = E \begin{bmatrix} x_0(n)x_0(n) & x_0(n)x_1(n) \\ x_1(n)x_0(n) & x_1(n)x_1(n) \end{bmatrix}$$

$$= E \begin{bmatrix} A^2 \cos^2(\omega_0 n) & A^2 \cos(\omega_0 n)\sin(\omega_0 n) \\ A^2 \sin(\omega_0 n)\cos(\omega_0 n) & A^2 \sin^2(\omega_0 n) \end{bmatrix} = \begin{bmatrix} \dfrac{A^2}{2} & 0 \\ 0 & \dfrac{A^2}{2} \end{bmatrix}.$$

This equation shows that because of the 90° phase shift, $x_0(n)$ is orthogonal to $x_1(n)$ and the off-diagonal terms in the **R** matrix are 0. The eigenvalues λ_1 and λ_2 of the **R** matrix are identical and equal to $A^2/2$. Therefore the system has very fast convergence since the eigenvalue spread equals 1. The time constant of the adaptation is approximated as

$$\tau_{mse} \le \frac{1}{\mu\lambda} = \frac{2}{\mu A^2},$$

which is determined by the power of the reference sinewave and the step size μ.

8.5.5 Adaptive Channel Equalization

In digital communications, considerable effort has been devoted to the development of data-transmission systems that utilize the available telephone channel bandwidth efficiently. The transmission of high-speed data through a channel is limited by intersymbol interference (ISI) caused by distortion in the transmission channel. High-speed data transmission through channels with severe distortion can be achieved in several ways, such as (1) by designing the transmit and receive filters so that the combination of filters and channel results in an acceptable error from the combination of ISI and noise; and (2) by designing an equalizer in the receiver that counteracts the channel distortion. The second method is the most commonly used.

As illustrated in Figure 8.11, the received signal $y(n)$ is different from the original signal $x(n)$ because it was distorted by the overall channel transfer function $C(z)$, which includes the transmit filter, the transmission medium, and the receive filter. To recover the original signal, $x(n)$, we need to process $y(n)$ using the equalizer $W(z)$, which is the inverse of the channel's transfer function $C(z)$ to compensate for the channel distortion. That is, we have to design the equalizer

$$W(z) = \frac{1}{C(z)}, \qquad (8.5.11)$$

i.e., $C(z)W(z) = 1$ such that $\hat{x}(n) = x(n)$.

In practice, the telephone channel is time varying and is unknown in the design stage due to variations in the transmission medium. The transmit and receive filters that are

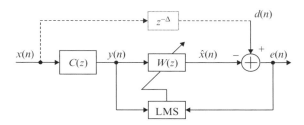

Figure 8.11 Cascade of channel with an ideal adaptive channel equalizer

designed based on the average channel characteristics may not adequately reduce intersymbol interference. Thus we need an adaptive equalizer that provides precise compensation over the time-varying channel. As illustrated in Figure 8.11, the adaptive channel equalizer is an adaptive filter with coefficients that are adjusted using the LMS algorithm.

As shown in Figure 8.11, an adaptive filter requires the desired signal $d(n)$ for computing the error signal $e(n)$ for the LMS algorithm. In theory, the delayed version of the transmitted signal, $x(n - \Delta)$, is the desired response for the adaptive equalizer $W(z)$. However, with the adaptive filter located in the receiver, the desired signal generated by the transmitter is not available at the receiver. The desired signal may be generated locally in the receiver using two methods. A decision-directed algorithm, in which the equalized signal $\hat{x}(n)$ is sliced to form the desired signal, is the simplest and can be used for channels that have only a moderate amount of distortion. However, if the error rate of the data derived by slicing is too high, the convergence may be seriously impaired. In this case, the training of an equalizer by using that sequence is agreed on beforehand by the transmitter and the receiver.

During the training stage, the adaptive equalizer coefficients are adjusted by transmitting a short training sequence. This known transmitted sequence is also generated in the receiver and is used as the desired signal $d(n)$ for the LMS algorithm. A widely used training signal consists of pseudo-random noise (will be introduced in Section 9.2) with a broad and flat power spectrum. After the short training period, the transmitter begins to transmit the data sequence. In order to track the possible slow time variations in the channel, the equalizer coefficients must continue to be adjusted while receiving data. In this data mode, the output of the equalizer, $\hat{x}(n)$, is used by a decision device (slicer) to produce binary data. Assuming the output of the decision device is correct, the binary sequence can be used as the desired signal $d(n)$ to generate the error signal for the LMS algorithm.

An equalizer for a one-dimensional baseband system has real input signals and filter coefficients. However, for a two-dimensional quadrature amplitude modulation (QAM) system, both signals and coefficients are complex. All operations must use complex arithmetic and the complex LMS algorithm expressed as

$$\mathbf{w}(n + 1) = \mathbf{w}(n) + \mu \mathbf{x}^*(n)e(n), \tag{8.5.12}$$

where * denotes complex conjugate and

$$\mathbf{x}(n) = \mathbf{x}_R(n) + j\mathbf{x}_I(n) \tag{8.5.13}$$

and

$$\mathbf{w}(n) = \mathbf{w}_R(n) + j\mathbf{w}_I(n). \tag{8.5.14}$$

In (8.5.13) and (8.5.14), the subscript R and I represent real and imaginary parts of complex numbers. The complex output $\hat{x}(n)$ is given by

$$\hat{x}(n) = \mathbf{w}^T(n)\mathbf{x}(n). \tag{8.5.15}$$

Since all multiplications are complex, the equalizer usually requires four times as many multiplications.

8.6 Implementation Considerations

The adaptive algorithms introduced in previous sections assume the use of infinite precision for the signal samples and filter coefficients. In practice, an adaptive algorithm is implemented on finite-precision hardware. It is important to understand the finite-wordlength effects of adaptive algorithms in meeting design specifications.

8.6.1 Computational Issues

The coefficient update defined in (8.2.21) requires $L + 1$ multiplications and L additions by multiplying $\mu^* e(n)$ outside the loop. Given the input vector $\mathbf{x}(n)$ stored in the array x[], the error signal en, the weight vector w[], the step size mu, and the filter order L, Equation (8.2.21) can be implemented in C as follows:

```
uen = mu* en;              /* u* e(n)               */
for (l = 0, l < L, l++)    /* l = 0, 1, ..., L−1   */
{
    w[l] += uen* x[l];     /* LMS update            */
}
```

The architecture of most commercially available DSP chips has been optimized for convolution operations to compute output $y(n)$ given in (8.2.19) in L instruction cycles. However, the weight update operations in (8.2.21) cannot take advantage of this special architecture because each update cycle involves loading the weight value into the accumulator, performing a multiply–add operation, and storing the result back into memory. With typical DSP architecture, the memory transfer operations take as many machine cycles as the multiply–add, thus resulting in a dominant computational burden involved in the weight update.

Let T_p denote the total processing time for each input sample. Real-time processing requires that $T_p < T$, where T is the sampling period. This means that if the algorithm is too complicated, we need to reduce the complexity of computation to allow $T_p < T$. Because the output convolution is more important and can be implemented very efficiently using DSP chips, skipping part of the weight update is suggested. That is, the simplest solution is to update only a fraction of the coefficients each sample period. The principal cost is a somewhat slower convergence rate.

In a worst-case scenario, we might update only one weight during a sample period. The next weight would be updated for the next sample period, and so on. When the computing power permits, a group of samples can be updated during each sample period.

8.6.2 Finite-Precision Effects

In the digital implementation of adaptive algorithms, both the signals and the internal algorithmic quantities are carried to a certain limited precision. Therefore an adaptive filter implementation with limited hardware precision requires special attention because of the potential accumulation of quantization and arithmetic errors to unacceptable levels as well as the possibility of overflow. This section analyzes finite-precision effects in adaptive filters using fixed-point arithmetic and presents methods for confining these effects to acceptable levels.

We assume that the input data samples are properly scaled so that their values lie between -1 and 1. Each data sample and filter coefficient is represented by B bits (M magnitude bits and one sign bit). For the addition of digital variables, the sum may become larger than 1. This is known as overflow. As introduced in Section 3.6, the techniques used to inhibit the probability of overflow are scaling, saturation arithmetic, and guard bits. For adaptive filters, the feedback path makes scaling far more compli- cated. The dynamic range of the filter output is determined by the time-varying filter coefficients, which are unknown at the design stage.

For the adaptive FIR filter with the LMS algorithm, the scaling of the filter output and coefficients can be achieved by scaling the 'desired' signal, $d(n)$. The scale factor α, where $0 < \alpha \leq 1$, is implemented by right-shifting the bits of the desired signal to prevent overflow of the filter coefficients during the weight update. Reducing the magnitude of $d(n)$ reduces the gain demand on the filter, thereby reducing the magni- tude of the weight values. Usually, the required value of α is not expected to be very small. Since α only scales the desired signal, it does not affect the rate of convergence, which depends on the reference signal $x(n)$. An alternative method for preventing overflow is to use the leaky LMS algorithm described in Section 8.4.2.

With rounding operations, the finite-precision LMS algorithm can be described as follows:

$$y(n) = R\left[\sum_{l=0}^{L-1} w_l(n)x(n-l)\right], \tag{8.6.1}$$

$$e(n) = R[\alpha d(n) - y(n)], \tag{8.6.2}$$

$$w_l(n+1) = R[w_l(n) + \mu x(n-l)e(n)], \quad l = 0, 1, \ldots, L-1, \tag{8.6.3}$$

where $R[x]$ denotes the fixed-point rounding of the quantity x.

When the convolution sum in (8.6.1) is calculated using a multiplier with an internal double-precision accumulator, internal roundoff noise is avoided. Therefore roundoff error is only introduced when the product is transferred out of the accumulator and the result is rounded to single precision. When updating weights according to (8.6.3), the product $\mu x(n-l)e(n)$ produces a double-precision number, which is added to the original stored weight value, $w_l(n)$, then is rounded to form the updated value, $w_l(n+1)$. Insufficient precision provided in the weight value will cause problems such as coefficient bias or stalling of convergence, and will then be responsible for excess error in the output of the filter.

By using the assumptions that quantization and roundoff errors are zero-mean white noise independent of the signals and each other, that the same wordlength is used for both signal samples and coefficients, and that μ is sufficiently small, the total output MSE is expressed as

$$\xi = \xi_{\min} + \frac{\mu}{2} L\sigma_x^2 \xi_{\min} + \frac{L\sigma_e^2}{2\alpha^2 \mu} + \frac{1}{\alpha^2}\left[|\mathbf{w}^\circ|^2 + k\right]\sigma_e^2, \tag{8.6.4}$$

where $\xi_{\min}$ is the minimum MSE defined in (8.2.13), σ_x^2 is the variance of the input signal $x(n)$, σ_e^2 is as defined in (3.5.6), k is the number of rounding operations in (8.6.1) ($k = 1$ if a double-precision accumulator is used), and $\mathbf{w}^\circ$ is the optimum weight vector.

The second term in (8.6.4) represents excess MSE due to algorithmic weight fluctuation and is proportional to the step size μ. For fixed-point arithmetic, the finite-precision error given in (8.6.4) is dominated by the third term, which reflects the error in the quantized weight vector, and is inversely proportional to the step size μ. The last term in (8.6.4) arises because of two quantization errors – the error in the quantized input vector and the error in the quantized filter output $y(n)$.

Whereas the excess MSE given in the second term of (8.6.4) is proportional to μ, the power of the roundoff noise in the third term is inversely proportional to μ. Although a small value of μ reduces the excess MSE, it may result in a large quantization error. There will be an optimum step size that achieves a compromise between these competing goals. The total error for a fixed-point implementation of the LMS algorithm is minimized using the optimum μ° expressed as

$$\mu^\circ = \frac{2^{-M}}{2\alpha}\sqrt{\frac{1}{3\xi_{\min}\sigma_x^2}}. \tag{8.6.5}$$

In order to stabilize the digital implementation of the LMS algorithm, we may use the leaky LMS algorithm to reduce numeric errors accumulated in the filter coefficients. As discussed in Section 8.4.2, the leaky LMS algorithm prevents overflow in a finite-precision implementation by providing a compromise between minimizing the MSE and constraining the energy of the adaptive filter impulse response.

There is still another factor to consider in the selection of step size μ. As mentioned in Section 8.2, the adaptive algorithm is aimed at minimizing the error signal, $e(n)$. As the weight vector converges, the error term decreases. At some point, the update term will be rounded to 0. Since $\mu x(n-l)e(n)$ is a gradually decreasing random quantity, fewer and fewer values will exceed the rounding threshold level, and eventually the weight will stop changing almost completely. The step size value μ° given in (8.6.5) is shown to be too small to allow the adaptive algorithm to converge completely. Thus the 'optimal' value in (8.6.5) may not be the best choice from this standpoint.

From (8.6.1)–(8.6.3), the digital filter coefficients, as well as all internal signals, are quantized to within the least significant bit LSB $\equiv 2^{-B}$. From (8.6.3), the LMS algorithm modifies the current parameter settings by adding a correction term, $R[\mu x(n-l)e(n)]$. Adaptation stops when the correction term is smaller in magnitude than the LSB. At this point, the adaptation of the filter virtually stops. Roundoff precludes the tap weights reaching the optimum (infinite-precision) value. This phenomenon is known as 'stalling' or 'lockup'.

The condition for the lth component of the weight vector $w_l(n)$ not to be updated is whenever the corresponding correction term for $w_l(n)$ in the update equation is smaller in magnitude than the least significant bit of the weight:

$$|\mu x(n - l)e(n)| < 2^{-M}. \tag{8.6.6}$$

Suppose that this equation is first satisfied for $l = 0$ at time n. As the particular input sample $x(n)$ propagates down the tapped-delay-line, the error will further decrease in magnitude, and thus this sample will turn off all weight adaptation beyond this point.

To get an approximate condition for the overall algorithm to stop adapting, we can replace $|x(n - l)|$ and $|e(n)|$ with their standard deviation values, σ_x and σ_e, respectively. The condition for the adaptation to stop becomes

$$\mu\sigma_x\sigma_e < 2^{-M}. \tag{8.6.7}$$

We have to select the step size μ to satisfy

$$\mu \geq \mu_{\min} \equiv \frac{2^{-M}}{\sigma_x\sigma_e} \tag{8.6.8}$$

to prevent early termination of adaptation. When adaptation is prematurely terminated by quantization effects, the total output MSE can be decreased by increasing the step size μ.

In steady state, the optimum step size μ^o in (8.6.5) is usually smaller than the $\mu_{\min}$ specified in (8.6.8). Therefore if the value $\mu \leq \mu^o < \mu_{\min}$ is used, the adaptation essentially stops before complete convergence of the algorithm is attained. In order to prevent this early termination of the adaptation, some value of

$$\mu \geq \mu_{\min} > \mu^o \tag{8.6.9}$$

is selected. In this case, the excess MSE due to misadjustment is larger than the finite-precision error.

In conclusion, the most important design issue is to find the best value of μ that satisfies

$$\mu_{\min} < \mu < \frac{1}{L\sigma_x^2}. \tag{8.6.10}$$

To prevent algorithm stalling due to finite-precision effects, the design must allow the residual error to reach small non-zero values. This can be achieved by using a sufficiently large number of bits, and/or using a large step size μ, while still guaranteeing convergence of the algorithm. However, this will increase excess MSE as shown in (8.6.4).

8.7 Experiments Using the TMS320C55x

The adaptive filtering experiments we will conduct in this section are filters that adjust their coefficients based on a given adaptive algorithm. Unlike the time-invariant filters introduced in Chapters 5 and 6, the adaptation algorithms will optimize the filter response based on predetermined performance criteria. The adaptive filters can be realized as either FIR or IIR filters. As explained in Section 8.2.1, the FIR filters are always stable and provide linear phase responses. We will conduct our experiments using FIR filters in the subsequence sections.

As introduced in Section 8.2, there are many adaptive algorithms that can be used for DSP applications. Among them, the LMS algorithm has been widely used in real-time applications, such as adaptive system identification, adaptive noise cancellation, channel equalization, and echo cancellation. In this section, we will use the TMS320C55x to implement an adaptive identification system and an adaptive linear predictor.

8.7.1 Experiment 8A – Adaptive System Identification

The block diagram of adaptive system identification is shown in Figure 8.6. The input sample $x(n)$ is fed to both the unknown system and the adaptive filter. The output of the unknown system is used by the adaptive filter as the desired signal $d(n)$. The adaptive algorithm minimizes the differences between the outputs of the unknown system and adaptive filter. The filter coefficients are continuously adjusted until the error signal has been minimized. When the adaptive filter has converged, the coefficients of the filter describe the characteristics of the unknown system.

As shown in Figure 8.6, the system identification consists of three basic elements – a signal generator, an adaptive filter, and an unknown system that needs to be modeled. The input signal $x(n)$ should have a broad spectrum to excite all the poles and zeros of the unknown system. Both the white noise and the chirp signal are widely used for system identification. The signal generation algorithms will be introduced in Sections 9.1 and 9.2.

For the adaptive system identification experiment, we use the LMS algorithm in conjunction with an FIR filter as shown in Figure 8.6. In practical applications, the unknown system is a physical plant with both the input and output connected to the adaptive filter. However, for experimental purposes and to better understand the properties of adaptive algorithms, we simulate the unknown system in the same program. The adaptive system identification operations can be expressed as:

1. Place the current input sample $x(n)$ generated by the signal generator into x[0] of the signal buffer.

2. Compute the FIR filter output

$$y(n) = \sum_{l=0}^{L-1} w_l(n)x(n-l).$$

(8.7.1)

3. Calculate the error signal

$$e(n) = d(n) - y(n). \tag{8.7.2}$$

4. Update the adaptive filter coefficients

$$w_l(n + 1) = w_l(n) + \mu e(n)x(n - l), \quad l = 0, 1, \ldots, L - 1. \tag{8.7.3}$$

5. Update the signal buffer

$$x(n - l - 1) = x(n - l), \quad l = L - 2, \ L - 1, \ldots, 1, 0. \tag{8.7.4}$$

The adaptive system identification shown in Figure 8.6 can be implemented in C language as follows:

```
/* Simulate unknown system                                          */
    x1[0] = input;                /* Get input signal x(n)  */
    d = 0.0;
    for (i = 0; i < N1; i++)      /* Compute d(n)           */
      d += (coef[i] * x1[i]);
    for (i = N1-1; i > 0; i--)    /* Update signal buffer   */
      x1[i] = x1[i-1];            /*    of unknown system    */

/* Adaptive system identification operation                         */
    x[0] = input;                 /* Get input signal x(n)  */
    y = 0.0;
    for (i = 0; i < N0; i++)      /* Compute output y(n)    */
      y += (w[i] * x[i]);
    e = d - y;                    /* Calculate error e(n)   */
    uen = twomu* e;               /* uen = mu* e(n)          */
    for (i = 0; i < N0; i++)      /* Update coefficients     */
      w[i] += (uen* x[i]);
    for (i = N0-1; i > 0; i--)    /* Update signal buffer   */
      x[i] = x[i-1];              /*    of adaptive filter   */
```

The unknown system for this example is an FIR filter with the filter coefficients given by coef[]. The input is a zero-mean random noise. The unknown system output d is used as the desired signal for the adaptive filter, and the adaptive filter coefficients w[i], i=0, 1, ...NO, will match closely to the unknown system response after the adaptive filter reaches its steady-state response.

Experiment 8A consists of the following modules – an adaptive FIR filter using the LMS algorithm implemented using the C55x assembly language, a random noise generator, an initialization function, and a C program for testing the adaptive system identification experiment. These programs are listed in Table 8.1 to Table 8.4.

The assembly routine listed in Table 8.1 implements the adaptive FIR filter using the LMS algorithm. The input signal is pointed by the auxiliary register AR0, and the desired signal is ponted by AR1. The auxiliary registers AR3 and AR4 are used as circular pointers for the signal buffer and coefficient buffer, respectively. The outer

Table 8.1   Implementation of adaptive filter using the C55x assembly code

```
; AR0 -> in[] is input signal buffer
; AR1 -> d[] is desired signal buffer
; AR3 -> x[] is circular buffer
; AR4 -> w[] is circular buffer

        rptblocal loop-1              ; for (n = 0; n < Ns; n++)
        mov  *AR0+, *AR3              ; x[n] = in[n]
        mpym *AR3+, *AR4+, AC0        ; temp = w[0] * x[0]
||      rpt  CSR                     ; for (i = 0; i < N-1; i++)
        macm *AR3+, *AR4+, AC0        ;    y += w[i] * x[i]
        sub  *AR1+ << #16, AC0        ; AC0 = -e = y-d[n] , AR1 -> d[n]
        mpyk #-TWOMU, AC0             ; AC0 = mu* e[n]
        mov  rnd(hi(AC0)), mmap(T1)   ; T1 = uen = mu* e[n]
        rptblocal lms_loop-1          ; for (j = 0; i < N-2; i++)
        mpym *AR3+, T1, AC0           ;    AC0 = uen* x[i]
        add  *AR4 << #16, AC0         ;    w[i] += uen* x[i]
        mov  rnd(hi(AC0)), *AR4+
lms_loop
        mpym *AR3, T1, AC0            ; w[N-1] += uen* x[N-1]
        add  *AR4 << #16, AC0
        mov  rnd(hi(AC0)), *AR4+      ; Store the last w[N-1]
loop
```

block-repeat loop controls the process of signal samples in blocks, while the two inner repeat loops perform the adaptive filtering sample-by-sample. The repeat instruction

```
rpt   CSR
macm  *AR3+, *AR4+, AC0        ; y += w[i] * x[i]
```

performs the FIR filtering, and the inner block-repeat loop, `lms_loop`, updates the adaptive filter coefficients.

The zero-mean random noise generator given in Table 8.2 is used to generate testing data for both the unknown system and the adaptive filter. The function `rand()` will generate a 16-bit unsigned integer number between 0 and 65 536. We subtract 0x4000 from it to obtain the zero-mean pseudo-random number from −32 768 to 32 767.

The signal buffers and the adaptive filter coefficient buffer are initialized to 0 by the function `init.c` listed in Table 8.3. For assembly language implementation, we apply the block processing structure as we did for the FIR filter experiments in Chapter 5. To use the circular buffer scheme, we pass the signal buffer index as an argument to the adaptive filter subroutine. After a block of samples are processed, the subroutine returns the index for the adaptive filter to use in the next iteration.

The adaptive system identification is tested by the C function `exp8a.c` given by Table 8.4. The signal and coefficient buffers are initialized to 0 first. The random signal generator is then used to generate Ns samples of white noise. The FIR filter used to

Table 8.2   List of random noise generator in C

```
/*
  random.c — Zero-mean random noise generator
*/
#include <math.h>

void random(int *x, unsigned int N)
{
  unsigned int t;
  for(t = N; t > 0; t--)
    *x++ = rand() — 0x4000;    /* Zero-mean */
}
```

Table 8.3   List of buffer initialization function

```
/*
  init.c — Initialize an array to zero
*/

void init(int *ptr, unsigned int N)
{
  unsigned int i;

  for(i = N; i > 0; i--)
    *ptr++ = 0;
}
```

Table 8.4   List of C program for Experiment 8A

```
/*
  exp8a.c — C program for Experiment 8A
    Adaptive system identification using the LMS algorithm
*/

#include "LP_coef.dat"

#define N0 48       /* Adaptive filter order      */
#define N1 48       /* Unknown system order       */
#define Ns 128      /* Number of input signal     */
extern unsigned int fir_filt(int *, unsigned int, int *,
                            unsigned int, int *, int *, unsigned int);
extern unsigned int adaptive(int *, int *, int *, int *,
                            unsigned int, unsigned int, unsigned int);
extern void init(int *, unsigned int);
extern void random(int *, unsigned int);
```

Table 8.4 (*continued*)

```
int w[N0],          /* Adaptive filter coefficients  */
    d_sys[N0],      /* Adaptive filter signal buffer */
    d_fir[N1],      /* Unknown system signal buffer  */
    in[Ns],         /* Input sample buffer           */
    d[Ns];          /* Unknown system output buffer  */

unsigned int fir_index, sys_index;

void main()
{
  init(w, N0);      /* Initialize coefficients       */
  init(d_sys, N0);  /* Initialize x(n) signal buffer */
  init(d_fir, N1);  /* Initialize d(n) signal buffer */
  fir_index = 0;    /* Initialize d(n) buffer index  */
  sys_index = 0;    /* Initialize x(n) buffer index  */
  for(;;)           /* Generate samples and          */
  {                 /*    identify unknown system    */
    random(x, Ns);
    fir_index = fir_filt(in, Ns, LP_coef, N1, d, d_fir, fir_index);
    sys_index = adaptive(in, d, d_sys, w, Ns, N0, sys_index);
  }
}
```

simulate the unknown system is implemented in Experiment 5A. The adaptive filter uses the unknown FIR filter output $d(n)$ as the desired signal to produce the error signal that is used for the LMS algorithm. After several iterations, the adaptive filter converges and its coefficient vector w[] contains N coefficients that can be used to describe the unknown system in the form of an FIR filter. The results of the system identification are plotted in Figure 8.12. The impulse responses (left) and the frequency responses (right) of the unknown system (top) and the adaptive model (bottom) are almost identical.

Complete the following steps for Experiment 8A:

1. This experiment uses the following files: exp8.cmd, expt8a.c, init.c, random.c, adaptive.asm, fir_flt.asm, LP_coef.dat, and randdata.dat, where the assembly routine fir_flt.asm and its coefficients LP_coef.dat are identical to those used for the experiments in Chapter 5.

2. Create the project epx8a, add files exp8.cmd, expt8a.c, init.c, random.c, adaptive.asm, and fir_flt.asm into the project. Build, debug, and run the experiment using the CCS.

3. Configure the CCS, and set the animation option for viewing the coefficient buffer w[] of the adaptive filter, and LP_coef[] of the unknown the system in both the time domain and frequency domain.

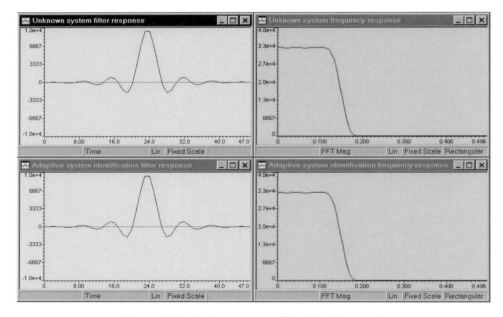

Figure 8.12   Adaptive system identification results

4. Verify the adaptation process by viewing how the adaptive filter coefficients are adjusted. Record the steady-state values of w[], and plot the magnitude responses of the adaptive filter and the unknown system. Save the adaptive filter coefficients, and compare them with the unknown system coefficients given in the file LP_coef.dat.

5. Adjust step size μ, and repeat the adaptive system identification process. Observe the change of the system performance.

6. Increase the number of the adaptive filter coefficients to NO = 64, and observe the system performance.

7. Reduce the number of the adaptive filter coefficients to NO = 32, and observe the system performance.

8.7.2   Experiment 8B – Adaptive Predictor Using the Leaky LMS Algorithm

As shown in Figure 8.7, an adaptive predictor receives the primary signal that consists of the broadband components $v(n)$ and the narrowband components $s(n)$. An adaptive system can separate the narrowband signal from the broadband signal. The output of the adaptive filter is the narrowband signal $y(n) \approx s(n)$. For applications such as spread spectrum communications, the narrowband interference can be tracked and removed by the adaptive filter. The error signal, $e(n) \approx v(n)$, contains the desired broadband signal.

We use a fixed delay Δ in between the primary input signal and the reference input as shown in Figure 8.7. If we choose a long enough delay, we can de-correlate the broadband components at the reference input from those at the primary signal. The adaptive filter output $y(n)$ will be the narrowband signal because its periodic nature still keeps them correlated. If the narrowband components are desired, the filter output $y(n)$ is used as the system output. On the other hand, if the broadband signal is corrupted by a narrowband noise, the adaptive filter will reduce the narrowband interference by subtracting the estimated narrowband components from the primary signal. Thus the error output $e(n)$ is used as the system output that consists of broadband signal.

In the experiment, we use the white noise as the broadband signal. Since the white noise is uncorrelated, the delay $\Delta = 1$ is chosen. The adaptive predictor operation is implemented as follows:

1. Compute the FIR filter output

$$y(n) = \sum_{l=0}^{L-1} w_l(n) x(n - l - 1). \tag{8.7.5}$$

2. Calculate the error signal

$$e(n) = x(n) - y(n). \tag{8.7.6}$$

3. Update the adaptive filter coefficients

$$w_l(n+1) = w_l(n) + \mu e(n) x(n - l - 1), \quad l = 0, 1, \dots, L-1. \tag{8.7.7}$$

4. Update the signal buffer for adaptive filter and place the new sample into the buffer

$$x(n - l - 1) = x(n - l), \quad l = L - 1, \dots, 1, \tag{8.7.8a}$$

$$x(n) = input. \tag{8.7.8b}$$

The adaptive predictor written in floating-point C is given in Table 8.5. The fixed-point implementation using the intrinsics can be implemented and compared against the floating-point implementation. Finally, the assembly routine can be written to maximize the run-time efficiency and minimize the program memory space usage. The adaptive predictor using the leaky LMS algorithm written in the C55x assmebly language is listed in Table 8.6.

In practice, it is preferred to initialize the adaptive filter coefficients to a known state. The initialization can be done in two ways. If we know statistical characteristics of the system, we can preset several adaptive filter coefficients to some predetermined value. Using the preset values, the adaptation process usually converges to the steady state at a faster rate. However, if we do not have any prior knowledge of the system, a common practice is to start the adaptive process by initializing the coefficients to 0. The function init.c listed in Table 8.3 is used to set both the coefficient and signal buffers to 0 at the beginning of the adaptive process.

Table 8.5 Implementation of adaptive linear predictor in C

```
/*
  alp.c — Adaptive linear predictor
*/

#define twomu(96.0/32768.0)    /* Step size mu            */

void alp(float *in, float *y, float *e, float *x, float *w,
         unsigned int Ns, unsigned int N)
{
  unsigned int n;
  int i;
  float temp;
  float uen;

  for(n = 0; n < Ns; n++)
  {
    temp = 0.0;
    for(i = N−1; i >= 0; i−−)  /* FIR filtering           */
      temp += (w[i] * x[i]);
    y[n] = temp;
    e[n] = in[n] − y[n];        /* Calculate error         */
    uen = twomu* e[n];          /* uen = mu* e(n)          */
    for(i = N−1; i >= 0; i−−)  /* Update coefficients     */
      w[i] += uen* x[i];
    for(i = N−1; i > 0; i−−)   /* Update signal buffer    */
      x[i] = x[i−1];
    x[0]  = in[n];
  }
}
```

Table 8.6 Assembly program implementation of adaptive linear predictor

```
; AR0 -> in[] is the input buffer
; AR1 -> y[] is the output buffer
; AR2 -> e[] is the error buffer
; AR3 -> x[] is circular buffer
; AR4 -> w[] is circular buffer

      mov   #ALPHA,T0                ; T0 = leaky factor α
||    rptblocal loop-1              ; for(n = 0; n < Ns; n++)
      mpym  *AR3+, *AR4+, AC0        ;   temp = w[0] * x[0]
||    rpt   CSR                      ; for(i = 0; i < N-1; i++)
      macm  *AR3+, *AR4+, AC0        ;   temp = w[i] * x[i]
      mov   rnd(hi(AC0)),*AR1        ; y[t] = temp;
      sub   *AR0, *AR1+, AC0
      mov   rnd(hi(AC0)), *AR2+      ; e[n] = in[n] − y[n]
```

Table 8.6 (*continued*)

```
||    mpyk  #TWOMU, AC0
      mov   rnd(hi(AC0)), mmap(T1)     ; T1= mu* e[n] = uen
      mpym  *AR4, T0, AC0
||    rptblocal lms_loop-1             ; for(j = 0; i < N-2; i++)
      macm  *AR3+, T1, AC0             ;    w[i] = alpha*w[i] + uen*x[i]
      mov   rnd(hi(AC0)), *AR4+
      mpym  *AR4, T0, AC0
lms_loop
      macm  *AR3,T1, AC0               ; w[N-1] = α*w[N-1] + uen*x[N-1]
      mov   rnd(hi(AC0)), *AR4+        ; Store the last w[i]
      mov   *AR0+, *AR3                ; x[n] = in[n]
loop
```

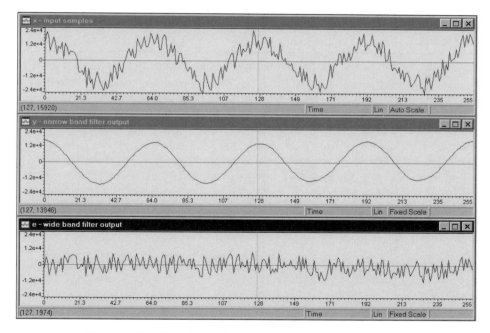

Figure 8.13 The signal plots of the adaptive linear predictor

The experiment results are shown in Figure 8.13. The input signal $x(n)$ shown in the top window contains both the broadband random noise and the narrowband sinusoid signal. The adaptive filter output $y(n)$ consisting of the narrowband sinusoid signal is shown in the middle window. The adaptive linear predictor output $e(n)$ shown in the bottom window contains the broadband noise.

The signal generator `signal.c` listed in Table 8.7 is used for the experiment to produce a sinusoidal signal embedded in random noise.

The C program `exp8b.c` is listed in Table 8.8. The block size is chosen as 256. The adaptive FIR filter order is 48. The initialization is performed once at the beginning of the experiment. The adaptation step size is set to $\mu = 96/32\,768$. The system uses the leaky LMS algorithm with leaky factor set to $\alpha = 32\,704/32\,768 = 0.998$.

Go through the following steps for Experiment 8B:

1. Create the project `epx8b`, include `exp8.cmd`, `exp8b.c`, `init.c`, `alp.asm`, `singal.c` and `noise.dat`. These files can be found in the software package.

2. Build, debug, and run the experiment using the CCS.

3. Configure the CCS, and set the animation option for viewing the output of the adaptive filter `y[]`, the output of the system `e[]`, the input signal `in[]`, and the adaptive filter coefficients `w[]` at a block-by-block basis.

Table 8.7   List of C program for generating sinewave embedded in random noise

```
/*
   signal.c — Sinewave plus zero-mean random noise
*/
#include <math.h>
#include <intrindefs.h>

#define PI 3.1415962
#define K (Ns ≫ 6)
#define a1 0x4000
#define a2 0x4000
static unsigned int i = 0;

void cos_rand(int *x, unsigned int Ns)
{
   unsigned int t;
   float     two_pi_K_Ns;
   int       temp;
   long      ltemp;

   two_pi_K_Ns = 2.0* PI* K/Ns;
   for (t = Ns; t > 0; t--)
   {
      temp = (int)(0x7fff* cos(two_pi_K_Ns* i));
      ltemp = _lsmpy(a1, temp);
      temp = rand() −0x4000;
      *x++ = _smac(ltemp, a2, temp) ≫ 16;
      i++;
      i %= (Ns ≫ 1);
   }
}
```

Table 8.8 List of C program for Experiment 8B

```
/*
  exp8b.c — Experiment 8B, Adaptive linear predictor
*/

#define N  48        /* Adaptive FIR filter order      */
#define Ns 256       /* Number of input signal per block */

#pragma DATA_SECTION(e, "lms_err");
#pragma DATA_SECTION(y, "lms_out");
#pragma DATA_SECTION(x, "lms_in");
#pragma DATA_SECTION(d, "lms_data");
#pragma DATA_SECTION(w, "lms_coef");
#pragma DATA_SECTION(index, "lms_data");
#pragma CODE_SECTION(main, "lms_code");

int e[Ns],              /* Error signal buffer       */
    y[Ns],              /* Output signal buffer      */
    in[Ns],             /* Input signal buffer       */
    w[N],               /* Filter coefficient buffer */
    x[N],               /* Filter signal buffer      */
    index;

extern void init(int *, unsigned int);
extern unsigned int alp(int *, int *, int *, int *, int *,
                        unsigned int, unsigned int, unsigned int);
extern void cos_rand(int *, unsigned int);

void main(void)
{
  init(x,N);                          /* Initialize x[] to zero */
  init(w,N);                          /* Initialize w[] to zero */
  index = 0;
  for(;;)
  {
    cos_rand(x, Ns);                  /* Generate testing signal */
    index = alp(in, y, e, x, w, Ns, N, index);  /* Adaptive predictor */
  }
}
```

4. Verify the adaptive linear predictor and compare the results with Figure 8.13.

5. Verify the adaptation process by viewing how the adaptive coefficients w[] are adjusted. Record the steady-state values of w[], and plot the magnitude response of the adaptive filter.

6. Change the order of the adaptive filter and observe the system performance.

7. Adjust adaptation step size and observe the system performance.

8. Change the leaky factor value and observe the system performance.

9. Can we obtain a similar result without using the leaky LMS (by setting leaky factor to 0x7fff)? Find the steady-state adaptive filter coefficients w[] by running the adaptive predictor for a period of time, and compare the magnitude response with the one obtained in step 5.

References

[1] S. T. Alexander, *Adaptive Signal Processing*, New York: Springer-Verlag, 1986.

[2] M. Bellanger, *Adaptive Digital Filters and Signal Analysis*, New York: Marcel Dekker, 1987.

[3] P. M. Clarkson, *Optimal and Adaptive Signal Processing*, Boca Raton, FL: CRC Press, 1993.

[4] C. F. N. Cowan and P. M. Grant, *Adaptive Filters*, Englewood Cliffs, NJ: Prentice-Hall, 1985.

[5] J. R. Glover, Jr., 'Adaptive noise canceling applied to sinusoidal interferences,' *IEEE Trans. Acoust., Speech, Signal Processing*, ASSP-25, Dec. 1997, pp. 484–491.

[6] S. Haykin, *Adaptive Filter Theory*, 2nd Ed., Englewood Cliffs, NJ: Prentice-Hall, 1991.

[7] S. M. Kuo and C. Chen, 'Implementation of adaptive filters with the TMS320C25 or the TMS320C30,' in *Digital Signal Processing Applications with the TMS320 Family*, vol. 3, P. Papamichalis, Ed., Englewood Cliffs, NJ: Prentice-Hall, 1990, pp. 191–271, Chap. 7.

[8] S. M. Kuo and D. R. Morgan, *Active Noise Control Systems – Algorithms and DSP Implementations*, New York: Wiley, 1996.

[9] L. Ljung, *System Identification: Theory for the User*, Englewood Cliffs, NJ: Prentice-Hall, 1987.

[10] J. Makhoul, 'Linear prediction: A tutorial review,' *Proc. IEEE*, vol. 63, Apr. 1975, pp. 561–580.

[11] J. R. Treichler, C. R. Johnson, Jr., and M. G. Larimore, *Theory and Design of Adaptive Filters*, New York: Wiley, 1987.

[12] B. Widrow, J. R. Glover, J. M. McCool, J. Kaunitz, C. S. Williams, R. H. Hern, J. R. Zeidler, E. Dong, and R. C. Goodlin, 'Adaptive noise canceling: principles and applications,' *Proc. IEEE*, vol. 63, Dec. 1975, pp. 1692–1716.

[13] B. Widrow and S. D. Stearns, *Adaptive Signal Processing*, Englewood Cliffs, NJ: Prentice-Hall, 1985.

[14] M. L. Honig and D. G. Messerschmitt, *Adaptive Filters: Structures, Algorithms, and Applications*, Boston, MA: Kluwer Academic Publishers, 1986.

Exercises

Part A

1. Determine the autocorrelation function of the following signals:

   (a) $x(n) = A \sin(2\pi n/N)$, and

   (b) $y(n) = A \cos(2\pi n/N)$.

2. Find the crosscorrelation functions $r_{xy}(k)$ and $r_{yx}(k)$, where $x(n)$ and $y(n)$ are defined in the Problem 1.

3. Let $x(n)$ and $y(n)$ be two independent zero-mean WSS random signals. The random signal $w(n)$ is obtained by using

$$w(n) = ax(n) + by(n),$$

where a and b are constants. Express $r_{ww}(k)$, $r_{wx}(k)$, and $r_{wy}(k)$ in terms of $r_{xx}(k)$ and $r_{yy}(k)$.

4. An estimator $\bar{x}$ of random process x is unbiased if

$$E[\bar{x}] = x.$$

Show that the sample mean estimator given in (8.1.13) is unbiased, but the sample variance estimator given in (8.1.14) is biased. That is, show that

$$E[\bar{\sigma}_x^2] = \sigma_x^2\left(1 - \frac{1}{N}\right) \neq \sigma_x^2.$$

5. Show that the PDS $P_{xx}(\omega)$ of a WSS signal $x(n)$ is a real-valued function of ω.

6. If $x(n)$ is a real-valued random variable, show that

   (a) $P_{xx}(\omega) = P_{xx}(-\omega)$, and

   (b) $P_{xx}(\omega) \geq 0$.

7. Find the power density spectrum $P_{xx}(z)$ of a random signal with the following autocorrelation function:

$$r_{xx}(k) = 0.8^{|k|}, \quad -\infty < k < \infty$$

8. Consider a second-order autoregressive (AR) process defined by

$$d(n) = v(n) - a_1 d(n-1) - a_2 d(n-2),$$

where $v(n)$ is a white noise of zero mean and variance σ_v^2. This AR process is generated by filtering $v(n)$ using the second-order IIR filter $H(z)$.

   (a) Derive the IIR filter transfer function $H(z)$.

   (b) Consider a second-order optimum FIR filter shown in Figure 8.5. If the desired signal is $d(n)$, the primary input $x(n) = d(n-1)$. Find the optimum weight vector $\mathbf{w}^o$ and the minimum mean-squared error $\xi_{\min}$.

Part B

9. Given the two finite-length sequences

   $x(n) = \{1\ 3\ -2\ 1\ 2\ -1\ 4\ 4\ 2\}$, and

   $y(n) = \{2\ -1\ 4\ 1\ -2\ 3\}$.

   Using MATLAB to compute and plot the crosscorrelation function $r_{xy}(k)$ and the autocorrelation function $r_{xx}(k)$.

10. Write a MATLAB script to generate the length 1024 signal defined as

$$x(n) = 0.8\sin(\omega_0 n) + v(n),$$

where $\omega_0 = 0.1\pi$, $v(n)$ is a zero-mean random noise with variance $\sigma_v^2 = 1$ (see Section 3.3 for details). Compute and plot, $r_{xx}(k)$, $k = 0, 1, \ldots, 127$ using MATLAB.

11. Consider the Example 8(b). The digital filter is a second-order FIR filter using the LMS algorithm. The AR parameters are $a_1 = -0.195$, $a_2 = 0.95$, and $\sigma_v^2 = 0.0965$. Simulate the operation of the adaptive filter using either MATLAB or C program. After the convergence of filter.

(a) Plot the learning curve $E[e^2(n)]$, which can be approximated by the smoothed $e^2(n)$ using the first-order IIR filter.

(b) Repeat (a) using different values of step size μ. Discuss the convergence speed and the excess MSE related to μ.

(c) Repeat the problem (a) using the parameters $a_1 = -1.9114$, $a_2 = 0.95$, and $\sigma_v^2 = 0.038$. Explain why the convergence is much slower than the problem (a) by analyzing the eigenvalue spread given in (8.3.7).

(d) Plot the coefficient tracks $w_0(n)$ and $w_1(n)$, and show the coefficients converge to the optimum values derived in Example 8(b).

12. Implement the adaptive system identification technique illustrated in Figure 8.5 using MATLAB or C program. The input signal is a zero-mean, unit-variance white noise. The unknown system is defined by the room impulse response used in Chapter 4.

13. Implement the adaptive line enhancer illustrated in Figure 8.7 using MATLAB or C program. The desired signal is given by

$$x(n) = \sqrt{2}\sin(\omega n) + v(n),$$

where frequency $\omega = 0.2\pi$ and $v(n)$ is the zero-mean white noise with unit variance. The decorrelation delay $\Delta = 1$. Plot both $e(n)$ and $y(n)$.

14. Implement the adaptive noise cancellation illustrated in Figure 8.8 using MATLAB or C program. The primary signal is given by

$$d(n) = \sin(\omega n) + 0.8v(n) + 1.2v(n-1) + 0.25v(n-2),$$

where $v(n)$ is defined by Problem 13. The reference signal is $v(n)$. Plot $e(n)$.

15. Implement the single-frequency adaptive notch filter illustrated in Figure 8.10 using MATLAB or C program. The desired signal $d(n)$ is given in Problem 14, and $x(n)$ is given by

$$x(n) = \sqrt{2}\sin(\omega n).$$

Plot $e(n)$ and the magnitude response of second-order FIR filter after convergence.

Part C

16. Replace the unknown system in the Experiment 8A with the IIR filter `iirform2.asm` from Chapter 6. Adjust the adaptive filter order to find the FIR filter coefficients that are the best to identify the unknown IIR filter. Verify the system identification by comparing the adaptive FIR filter magnitude response with the IIR filter response.

17. Given a corrupted primary input $d(n) = 0.25\cos(2\pi nf_1/f_s) + 0.25\sin(2\pi nf_2/f_s)$, and the reference signal $x(n) = 0.125\cos(2\pi nf_2/f_s)$, where f_s is sampling frequency, f_1 and f_2 are the frequencies of the desired signal and interference, respectively. Implement the adaptive noise canceler that removed the interference signal.

18. Implement the adaptive linear predictor using the normalized LMS algorithm in real-time using an EVM or DSK. Use signal generators to generate a sinusoid and white noise. Connect both signals to the EVM input with a coupler. Run an adaptive linear predictor and display both the input and the adaptive filter output on an oscilloscope.

9

Practical DSP Applications in Communications

There are many DSP applications that are used in our daily lives, some of which have been introduced in previous chapters. DSP algorithms, such as random number generation, tone generation and detection, echo cancellation, channel equalization, noise reduction, speech and image coding, and many others can be found in a variety of communication systems. In this chapter, we will introduce some selected DSP applications in communications that played an important role in the realization of the systems.

9.1 Sinewave Generators and Applications

When designing algorithms for a sinewave (sine or cosine function) generation, several characteristics should be considered. These issues include total harmonic distortion, frequency and phase control, memory usage, execution time, and accuracy. The total harmonic distortion (THD) determines the purity of a sinewave and is defined as

$$\text{THD} = \frac{\text{spurious harmonic power}}{\text{total waveform power}}, \tag{9.1.1}$$

where the spurious harmonic power relates to the unwanted harmonic components of the waveform. For example, a sinewave generator with a THD of 0.1 percent has a distortion power level approximately 30 dB below the fundamental component. This is the most important characteristic from the standpoint of performance. The other characteristics are closely related to details of the implementation.

Polynomials can be used to express or approximate some trigonometric functions. However, the sine or cosine function cannot be expressed as a finite number of additions and multiplications. We must depend on approximation. Because polynomial approximations can be computed with multiplications and additions, they are ready to be implemented on DSP devices. For example, the sine function can be approximated by (3.8.1). The implementation of a sinewave generation using polynomial approximation is given in Section 3.8.5. As discussed in Chapter 6, another approach of generating sinusoidal signals is to design a filter $H(z)$ whose impulse response $h(n)$ is the desired sinusoidal waveform. With an impulse function $\delta(n)$ used as input, the IIR filter will generate the desired impulse response (sinewave) at the output. In this section, we will discuss the lookup-table method for generating sinewaves.

9.1.1 Lookup-Table Method

The lookup-table method or wavetable generator is probably the most flexible and conceptually simple method for generating sinusoidal waveforms. The technique simply involves the readout of a series of stored data values representing discrete samples of the waveform to be generated. The data values can be obtained either by sampling the appropriate analog waveform, or more commonly, by computing the desired values using MATLAB or C programs. Enough samples are generated and stored to accurately represent one complete period of the waveform. The periodic signal is then generated by repeatedly cycling through the data memory locations using a circular pointer. This technique is also used for generating computer music.

A sinewave table contains equally spaced sample values over one period of the waveform. An N-point sinewave table can be computed by evaluating the function

$$x(n) = \sin\left(\frac{2\pi n}{N}\right), \quad n = 0, 1, \ldots, N - 1. \tag{9.1.2}$$

These sample values must be represented in binary form. The accuracy of the sine function is determined by the wordlength used to represent data and the table length. The desired sinewave is generated by reading the stored values in the table at a constant (sampling) rate of step Δ, wrapping around at the end of the table whenever the pointer exceeds $N - 1$. The frequency of the generated sinewave depends on the sampling period T, table length N, and the sinewave table address increment Δ:

$$f = \frac{\Delta}{NT} \text{ Hz.} \tag{9.1.3}$$

For the designed sinewave table of length N, a sinewave of frequency f with sampling rate f_s can be generated by using the pointer address increment

$$\Delta = \frac{Nf}{f_s}, \quad \Delta \leq \frac{N}{2}. \tag{9.1.4}$$

To generate L samples of sinewave $x(l)$, $l = 0, 1, \ldots, L - 1$, we use a circular pointer k such that

$$k = (m + l\Delta)_{\text{mod } N}, \tag{9.1.5}$$

where m determines the initial phase of sinewave. It is important to note that the step Δ given in (9.1.4) may be a non-integer, thus $(m + l\Delta)$ in (9.1.5) is a real number. That is, a number consisting of an integer and a fractional part. When fractional values of Δ are used, samples of points between table entries must be estimated using the table values.

The easy solution is to round this non-integer index to the nearest integer. However, the better but more complex solution is to interpolate the two adjacent samples.

The lookup-table method is subject to the constraints imposed by aliasing, requiring at least two samples per period in the generated waveform. Two sources of error in the lookup-table algorithm cause harmonic distortion:

1. An amplitude quantization error is introduced by representing the sinewave table values with finite-precision numbers.

2. Time-quantization errors are introduced when points between table entries are sampled, which increase with the address increment Δ.

The longer the table is, the less significant the second error will be. To reduce the memory requirement for generating a high accuracy sinewave, we can take advantage of waveform symmetry, which in effect results in a duplication of stored values. For example, the values are repeated (regardless of sign change) four times every period. Thus only a quarter of the memory is needed to represent the waveform. However, the cost is a greater complexity of algorithm to keep track of which quadrant of the waveform is to be generated and with the correct sign. The best compromise will be determined by the available memory and computation power for a given application on the target DSP hardware.

To decrease the harmonic distortion for a given table size N, an interpolation scheme can be used to more accurately compute the values between table entries. Linear interpolation is the simplest method for implementation. For linear interpolation, the sine value for a point between successive table entries is assumed to lie on the straight line between the two values. Suppose the integer part of the pointer is i ($0 \leq i < N$) and the fractional part of the pointer is f ($0 < f < 1$), the sine value is computed as

$$x(n) = s(i) + f \cdot [s(i+1) - s(i)], \tag{9.1.6}$$

where $[s(i+1) - s(i)]$ is the slope of the line segment between successive table entries i and $i + 1$.

Example 9.1: A cosine/sine function generator using table-lookup method with 1024 points cosine table can be implemented using the TMS320C55x assembly code as follows:

```
;   cos_sin.asm — Table lookup sinewave generator
;                  with 1024-point cosine table range (0—π)
;
;   Prototype: void cos_sin(int, int *, int *)
;   Entry: arg0: T0 — a (alpha)
;          arg1: AR0 — pointer to cosine
;          arg2: AR1 — pointer to sine

     .def   _cos_sin
     .ref   tab_0_PI
     .sect  "cos_sin"
```

```
_cos_sin
        mov    T0, AC0            ; T0 = a
        sfts   AC0, #11           ; Size of lookup table
        mov    #tab_0_PI, T0      ; Table based address
||      mov    hi(AC0), AR2
        mov    AR2, AR3
        abs    AR2                ; cos(-a) = cos(a)
        add    #0x200, AR3        ; 90 degree offset for sine
        and    #0x7ff, AR3        ; Modulo 0x800 for 11-bit
        sub    #0x400, AR3        ; Offset 180 degree for sine
        abs    AR3                ; sin(-a) = sin(a)
||      mov    *AR2(T0), *AR0     ; *AR0 = cos(a)
        mov    *AR3(T0), *AR1     ; *AR1 = sin(a)
        ret
        .end
```

In this example, we use a half table $(0 \rightarrow \pi)$. Obviously, a sine (or cosine) function generator using the complete table $(0 \rightarrow 2\pi)$ can be easily implemented using only a few lines of assembly code, while a function generator using a quarter table $(0 \rightarrow \pi/2)$ will be more challenging to implement efficiently. The assembly program `cos_sin.asm` used in this example is available in the software package.

9.1.2 Linear Chirp Signal

A linear chirp signal is a waveform whose instantaneous frequency increases linearly with time between two specified frequencies. It is a broadband waveform with the lowest possible peak to root-mean-square amplitude ratio in the desired frequency band. The digital chirp waveform is expressed as

$$c(n) = A \sin[\phi(n)], \tag{9.1.7}$$

where A is a constant amplitude and $\phi(n)$ is a quadratic phase of the form

$$\phi(n) = 2\pi \left[f_L n + \left(\frac{f_U - f_L}{2(N-1)} \right) n^2 \right] + \alpha, \quad 0 \leq n \leq N-1, \tag{9.1.8}$$

where N is the total number of points in a single chirp. In (9.1.8), α is an arbitrary constant phase factor, f_L and f_U, are the normalized lower and upper band limits, respectively, which are in the range $0 \leq f \leq 0.5$. The waveform periodically repeats with

$$\phi(n + kN) = \phi(n), \quad k = 1, 2, \ldots \tag{9.1.9}$$

The instantaneous normalized frequency is defined as

$$f(n) = f_L + \left(\frac{f_U - f_L}{N-1} \right) n, \quad 0 \leq n \leq N-1. \tag{9.1.10}$$

This expression shows that the instantaneous frequency goes from $f(0) = f_L$ at time $n = 0$ to $f(N-1) = f_U$ at time $n = N-1$.

Because of the complexity of the linear chirp signal generator, it is more convenient for real-time applications to generate such a sequence by a general-purpose computer and store it in a lookup table. Then the lookup-table method introduced in Section 9.1.1 can be used to generate the desired signal.

An interesting application of chirp signal generator is generating sirens. The electronic sirens are often created by a small generator system inside the vehicle compartment. This generator drives either a 60 or 100 Watt loudspeaker system present in the light bar mounted on the vehicle roof or alternatively inside the vehicle radiator grill. The actual siren characteristics (bandwidth and duration) vary slightly from manufacturers. The wail type of siren sweeps between 800 Hz and 1700 Hz with a sweep period of approximately 4.92 seconds. The yelp siren has similar characteristics to the wail but with a period of 0.32 seconds.

9.1.3   DTMF Tone Generator

A common application of sinewave generator is the all-digital touch-tone phone that uses a dual-tone multi-frequency (DTMF) transmitter and receiver. DTMF also finds widespread use in electronic mail systems and automated telephone servicing systems in which the user can select options from a menu by sending DTMF signals from a telephone.

Each key-press on the telephone keypad generates the sum of two tones expressed as

$$x(n) = \cos(2\pi f_L nT) + \cos(2\pi f_H nT), \tag{9.1.11}$$

where T is the sampling period and the two frequencies f_L and f_H uniquely define the key that was pressed. Figure 9.1 shows the matrix of sinewave frequencies used to encode the 16 DTMF symbols. The values of the eight frequencies have been chosen carefully so that they do not interfere with speech.

The low-frequency group (697, 770, 852, and 941 Hz) selects the four rows frequencies of the 4×4 keypad, and the high-frequency group (1209, 1336, 1477, and 1633 Hz)

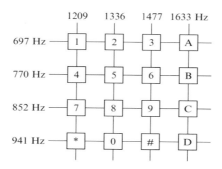

Figure 9.1   Matrix telephone keypad

selects the columns frequencies. A pair of sinusoidal signals with f_L from the low-frequency group and f_H from the high-frequency group will represent a particular key. For example, the digit '3' is represented by two sinewaves at frequencies 697 Hz and 1477 Hz. The row frequencies are in the low-frequency range below 1 kHz, and the column frequencies are in the high-frequency between 1 kHz and 2 kHz. The digits are displayed as they appear on a telephone's 4 × 4 matrix keypad, where the fourth column is omitted on standard telephone sets.

The generation of dual tones can be implemented by using two sinewave generators connected in parallel. Each sinewave generator can be realized using the polynomial approximation technique introduced in Section 3.8.5, the recursive oscillator introduced in Section 6.6.4, or the lookup-table method discussed in Section 9.1.1. Usually, DTMF signals are interfaces to the analog world via a CODEC (coder/decoder) chip with an 8 kHz sampling rate.

The DTMF signal must meet timing requirements for duration and spacing of digit tones. Digits are required to be transmitted at a rate of less than 10 per second. A minimum spacing of 50 ms between tones is required, and the tones must be present for a minimum of 40 ms. A tone-detection scheme used to implement a DTMF receiver must have sufficient time resolution to verify correct digit timing. The issues of tone detection will be discussed later in Section 9.3.

9.2   Noise Generators and Applications

Random numbers are useful in simulating noise and are used in many practical applications. Because we are using digital hardware to generate numbers, we cannot produce perfect random numbers. However, it is possible to generate a sequence of numbers that are unrelated to each other. Such numbers are called pseudo-random numbers (PN sequence).

Two basic techniques can be used for pseudo-random number generation. The lookup-table method uses a set of stored random samples, and the other is based on random number generation algorithms. Both techniques obtain a pseudo-random sequence that repeats itself after a finite period, and therefore is not truly random at all time. The number of stored samples determines the length of a sequence generated by the lookup-table method. The random number generation algorithm by computation is determined by the register size. In this section, two random number generation algorithms will be introduced.

9.2.1   Linear Congruential Sequence Generator

The linear congruential method is probably the most widely used random number generator. It requires a single multiplication, addition, and modulo division. Thus it is simple to implement on DSP chips. The linear congruential algorithm can be expressed as

$$x(n) = [ax(n-1) + b]_{\mathrm{mod}\ M},\tag{9.2.1}$$

where the modulo operation (mod) returns the remainder after division by M. The constants a, b, and M are chosen to produce both a long period and good statistical characteristics of the sequences. These constants can be chosen as

$$a = 4K + 1, \tag{9.2.2}$$

where K is an odd number such that a is less than M, and

$$M = 2^L \tag{9.2.3}$$

is a power of 2, and b can be any odd number. Equations (9.2.2) and (9.2.3) guarantee that the period of the sequence in (9.2.1) is of full-length M.

A good choice of parameters are $M = 2^{20} = 10\,48\,576$, $a = 4(511) + 1 = 2045$, and $x(0) = 12\,357$. Since a random number routine usually produces samples between 0 and 1, we can normalize the nth random sample as

$$r(n) = \frac{x(n) + 1}{M + 1} \tag{9.2.4}$$

so that the random samples are greater than 0 and less than 1. Note that the random numbers $r(n)$ can be generated by performing Equations (9.2.1) and (9.2.4) in real time. A C function (`uran.c` in the software package) that implements this random number generator is listed in Table 9.1.

Example 9.2: Most of the fixed-point DSP processors are 16-bit. The following TMS320C55x assembly code implements an $M = 2^{16}$ (65 536) random number generator.

Table 9.1 C program for generating linear congruential sequence

```
/***********************************************************************
*   URAN — This function generates pseudo-random numbers             *
***********************************************************************/
static long n = (long)12357;              // Seed x(0)= 12357
float uran()
{
    float ran;                            // Random noise r(n)
    n = (long)2045*n+1L;                  // x(n)= 2045*x(n−1)+1
    n -= (n/1048576L)*1048576L;           // x(n)= x(n) − INT[x(n)/
                                          //          1048576] *1048576
    ran = (float)(n+1L)/(float)1048577;   // r(n)= FLOAT[x(n)+1] /
                                          //          1048577
    return(ran);                          // Return r(n) to the main
}                                         //    function
```

```
;   rand16_gen.asm — 16-bit zero-mean random number generator
;
;   Prototype: int rand16_gen(int *)
;
;   Entry: arg0 — AR0 pointer to seed value
;   Return: T0 — random number
C1   .equ   0x6255
C2   .equ   0x3619
     .def   _rand16_gen
     .sect "rand_gen"
_rand16_gen
     mov    #C1, T0
     mpym   *AR0, T0, AC0   ; seed = (C1* seed+C2)
     add    #C2, AC0
     and    #0xffff, AC0    ; seed %= 0x8000
     mov    AC0, *AR0
     sub    #0x4000, AC0    ; Zero-mean random number
     mov    AC0, T0
     ret
     .end
```

The assembly program `rand16_gen.asm` used for this example is available in the software package.

9.2.2 Pseudo-Random Binary Sequence Generator

A shift register with feedback from specific bits can also generate a repetitive pseudo-random sequence. A schematic of the 16-bit generator is shown in Figure 9.2, where the functional circle labeled 'XOR' performs the exclusive-OR function of its two binary inputs. The sequence itself is determined by the position of the feedback bits on the shift register. In Figure 9.2, x_1 is the output of b_0 XOR with b_2, x_2 is the output of b_{11} XOR with b_{15}, and x is the output of x_1 XOR with x_2.

An output from the sequence generator is the entire 16-bit word. After the random number is generated, every bit in the register is shifted left 1 bit (b_{15} is lost), and then x is shift to b_0 to generate the next random number. A shift register length of 16 bits can

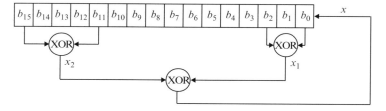

Figure 9.2   A 16-bit pseudo random number generator

readily be accommodated by a single word on the many 16-bit DSP devices. Thus memory usage is minimum. It is important to recognize, however, that sequential words formed by this process will be correlated. The maximum sequence length before repetition is

$$L = 2^M - 1,$$ (9.2.5)

where M is the number of bits of the shift register.

Example 9.3: The PN sequence generator given in Table 9.2 uses many Boolean operations. The C program requires at least 11 operations to complete the computation. The following TMS320C55x assembly program computes the same PN sequence in 11 cycles:

```
;   pn_gen.asm — 16-bit zero-mean PN sequence generator
;
;   Prototype: int pn_gen(int *)
;
;   Entry: arg0 — AR0 pointer to the shift register
;   Return: T0 — random number

BIT15    .equ 0x8000                   ; b15
BIT11    .equ 0x0800                   ; b11
```

Table 9.2 C program for generating PN sequence

```
/***************************************************************
*    PN Sequence generator                                     *
****************************************************************/
static int shift_reg;

int pn_sequence(int * sreg)
{
    int b2, b11, b15;
    int x1, x2;              /* x2 also used for x          */

    b15 = * sreg >> 15;
    b11 = * sreg >> 11;
    x2 = b15^b11;            /* First XOR bit15 and bit11 */
    b2 = * sreg >> 2;
    x1 = * sreg ^b2;         /* Second XOR bit2 and bit0   */
    x2 = x1^x2;              /* Final XOR of x1 and x2     */
    x2 &= 1;
    * sreg = * sreg << 1;
    * sreg = * sreg | x2;    /* Update the shift register */
    x2 = * sreg−0x4000;      /* Zero-mean random number    */
    return x2;
}
```

```
BIT2    .equ 0x0004                     ; b2
BIT0    .equ 0x0001                     ; b0
        .def _pn_gen
        .sect "rand_gen"
_pn_gen
        mov   *AR0, AC0                 ; Get register value
        bfxtr #(BIT15|BIT2), AC0, T0    ; Get b15 and b2
        bfxtr #(BIT11|BIT0), AC0, T1    ; Get b11 and b0
        sfts  AC0, #1
||      xor   T0, T1                    ; XOR all 4 bits
        mov   T1, T0
        sfts  T1, #-1
        xor   T0, T1                    ; Final XOR
        and   #1, T1
        or    T1, AC0
        mov   AC0, *AR0                 ; Update register
        sub   #0x4000, AC0, T0          ; Zero-mean random number
||      ret
        .end
```

The C program pn_sequence.c and the TMS320C55x assembly program pn_gen.asm for this example are available in the software package.

9.2.3 Comfort Noise in Communication Systems

In voice-communication systems, the complete suppression of a signal using residual echo suppressor (will be discussed later in Section 9.4) has an adverse subjective effect. This problem can be solved by adding a low-level comfort noise, when the signal is suppressed by a center clipper. As illustrated in Figure 9.3, the output of residual echo suppressor is expressed as

$$y(n) = \begin{cases} \alpha v(n), & |x(n)| \leq \beta \\ x(n), & |x(n)| > \beta, \end{cases} \qquad (9.2.6)$$

where $v(n)$ is an internally generated zero-mean pseudo-random noise and $x(n)$ is the input applied to the center clipper with the clipping threshold β.

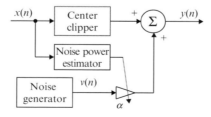

Figure 9.3 Injection of comfort noise with active center clipper

The power of the comfort noise should match the background noise when neither talker is active. Therefore the algorithm shown in Figure 9.3 is the process of estimating the power of the background noise in $x(n)$ and generating the comfort noise of the same power to replace signals suppressed by the center clipper.

9.2.4 Off-Line System Modeling

As discussed in Section 8.5, several applications require knowledge of the transfer function $H(z)$ of an unknown system. Assuming that the characteristics of the system are time-invariant, off-line modeling can be used to estimate $H(z)$ during an initial training stage. White noise is an ideal broadband training signal in system identification because it has a constant spectral density at all frequencies. A repeated linear chirp signal introduced in Section 9.1.2 can also be used because it has the lowest peak factor with the most concentrated power distribution over the required frequency range. When such a waveform is used for system identification, the required measurement time can be quite short relative to the time required for repetitive measurements using other waveforms.

The block diagram of the off-line system modeling is shown in Figure 9.4, where uncorrelated random noise $x(n)$ is internally generated by the DSP system. Detailed noise generation methods are given in Section 9.2. As illustrated in Figure 9.4, the random noise is used as the input to an unknown system $H(z)$ and an adaptive filter $\hat{H}(z)$. The off-line system modeling procedure is summarized as follows:

1. Generate the random noise $x(n)$. In the acoustic echo canceler (will be discussed in Section 9.5), $x(n)$ is converted to an analog signal, amplified, and then used to drive a loudspeaker.

2. Obtain the desired signal $d(n)$. In the acoustic echo canceler, $d(n)$ is the digital signal picked up by a microphone.

3. Apply an adaptive algorithm as follows:

   a. Compute the filter output

   $$y(n) = \sum_{l=0}^{L-1} \hat{h}_l(n)x(n-l),\qquad (9.2.7)$$

   where $\hat{h}_l(n)$ is the lth coefficient of the adaptive filter $\hat{H}(z)$ at time n.

   b. Compute the error signal

   $$e(n) = d(n) - y(n).\qquad (9.2.8)$$

   c. Update the filter coefficients using the LMS algorithm

   $$\hat{h}_l(n+1) = \hat{h}_l(n) + \mu x(n-l)e(n),\quad l = 0, 1, \ldots, L-1.\qquad (9.2.9)$$

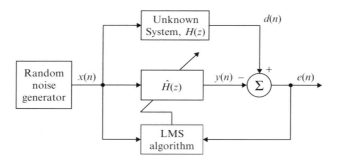

Figure 9.4   Off-line modeling of an unknown system using adaptive filter

4. Go to step 1 for the next iteration until the adaptive filter $\hat{H}(z)$ converges to the optimum solution. That is, the power of $e(n)$ is minimized.

   After convergence of the algorithm, the adaptation is stopped and coefficients $\hat{h}_l$, $l = 0, 1, \ldots, L - 1$ are fixed. It is important to note that an averaging technique can be used to obtain better results. If the algorithm converges at time $n = N$, the coefficients are averaged over the next M samples as

$$\hat{h}_l = \frac{1}{M} \sum_{n=N}^{N+M-1} \hat{h}_l(n), \quad l = 0, 1, \ldots, L - 1. \tag{9.2.10}$$

9.3   DTMF Tone Detection

This section introduces detection methods for DTMF tones used in the communication networks. The correct detection of a digit requires both a valid tone pair and the correct timing intervals. DTMF signaling is used both to set up a call and to control features such as call forwarding and teleconferencing calling. In some applications, it is necessary to detect DTMF signaling in the presence of speech, so it is important that the speech waveform is not interpreted as valid signaling tones.

9.3.1   Specifications

The implementation of a DTMF receiver involves the detection of the signaling tones, validation of a correct tone pair, and the timing to determine that a digit is present for the correct amount of time and with the correct spacing between tones. In addition, it is necessary to perform additional tests to improve the performance of the decoder in the presence of speech. A DSP implementation is useful in applications in which the digitized signal is available and several channels need to be processed such as in a private branch exchange.

DTMF receivers are required to detect frequencies with a tolerance of ±1.5 percent as valid tones. Tones that are offset by ±3.5 percent or greater must not be detected. This requirement is necessary to prevent the detector from falsely detecting speech and other signals as valid DTMF digits. The receiver is required to work with a worst-case signal-to-noise ratio of 15 dB and with a dynamic range of 26 dB.

Another requirement of the receiver is the ability to detect DTMF signals when two tones are received at different levels. The high-frequency tone may be received at a lower level than the low-frequency tone due to the magnitude response of the communication channel. This level difference is called twist, and the situation described above is called a forward (or standard) twist. Reverse twist occurs when the low-frequency tone is received at a lower level than the high-frequency tone. The receiver must operate with a maximum of 8 dB normal twist and 4 dB reverse twist. A final requirement for the receiver is that it operates in the presence of speech without incorrectly identifying the speech signal as valid DTMF tones. This is referred to as talk-off performance.

9.3.2 Goertzel Algorithm

The principle of DTMF detection is to examine the energy of the received signal at the DTMF frequencies (defined in Figure 9.1) to determine whether a valid DTMF tone pair has been received. The detection algorithm can be a DFT implementation using an FFT algorithm or a filter-bank implementation. An FFT can be used to calculate the energies of N evenly spaced frequencies. To achieve the frequency resolution required to detect the eight DTMF frequencies within ±1.5 percent frequency deviation, a 256-point FFT is needed for an 8 kHz sample rate. For the relatively small number of tones to be detected, the filter-bank implementation is more efficient.

Since only eight frequencies are of interest, it is more efficient to use the DFT directly to compute

$$X(k) = \sum_{n=0}^{N-1} x(n) W_N^{kn} \tag{9.3.1}$$

for eight different values of k that correspond to the DTMF frequencies defined in Figure 9.1. The DFT coefficients can be more efficiently calculated by using the Goertzel algorithm, which can be interpreted as a matched filter for each frequency k as illustrated in Figure 9.5. In this figure, $x(n)$ is the input signal of the system, $H_k(z)$ is the transfer function of the filter at kth frequency bin, and $X(k)$ is the corresponding filter output.

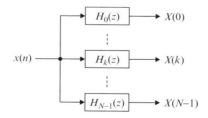

Figure 9.5   Flow graph of Goertzel filters

From (7.1.4), we have

$$W_N^{-kN} = e^{j(2\pi/N)kN} = e^{j2\pi k} = 1. \tag{9.3.2}$$

Multiplying the right-hand side of (9.3.1) by W_N^{-kN}, we have

$$X(k) = W_N^{-kN} \sum_{n=0}^{N-1} x(n) W_N^{kn} = \sum_{n=0}^{N-1} x(n) W_N^{-k(N-n)}. \tag{9.3.3}$$

Define the sequence

$$y_k(n) = \sum_{m=0}^{N-1} x(m) W_N^{-k(n-m)}, \tag{9.3.4}$$

this equation can be interpreted as a convolution of the finite-duration sequence $x(n)$, $0 \le n \le N - 1$, with the sequence $W_N^{-kn} u(n)$.

Consequently, $y_k(n)$ can be viewed as the output of a filter with impulse response $W_N^{-kn} u(n)$. That is, the filter with impulse response

$$h_k(n) = W_N^{-kn} u(n) \tag{9.3.5}$$

due to the finite-length input $x(n)$. Thus Equation (9.3.4) can be expressed as

$$y_k(n) = x(n) * W_N^{-kn} u(n). \tag{9.3.6}$$

From (9.3.3) and (9.3.4), and the fact that $x(n) = 0$ for $n < 0$ and $n \ge N$, we show that

$$X(k) = y_k(n)|_{n=N-1}. \tag{9.3.7}$$

That is, $X(k)$ is the output of filter $H_k(z)$ at time $n = N - 1$.

Taking the z-transform of (9.3.6) at both sides, we obtain

$$Y_k(z) = X(z) \frac{1}{1 - W_N^{-k} z^{-1}}. \tag{9.3.8}$$

The transfer function of the kth Goertzel filter is defined as

$$H_k(z) = \frac{Y_k(z)}{X(z)} = \frac{1}{1 - W_N^{-k} z^{-1}}, \quad k = 0, 1, \dots, N - 1. \tag{9.3.9}$$

This filter has a pole on the unit circle at the frequency $\omega_k = 2\pi k/N$. Thus the entire DFT can be computed by filtering the block of input data using a parallel bank of N filters defined by (9.3.9), where each filter has a pole at the corresponding frequency of the DFT. Since the Goertzel algorithm computes N DFT coefficients, the parameter N

must be chosen to make sure that $X(k)$ is close to the DTMF frequencies f_k. This can be accomplished by choosing N such that

$$\frac{f_k}{f_s} \cong \frac{k}{N}, \tag{9.3.10}$$

where the sampling frequency $f_s = 8\,\text{kHz}$ is used for most of telecommunication systems.

A signal-flow diagram of transfer function $H_k(z)$ is depicted in Figure 9.6. Since the coefficients W_N^{-k} are complex valued, the computation of each new value of $y_k(n)$ using Figure 9.6 requires four multiplications and additions. All the intermediary values $y_k(0)$, $y_k(1)$, ..., $y_k(N-1)$ must be computed in order to obtain the final output $y_k(N-1) = X(k)$. Therefore the computational algorithm shown in Figure 9.6 requires $4N$ complex multiplications and additions to compute $X(k)$ for each frequency index k.

The complex multiplications and additions can be avoided by combining the pair of filters that have complex-conjugated poles. By multiplying both the numerator and the denominator of $H_k(z)$ in (9.3.9) by the factor $(1 - W_N^k z^{-1})$, we have

$$H_k(z) = \frac{1 - W_N^k z^{-1}}{(1 - W_N^{-k} z^{-1})(1 - W_N^k z^{-1})} = \frac{1 - e^{j2\pi k/N} z^{-1}}{1 - 2\cos(2\pi k/N)z^{-1} + z^{-2}}. \tag{9.3.11}$$

The signal-flow graph of the transfer function defined by (9.3.11) is shown in Figure 9.7 using the direct-form II realization. The recursive part of the filter is on the left-hand side of the delay elements, and the non-recursive part is on the right-hand side. Since the

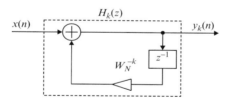

Figure 9.6   Flow graph of recursive computation of $X(k)$

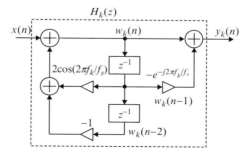

Figure 9.7   Detailed signal-flow diagram of Goertzel algorithm

output $y_k(n)$ is required only at time $N - 1$, we just need to compute the non-recursive part of the filter at the $(N - 1)$th iteration. The recursive part of algorithm can be expressed as

$$w_k(n) = x(n) + 2\cos(2\pi f_k/f_s)w_k(n - 1) - w_k(n - 2). \qquad (9.3.12)$$

The non-recursive calculation of $y_k(N - 1)$ is expressed as

$$X(k) = y_k(N - 1) = w_k(N - 1) - e^{-j2\pi f_k/f_s}w_k(N - 2). \qquad (9.3.13)$$

A further simplification of the algorithm is made by realizing that only the magnitude squared of $X(k)$ is needed for tone detection. From (9.3.13), the squared magnitude of $X(k)$ is computed as

$$|X(k)|^2 = w_k^2(N - 1) - 2\cos(2\pi f_k/f_s)w_k(N - 1)w_k(N - 2) + w_k^2(N - 2). \quad (9.3.14)$$

Therefore the complex arithmetic given in (9.3.13) is eliminated and (9.3.14) requires only one coefficient, $2\cos(2\pi f_k/f_s)$, for each $|X(k)|^2$ to be evaluated. Since there are eight possible tones to be detected, we need eight filters described by (9.3.12) and (9.3.14). Each filter is tuned to one of the eight frequencies defined in Figure 9.1. Note that Equation (9.3.12) is computed for $n = 0, 1, \ldots, N - 1$, but Equation (9.3.14) is computed only once at time $n = N - 1$.

9.3.3   Implementation Considerations

The flow chart of DTMF tone detection algorithm is illustrated in Figure 9.8. At the beginning of each frame of length N, the state variables $x(n)$, $w_k(n), w_k(n - 1)$, $w_k(n - 2)$, and $y_k(n)$ for each of the eight Goertzel filters and the energy are set to 0. For each sample, the recursive part of each filter defined in (9.3.12) is executed. At the end of each frame, i.e., $n = N - 1$, the squared magnitude $|X(k)|^2$ for each DTMF frequency is computed based on the (9.3.14). The following six tests are performed to determine if a valid DTMF digit has been detected.

Magnitude test

According to the International Telecommunication Union (ITU), previously the International Telegraph and Telephone Consultative Committee (CCITT), standard, the maximum signal level transmit to the public network shall not exceed $-9\,\text{dBm}$ $(+/- 1\,\text{dBm})$ (See Appendix A.6 for the definition of dBm). This leaves an average voice range of $-35\,\text{dBm}$ for a very weak long distance call, to $-10\,\text{dBm}$ for a local call. Generally, the DTMF receiver would be expected to operate at an average range of $-29\,\text{dBm}$ to $+1\,\text{dBm}$. The $+1\,\text{dBm}$ is an extreme, but could happen. Thus the largest magnitude in each band must be greater than a threshold of $-29\,\text{dBm}$, otherwise the DTMF signal should not be decoded. For this magnitude test, the squared

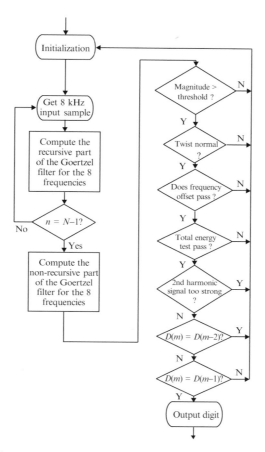

Figure 9.8 Flow chart for the DTMF tone detector

magnitude $|X(k)|^2$ defined in (9.3.14) for each DTMF frequency is computed. The largest magnitude in each group is obtained.

Twist test

Because of the frequency response of a telephone system, the tones may be attenuated according to the system's gains at the tonal frequencies. Consequently, we do not expect the high- and low-frequency tones to have exactly the same amplitude at the receiver, even though they were transmitted at the same strength. Twist is the difference, in decibels, between the low-frequency tone level and the high-frequency tone level. Forward twist exists when the high-frequency tone level is less than the low-frequency tone level. Generally, the DTMF digits are generated with some forward twist to compensate for greater losses at higher frequency within a long telephone cable. Different administrations recommend different amounts of allowable twist for a DTMF receiver. For example, Australia allows 10 dB, Japan allows only 5 dB, and AT&T recommends not more than 4 dB of forward twist or 8 dB of reverse twist.

Frequency offset test

This test is performed to prevent some broadband noises from being detected as effective tones. If the effective DTMF tones are present, the power levels at those two frequencies should be much higher than the power levels at the other frequencies. To perform this test, the largest magnitude in each group is compared to the magnitudes of other frequencies in that group. The difference must be greater than a predetermined threshold in each group.

Tone-to-total energy test

Similar to the frequency-offset test, the goal of this test is to reject some broad noises (such as speech) and further improve the robustness of the receiver. To perform this test, three different constants, $c1$, $c2$, and $c3$, are used. The energy of the detected tone in the low-frequency group is weighted by $c1$, the energy of the detected tone in the high-frequency group is weighted by $c2$, and the sum of the two energies is weighted by $c3$. Each of these terms must be greater than the summation of the energy of eight filter outputs. For this test, the total energy is computed as

$$E = \sum_{k=1}^{8} |X(k)|^2. \tag{9.3.15}$$

Second harmonic test

The objective of this test is to reject speech that has harmonics close to f_k so that they might be detected as DTMF tones. Since DTMF tones are pure sinusoids, they contain very little second harmonic energy. Speech, on the other hand, contains a significant amount of second harmonic energy. To test the level of second harmonic, the decoder must evaluate the second harmonic frequencies of all eight DTMF tones. These second harmonic frequencies (1394 Hz, 1540 Hz, 1704 Hz, 1882 Hz, 2418 Hz, 2672 Hz, 2954 Hz, and 3266 Hz) also can be detected using the Goertzel algorithm.

Digit decoder

Finally, if all five tests are passed, the tone pair is decoded as an integer between 1 and 16. Thus the digit decoder is implemented as

$$D(m) = C + 4(R - 1), \tag{9.3.16}$$

where $D(m)$ is the digit detected for frame m, $m = 0, 1, 2, \ldots$ is the frame index, C is the index of column frequencies which has been detected, and R is the index of row frequencies which has been detected. For example, if two frequencies 750 Hz and 1219 Hz are detected, the valid digit is computed as

$$D(m) = 2 + 4(2 - 1) = 6. \tag{9.3.17}$$

This value is placed in a memory location designated $D(m)$. If any of the tests fail, then '-1' representing 'no detection' is placed in $D(m)$. For a new valid digit to be declared, $D(m)$ must be the same for two successive frames, i.e., $D(m - 2) = D(m - 1)$. If the digit is valid for more than two successive frames, the receiver is detecting the continuation of a previously validated digit, and a third digit $D(m)$ is not the output.

There are two reasons for checking three successive digits at each pass. First, the check eliminates the need to generate hits every time a tone is present. As long as the tone is present, it can be ignored until it changes. Second, comparing digits $D(m - 2)$, $D(m - 1)$, and $D(m)$ improves noise and speech immunity.

9.4 Adaptive Echo Cancellation

Adaptive echo cancellation is an application of adaptive filtering to the attenuation of undesired echo in the telecommunication networks. This is accomplished by modeling the echo path using an adaptive filter and subtracting the estimated echo from the echo-path output. The development of echo canceling chips and advances in DSP processors have made the implementation of echo cancelers at commercially acceptable costs.

Beginning from canceling the voice echo in long-distance links and now being applied to control acoustic echo in hands-free telephones, adaptive echo cancelers have also found wide use in full-duplex data transmission over two-wire circuits such as high speed modems. In addition, echo canceling techniques are used in providing the digital data stream between the customer premise and serving central office. Since the requirements for voice and data echo cancelers are quite different, this section emphasizes on introducing voice echo cancelers for long-distance networks.

9.4.1 Line Echoes

One of the main problems associated with telephone communications is the generation of echoes due to impedance mismatches at various points in telecommunication networks. Such echoes are called line (or network) echoes. If the time delay between the speech and the echo is short, the echo is not noticeable. Distinct echoes are noticeable only if the delay exceeds tens of milliseconds, which are annoying and can disrupt a conversation under certain conditions. The deleterious effects of echoes depend upon their loudness, spectral distortion, and delay. In general, the longer the echo is delayed, the more echo attenuation is required. Echo is probably the most devastating degradation for long-distance telecommunications, especially if the two parties are separated by a great distance with a long transmission delay.

A simplified communication network is illustrated in Figure 9.9, where the local telephone set is connected to a central office by a two-wire line, in which both directions of transmission are carried on a single wire pair. The connection between two central offices is a four-wire facility, in which the two directions of transmission are segregated on physically different facilities. This is because long distance transmission requires amplification that is a one-way function and long-distance calls are multiplexed, which requires that signals in the two directions be sent over different slots. This four-wire transmission path may include various equipment, including switches, cross-connects, and multiplexers. The conversion between the two-wire and four-wire parts of the overall transmission link is done in the device called hybrid (H) located in the central office.

An ideal hybrid is a bridge circuit with the balancing impedance that is exactly equal to the impedance of the two-wire circuit at all frequencies. Therefore a hybrid circuit couples all energy on the incoming branch of the four-wire circuit into the two-wire circuit. Thus none of the incoming four-wire signal should be transferred to the outgoing branch of the four-wire circuit. In practice, the hybrid may be connected to any of the two-wire loops served by the central office. Thus the balancing network can only provide a compromise (fixed) impedance match. As a result, some of the incoming signals from the four-wire circuit that is supposed to go into the two-wire facility leak into the outgoing four-wire circuit, which is then returned to the source and is heard as an echo (shown in Figure 9.9). This echo requires special treatment if the round-trip delay exceeds 40 ms.

9.4.2 Adaptive Echo Canceler

In a telecommunication network using echo cancellation, an echo canceler is positioned in the four-wire section on the network near the origin of the echo source. The principle of the adaptive echo cancellation for one direction of transmission is illustrated in Figure 9.10. We show only one echo canceler located at the left end of network. To overcome the echoes in full-duplex communication network, it is desirable to cancel the echoes in both directions of the trunk. Thus an another echo canceler is symmetrically located at the other end. The reason for showing a telephone and two-wire line is to indicate that this side is called the near-end, while the other side is referred to as the far-end. The far-end talker is the talker who generates the echoes that will be canceled by the echo canceler.

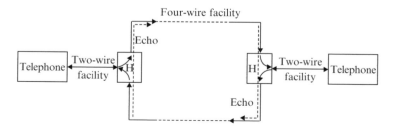

Figure 9.9   Long-distance telephone communication network

To explain the principle of echo cancellation, the function of the hybrid in Figure 9.10 can be illustrated in Figure 9.11, where the far-end signal $x(n)$ passing through the echo path $P(z)$ results in the undesired echo $r(n)$. The primary signal $d(n)$ is a combination of echo $r(n)$, near-end signal $u(n)$, and noise $v(n)$ which consists of the quantization noise from the A/D converter and other noises from the circuit. The adaptive filter $W(z)$ adaptively learns the response of the echo path $P(z)$ by using the far-end speech $x(n)$ as an excitation signal. The echo replica $y(n)$ is generated by $W(z)$, and is subtracted from the primary signal $d(n)$ to yield the error signal $e(n)$. Ideally, $y(n) \approx r(n)$ and the residual error $e(n)$ is substantially echo free.

A typical impulse response of echo path is shown in Figure 9.12. The time span over the impulse response of the echo path is significant (non-zero) and is typically about 4 ms. This portion is called the dispersive delay since it is associated with the frequency-dependent delay and loss through the echo path. Because of the existence of the four-wire circuit between the location of the echo canceler and the hybrid, the impulse response of echo path is delayed. Therefore the initial samples of $p(n)$ are all zeros, representing a flat delay between the canceler and the hybrid. The flat delay depends on the transmission delay and the delay through the sharp filters associated with frequency

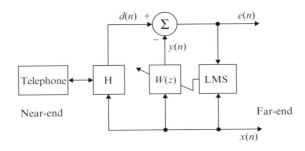

Figure 9.10 Block diagram of adaptive echo canceler

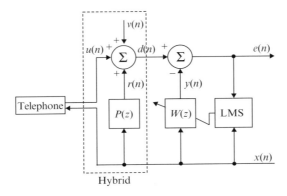

Figure 9.11 Equivalent diagram of echo canceler that show details of hybrid function

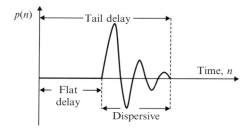

Figure 9.12   Typical impulse response of an echo path

division multiplex equipment. The sum of the flat delay and the dispersive delay is called the tail delay.

Assume that the echo path $P(z)$ is linear, time invariant, and with infinite impulse response $p(n)$, $n = 0, 1, \ldots, \infty$. As shown in Figure 9.11, the primary signal $d(n)$ can be expressed as

$$d(n) = r(n) + u(n) + v(n) = \sum_{l=0}^{\infty} p(l)x(n-l) + u(n) + v(n), \qquad (9.4.1)$$

where the additive noise $v(n)$ is assumed to be uncorrelated with the near-end speech $u(n)$ and the echo $r(n)$. The most widely used FIR filter generates the echo mimic

$$y(n) = \sum_{l=0}^{L-1} w_l(n)x(n-l) \qquad (9.4.2)$$

to cancel the echo signal $r(n)$. The estimation error is expressed as

$$
\begin{aligned}
e(n) &= d(n) - y(n) \\
&= \sum_{l=0}^{L-1} [p(l) - w_l(n)]x(n-l) + \sum_{l=L}^{\infty} p(l)x(n-l) + u(n) + v(n).
\end{aligned}
\qquad (9.4.3)
$$

As shown in (9.4.3), the adaptive filter $W(z)$ has to adjust its weights to mimic the response of echo path in order to cancel out the echo signal. The simple normalized LMS algorithm introduced in Section 8.4 is used for most voice echo cancellation applications. Assuming that disturbances $u(n)$ and $v(n)$ are uncorrelated with $x(n)$, we can show that $W(z)$ will converge to $P(z)$. Unfortunately this requires L to be quite large in many applications Echo cancellation is achieved if as $W(z) \approx P(z)$ shown in (9.4.3). Thus the residual error after the echo canceler has converged can be expressed as

$$e(n) \approx \sum_{l=L}^{\infty} p(l)x(n-l) + u(n) + v(n). \qquad (9.4.4)$$

By making the length L of $W(z)$ sufficiently long, this residual echo can be minimized. However, the excess MSE produced by the adaptive algorithm is also proportional to L.

Therefore there is an optimum order L that will minimize the MSE if an FIR filter is used.

The number of coefficients for the transversal filter is directly related to the tail delay (total delay) of the channel between the echo canceler and the hybrid. As mentioned earlier, the length of the impulse response of the hybrid (dispersive delay) is relatively short. However, the transmission delay (flat delay) from the echo canceler to the hybrid depends on the physical location of the echo canceler. As shown in Figure 9.13, the split echo canceler configuration is especially important for channels with particularly long delays, such as satellite channels. In the split configuration, the number of transversal filter coefficients need only compensate for the delay between the hybrid and the canceler and not the much longer delay through the satellite. Hence, the number of coefficients is minimized.

The design of an adaptive echo canceler involves many considerations, such as the speed of adaptation, the effect of near-end and far-end signals, the impact of signal levels and spectra, and the impact of nonlinearity. The echo canceler must accurately model the echo path and rapidly adapt to its variation. This involves the selection of an adaptive filter structure and an adaptation algorithm. Because the potential applications of echo cancellation are numerous, there have been considerable activities in the design of echo cancellation devices. The best selection depends on performance requirements for a particular application.

The effectiveness of an echo canceler is measured by the echo return loss enhancement (ERLE) defined as

$$\text{ERLE} = 10 \log \left\{ \frac{E[d^2(n)]}{E[e^2(n)]} \right\}. \tag{9.4.5}$$

For a given application, the ERLE depends on the step size μ, the filter length L, the signal-to-noise ratio (SNR), and the nature of signal in terms of power and spectral content. A larger value of step size provides a faster initial convergence, but the final ERLE is smaller due to the excess MSE. Provided the length is large enough to correct for the length of echo tail, increasing L further is detrimental since doubling L will reduce the ERLE by 3 dB.

Most echo cancelers aim at canceling echo components up to 30 dB. Further reduction of the residual echo can be achieved by using a residual echo suppressor that will be

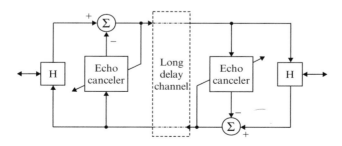

Figure 9.13 Split echo cancellation configuration

discussed later. Detailed requirements of an echo canceler are described in ITU recommendations G.165 and G.168, including the maximum residual echo level, the echo suppression effect on the hybrid, the convergence time must be less than 500 ms, the initial set-up time, and degradation in a double-talk situation.

The first special-purpose chip for echo cancellation implements a single 128-tap adaptive echo canceler [17]. Most echo cancelers were implemented using customized devices in order to handle the large amount of computation associated with it in real-time applications. Disadvantages of VLSI implementation include the high development cost and a lack of flexibility to meet application-specific requirements and improvements. There has been considerable activity in the design of devices using commercially available DSP chips.

9.4.3 Practical Considerations

There are some practical issues to be considered in designing adaptive echo canceler: (1) Adaptation should be stopped if the far-end signal $x(n)$ is absent. (2) We must also worry about the quality of adaptation over the large dynamic range of far-end signal power. (3) The adaptive process benefits when the far-end signal contains a well-distributed frequency component to persistently excite the adaptive system and the interfering signals $u(n)$ and $v(n)$ are small. When the reference $x(n)$ is a narrowband signal, the adaptive filter response cannot be controlled at frequencies other than that frequency band. If the reference signal later changes to a broadband signal, then the canceler may actually become an echo generator. Therefore a tone detector may be used to inhibit adaptation in this case.

As discussed in Section 9.4.2, the initial part of the impulse response of the echo path is all zeros, representing a flat transmission delay between the canceler and the hybrid. To take advantage of the flat delay, the structure illustrated in Figure 9.14 was developed, where Δ is a measure of flat delay and the order of shorter echo canceler $W(z)$ is $L - \Delta$. Estimation of the number of zero coefficients and using buffer indexing, one does not need to perform the actual adaptive filtering operation on the zero coefficients but simply index into the buffer appropriately. This technique can effectively reduce the real-time computational requirement. However, there are two difficulties: the multiple echoes and there has not been a good way to estimate the flat delay.

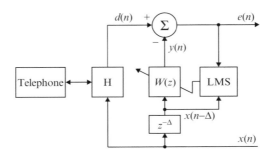

Figure 9.14 Adaptive echo canceler with effective flat-delay compensation

9.4.4 Double-Talk Effects and Solutions

Another extremely important problem of designing adaptive echo canceler is to handle double-talking, which is the simultaneous presence of both echo and near-end speech. An adaptive echo canceler estimates the impulse response of echo path using $x(n)$ and $d(n)$ as shown in Figure 9.11. For correctly identifying the characteristics of $P(z)$, $d(n)$ must originate solely from its input signal. During the double-talk periods, the error signal $e(n)$ described in (9.4.4) contains the residual echo, the uncorrelated noise $v(n)$, and the near-end speech $u(n)$. The effect of interpreting near-end speech as an error signal and making large corrections to the adaptive filter coefficients is a serious problem.

As long as the far-end signal $x(n)$ is uncorrelated with the near-end speech $u(n)$, this signal will not affect the asymptotic mean value of the filter coefficients. However, the variation in the filter coefficients about this mean will be increased substantially in the presence of the near-end talker due to the introduction of another large stochastic component in the adaptation. Thus the adaptive filter $W(z)$ is greatly disturbed in a very short time, resulting in performance degradation of the echo canceler. An unprotected algorithm may exhibit unacceptable behavior during double-talk periods and so some mechanisms to avoid its effects must be included. This problem may be solved by using a very small step size μ. However, this may result in slow adaptation.

An effective approach for the solving double-talk problem is to detect the occurrence of double-talking and to disable the adaptation of $W(z)$ during these periods. Note that nly the coefficient adaptation is disabled as illustrated in Figure 9.15, and the transmission channel remains open in both directions at all times. If the echo path does not change appreciably during the double-talk periods, the echo still can be canceled by the previously converged coefficients of $W(z)$ that are fixed during double-talk periods.

As shown in Figure 9.15, the speech detection and control block is used to control the adaptation of the adaptive filter $W(z)$ and the residual echo suppressor. The complexity of the double-talk detector (DTD), which detects the presence of near-end speech when the far-end speech is present, is much higher. A DTD is a very critical element in echo cancelers since an adaptive filter diverges quickly during double-talk unless the adaptation process is inhibited.

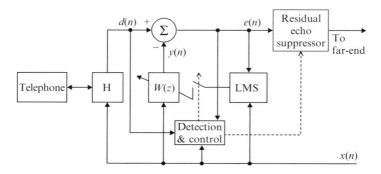

Figure 9.15 Adaptive echo canceler with speech detectors and residual echo suppressor

The conventional DTD is based on the echo return loss (ERL) or hybrid loss, which can be expressed as

$$\rho = 20 \log_{10} \left\{ \frac{E[|x(n)|]}{E[|d(n)|]} \right\}. \tag{9.4.6}$$

If the echo path is time-invariant, the ERL may be measured during the training period for some applications. In several adaptive echo cancelers, the value of ERL is assumed to be 6 dB. Based on this assumption, the near-end speech is present if

$$|d(n)| > \frac{1}{2} |x(n)|. \tag{9.4.7}$$

However, we cannot just compare the instantaneous absolute values $|d(n)|$ and $|x(n)|$ because of noise. Therefore the modified near-end speech detection algorithm declares the presence of near-end speech if

$$|d(n)| > \frac{1}{2} \max\{|x(n)|, \ldots, |x(n - L + 1)|\}. \tag{9.4.8}$$

This algorithm compares an instantaneous absolute value $|d(n)|$ with the maximum absolute value of $x(n)$ over a time window spanning the echo path delay range. The advantage of using an instantaneous power of $d(n)$ is fast response to the near-end speech. However, it will increase the probability of false alarm if noise exists in the network.

A more robust version of the algorithm uses short-term power estimates $P_x(n)$ and $P_d(n)$ to replace the instantaneous power $|x(n)|$ and $|d(n)|$. The short-term power estimates are implemented as the first-order IIR filter as follows:

$$P_x(n) = (1 - \alpha)P_x(n - 1) + \alpha|x(n)| \tag{9.4.9}$$

and

$$P_d(n) = (1 - \alpha)P_d(n - 1) + \alpha|d(n)|, \tag{9.4.10}$$

where $0 < \alpha \ll 1$. The use of a larger α results in robust detector in noise. However, it also results in slower response to the present of near-end speech. With these modified short-term power estimates, the near-end speech is detected if

$$P_d(n) > \frac{1}{2} \max\{P_x(n), P_x(n - 1), \ldots, P_x(n - L + 1)\}. \tag{9.4.11}$$

It is important to note that a considerable portion of the initial break-in near-end speech $u(n)$ may not be detected by this detector. Thus adaptation would proceed for a considerable amount of time in the presence of double-talking. Furthermore, the requirement of the buffer to store L power estimates increases the complexity of algorithm.

The assumption that the ERL is a constant of value 6 dB is usually incorrect in most applications. Even if the ERL is 6 dB, $p_d(n)$ can still be greater than the threshold without near-end speech because of the dispersive characteristics of the echo path and/ or far-end speech. If the ERL is higher than 6 dB, it will take longer to detect the presence of near-end speech. On the other hand, if the ERL is below 6 dB, most far-end speech will be falsely detected as near-end speech. For practical applications, it is better to dynamically estimate the time-varying threshold ρ by observing the signal level of $x(n)$ and $d(n)$ when the near-end speech $u(n)$ is absent.

9.4.5 Residual Echo Suppressor

Nonlinearities in the echo path of the telephone circuit and uncorrelated near-end speech limit the amount of achievable cancellation in a typical adaptive echo canceler from 30 to 35 dB. The residual echo suppressor shown in Figure 9.15 is used to remove the last vestiges of remaining echo. This device also effectively removes echo during the initial convergence of the echo canceler if off-line training stage is prohibited.

The most widely used residual echo suppressor is a center clipper with an input–output characteristic illustrated in Figure 9.16. The center clipper is used to remove the low-level echo signal caused by circuit noises, finite-precision errors, etc., which cannot be canceled by the echo canceler. This nonlinear operation is expressed as

$$y(n) = \begin{cases} 0, & |x(n)| \leq \beta \\ x(n), & |x(n)| > \beta, \end{cases} \tag{9.4.12}$$

where β is the clipping level. This center clipper completely eliminates signals below the clipping level, but leaves instantaneous signal values greater than the clipping level unaffected. Thus large signals go through unchanged, but small signals are eliminated. Since small signals are consistent with echo, the device achieves the function of residual echo suppression. The clipping threshold β determines how 'choppy' the speech will sound with respect to the echo level. A large value of β suppresses all the residual echoes but also deteriorates the quality of the near-end speech. Usually the threshold is set so as to equal or exceed the return echo peak amplitude.

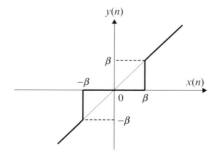

Figure 9.16   Input–output relationship of center clipper

As discussed in Section 9.2.3, the complete suppression of a signal has an undesired effect. This problem was solved by injecting a low-level comfort noise when the signal is suppressed. The comfort noise algorithm is described by (9.2.6) and is shown in Figure 9.3.

9.5   Acoustic Echo Cancellation

In recent years there has been a growing interest in acoustic echo cancellation for a hands-free cellular phone in mobile environments and a speakerphone in teleconferencing applications. There are three major components to form acoustic echoes: (1) Acoustic energy coupling between microphone and loudspeaker. (2) Multiple-path sound reflections known as the room reverberations of far-end speech signals. (3) The sound reflections of the near-end speech signal. These echoes disturb natural dialog and, at its worst, cause howling. Echo suppressors, earphone, and directional microphones have been conventional solutions to these problems, but have placed physical restrictions on the talkers. In this section, we are mainly concerned with the cancellation of first two components due to the fact that the near-end speech's echo is much less significant in terms of degrading the system performance.

9.5.1   Introduction

The speakerphone has become an important piece of office equipment because it provides the user the convenience of hands-free telephone conversation. When the telephone connection is between one or more hands-free telephones or between two conference rooms, a major source of echoes is the acoustic coupling between the loudspeaker and the microphone at each end. For reference purposes, the person using the speakerphone is the near-end talker and the person at the other end of the connection is the far-end talker. In the speakerphone illustrated in Figure 9.17, the far-end speech carried by telephone lines is put out through one or more loudspeakers. A power amplifier is used to produce a higher volume (usually 3 dB above normal speech level) of the incoming far-end signal so that it can be heard clearly by persons in the

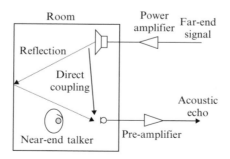

Figure 9.17   Acoustic echo in a room

conference room. Unfortunately, not only the direct coupling but also the sound bounces back and forth between the walls and the furniture will be picked up by the microphones and be transmitted back to the far-end. These acoustic echoes can be very annoying because they cause the far-end talker to hear a delayed version of his or her own speech.

The most effective technique to eliminate the acoustic echo is to use an adaptive echo cancellation discussed in the previous section. The basic concept of acoustic echo cancellation is similar to the line echo cancellation. However, the acoustic echo canceler requirements are different from those of line echo cancelers due to the fact that their functions are different and the different nature of the echo paths. Instead of the mismatch of the hybrid, a loudspeaker-room-microphone system needs to be modeled in these applications. The acoustic echo canceler controls the long echo using a high-order adaptive FIR filter. This full-band acoustic echo canceler will be discussed in this section. A more effective technique to cancel the acoustic echo is called the subband acoustic echo canceler, in which the input signal is split into several adjacent subbands and uses an independent low-order filter in each subband.

Compared with line echo cancellation, there are three major factors making the acoustic echo cancellation far more difficult. These factors are summarized as follows:

1. The reverberation of a room causes a long acoustic echo tail. As introduced in Section 4.5.2, the duration of the impulse response of the acoustic echo path is usually from 100 to 500 ms in a typical conference room. If an adaptive FIR filter is used to model this echo path, the order of the filter could be very high. For example, 3200 taps are needed to cancel 400 ms of echo at sampling rate 8 kHz.

2. The acoustic echo path is generally non-stationary and it may change rapidly due to the motion of people in the room, the position changes of the microphone and some other factors like temperature change, doors and/or windows opened or closed, etc. The canceler should trace these changes quickly enough to cancel the echoes, thus requiring a faster convergence algorithm.

3. The double-talk detection is much difficult since there is no guarantee of having a 6 dB acoustic loss such as the hybrid loss in the line echo case.

Therefore acoustic echo cancelers require more computation power, faster convergence adaptive algorithms, and more sophisticated double-talk detectors.

9.5.2 Acoustic Echo Canceler

A block diagram illustrating the use of an acoustic echo canceler in teleconferencing is given in Figure 9.18. The far-end speech $x(n)$ is diffused into the room by a loudspeaker and will be picked up by a microphone. The acoustic echoes occurred between the loudspeaker and the microphone can be modeled as an acoustic echo path $P(z)$, including the D/A converter, the smoothing lowpass filter, the power amplifier, the loudspeaker, the microphone, preamplifier, anti-aliasing lowpass filter, A/D converter, and the room transfer function from the loudspeaker to the microphone. The

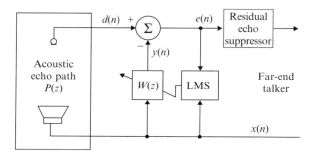

Figure 9.18   Block diagram of acoustic echo canceler

adaptive filter $W(z)$ models the acoustic echo path $P(z)$ and yields an echo replica $y(n)$, which is used to cancel acoustic echo components in the microphone signal $d(n)$.

An acoustic echo canceler removes the acoustic echoes by using the adaptive filter $W(z)$ to generate a replica of the echo expressed as

$$y(n) = \sum_{l=0}^{L-1} w_l(n)x(n - l). \qquad (9.5.1)$$

This replica is then subtracted from the microphone signal $d(n)$. The coefficients of $W(z)$ are updated by the normalized LMS algorithm expressed as

$$w_l(n + 1) = w_l(n) + \mu(n)e(n)x(n - l), \quad l = 0, 1, \ldots, L - 1, \qquad (9.5.2)$$

where $\mu(n)$ is the normalized step size by the power estimate of $x(n)$ and $e(n) = d(n) - y(n)$. This adaptation must be stopped if the near-end talker is speaking.

Acoustic echo cancelers usually operate in two modes. In the off-line training mode discussed in Section 9.2.4, the impulse response of the acoustic echo path is estimated with the use of white noise or chirp signals as the training signal $x(n)$. During the training mode, the correct length of the echo path response may be determined. In the subsequent on-line operating mode, an adaptive algorithm is used to track slight variations in the impulse response of echo path using the far-end speech $x(n)$.

9.5.3   Implementation Considerations

As shown in Figure 9.14, an effective technique to reduce filter order is to introduce a delay buffer of Δ samples at the input of adaptive filter. This compensates for delay in the echo path caused by the propagation delay from the loudspeaker to the microphone. Introducing this delay allows us to save on computation since it effectively forces these Δ coefficients to 0 without having to update their values. For example, with the distance between the loudspeaker and the microphone is around 10 cm, the measured time delay in the system is about 4.62 ms, which corresponds to $\Delta = 37$ at 8 kHz sampling rate.

For an adaptive FIR filter with the LMS algorithm, a large L requires a small step size μ, thus resulting in slow convergence. Therefore the filter is unable to track the relatively fast transient behavior of the acoustic echo path $P(z)$. Perhaps the number of taps could be reduced significantly by modeling the acoustic echo path as an IIR filter. However, there are difficulties such as the stability associated with the adaptive IIR structures.

As discussed in Chapter 8, if fixed-point arithmetic is used for implementation and μ is sufficiently small, the excess MSE is increased when a large L is used, and the numerical error (due to coefficient quantization and roundoff) is increased when a large L and a small are μ used. Furthermore, roundoff causes early termination of the adaptation when a small μ is used. In order to alleviate these problems, a higher dynamic range is required, which can be achieved by using floating-point arithmetic. However, this solution includes the added cost of more expensive hardware.

As mentioned earlier, the adaptation of coefficients must be stopped when the near-end talker is speaking. Most double-talk detectors for adaptive line echo cancelers discussed in the previous section are based on echo return loss (or acoustic loss) from the loudspeaker to the microphone. In acoustic echo canceler cases, this loss is very small or may be even a gain because of amplifiers used in the system. Therefore the higher level of acoustic echo than the near-end speech makes detection of weak near-end speech very difficult.

9.6 Speech Enhancement Techniques

In many speech communication settings, the presence of background noise degrades the quality or intelligibility of speech. This section discusses the design of single-channel speech enhancement (or noise reduction) algorithms, which use only one microphone to reduce background noise in the corrupted speech without an additional reference noise.

9.6.1 Noise Reduction Techniques

The wide spread use of cellular/wireless phones has significantly increased the use of communication systems in high noise environments. Intense background noise, however, often degrades the quality or intelligibility of speech, degrading the performance of many existing signal processing techniques such as speech coding, speech recognition, speaker identification, channel transmission, and echo cancellation. Since most voice coders and voice recognition units assume high signal-to-noise ratio (SNR), low SNR will deteriorate the performance dramatically. With the development of hands-free and voice-activated cellular phones, the noise reduction becomes increasingly important to improve voice quality in noisy environments.

The purpose of many speech enhancement algorithms is to reduce background noise, improve speech quality, or suppress undesired interference. There are three general classes of speech enhancement techniques: subtraction of interference, suppression of harmonic frequencies, and re-synthesis using vocoders. Each technique has its own set of assumptions, advantages, and limitations. The first class of techniques suppresses noise by subtracting a noise spectrum, which will be discussed in Section 9.6.2. The second type of speech enhancement is based on the periodicity of noise. These methods employ fundamental frequency tracking using adaptive comb filtering of the harmonic noise. The third class of techniques is based on speech modeling using iterative methods. These systems focus on estimating model parameters that characterize the speech signal, followed by re-synthesis of the noise-free signal. These techniques require a prior knowledge of noise and speech statistics and generally results in iterative enhancement schemes.

Noise subtraction algorithms can also be partitioned depending on whether a single-channel or dual-channel (or multiple-channel) approach is used. A dual-channel adaptive noise cancellation was discussed in Section 8.5. In this type of system, the primary channel contains speech with additive noise and the reference channel contains a reference noise that is correlated to the noise in the primary channel. In situations such as telephone or radio communications, only a single-channel system is available. A typical single-channel speech enhancement system is shown in Figure 9.19, where noisy speech $x(n)$ is the input signal of the system, which contains the speech signal $s(n)$ from the speech source and the noise $v(n)$ from the noise source. The output signal is the enhanced speech $\hat{s}(n)$. Characteristics of noise can only be estimated during silence periods between utterances, under the assumption that the background noise is stationary.

This section concentrates on the signal-channel speech enhancement system. Since only a single recording is available and the performance of the noise suppression system is based upon the accuracy of the background noise estimate, speech enhancement techniques must estimate noise characteristics during the non-speech periods when only background noise is present. Therefore an effective and robust voice activity detector (VAD) plays an important role in the single-channel noise suppression system.

Noise subtraction algorithms can be implemented in time-domain or frequency-domain. Based on the periodicity of voiced speech, the time-domain adaptive noise canceling technique can be utilized by generating a reference signal that is formed by delaying the primary signal by one period. Thus a complicated pitch estimation algorithm is required. Also, this technique can only be applied for voiced speech, but fails to process unvoiced speech. The frequency-domain implementation is based on short-time spectral amplitude estimation called the spectral subtraction. The basic idea was to obtain the short-term magnitude and phase spectra of the noisy speech during speech frames using the FFT, subtracting by an estimated noise magnitude spectrum, and inverse transforming this subtracted spectral amplitude using the phase of the original noisy speech. The enhancement procedure is performed frame-by-frame, thus data buffer requirements, block data handling, and time delay imposed by FFT complicate this technique for some real-time applications. Also, musical tone artifacts are often heard at frame boundaries in such reconstructed speech.

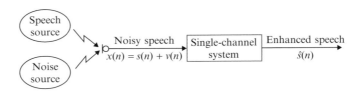

Figure 9.19 Single-channel speech enhancement system

Frequency-domain noise suppression also can be implemented in time-domain by first decomposing the corrupted speech signal into a different frequency band using bandpass filterbank. The noise power of each subband is then estimated during non-speech periods. Noise suppression is achieved through the use of the attenuation factor corresponding to the temporal signal power over estimated noise power ratio of each subband. Since the spectral subtraction algorithm provides the basic concept of filter-bank technique, it will be presented in detail in the next section.

9.6.2   Spectral Subtraction Techniques

Spectral subtraction offers a computationally efficient approach for reducing noise by using the FFT. Assume that a speech signal $s(n)$ has been degraded by an uncorrelated additive noise $v(n)$. As illustrated in Figure 9.20, this approach enhances speech by subtracting the estimate of the noise spectrum from the noisy speech spectrum. The noisy speech $x(n)$ is segmented and windowed. The FFT of each data window is taken and the magnitude spectrum is computed. A VAD is used to detect the speech and non-speech activities of the input signal. If the speech frame is detected, the system will perform the spectral subtraction and the enhanced speech signal $\hat{s}(n)$ will be generated. During the non-speech segment, the noise spectrum will be estimated and the data in the buffer will be attenuated to reduce noise.

There are two methods for generating the output during non-speech periods: (1) Attenuate the output by a fixed factor, and (2) set the output to 0. The experimental results show that having some residual noise (comfort noise) during non-speech frame will give higher speech quality. A possible reason for this is that noise present during speech frames is partially masked by the speech. Its perceived magnitude should be

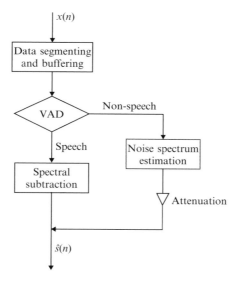

Figure 9.20   Block diagram of the spectral subtraction algorithm

balanced by the presence of the same amount of noise during non-speech segments. Setting the output to 0 has the effect of amplifying the noise during the speech segments. Therefore it is best to attenuate the noise by a fixed factor during the non-speech periods. A balance must be maintained between the magnitude and characteristics of the noise perceived during the speech segment and the noise that is perceived during the noise segment. A reasonable amount of attenuation was found to be about -30 dB. As a result, some undesirable audio effects such as clicking, fluttering, or even slurring of the speech signal are avoided.

As mentioned earlier, input signal from the A/D converter is segmented and windowed. To do this, the input sequence is separated into a half (50 percent) overlapped data buffer. Data at each buffer is then multiplied by the coefficients of the Hanning (or Hamming) window. After the noise subtraction, the time-domain enhanced speech waveform is reconstructed by the inverse FFT. These output segments are overlapped and added to produce the output signal. The processed data is stored in an output buffer.

Several assumptions were made for developing the algorithm. We assume that the background noise remains stationary such that its expected magnitude spectrum prior to speech segments unchanged during speech segments. If the environment changes, there is enough time to estimate a new magnitude spectrum of background noise before speech frame commences. For the slowly varying noise, the algorithm requires a VAD to determine that speech has ceased and a new noise spectrum could be estimated. The algorithm also assumes that significant noise reduction is possible by removing the effect of noise from the magnitude spectrum only.

Assuming that a speech signal $s(n)$ has been degraded by the uncorrelated additive signal $v(n)$, the corrupted noisy signal can be expressed as

$$x(n) = s(n) + v(n). \tag{9.6.1}$$

Taking the DFT of $x(n)$ gives

$$X(k) = S(k) + V(k). \tag{9.6.2}$$

Assuming that $v(n)$ is zero-mean and uncorrelated with $s(n)$, the estimate of $|S(k)|$ can be expressed as

$$|\hat{S}(k)| = |X(k)| - E|V(k)|, \tag{9.6.3}$$

where $E|V(k)|$ is the expected noise spectrum taken during the non-speech periods. Given the estimate $|\hat{S}(k)|$, the spectral estimate can be expressed as

$$\hat{S}(k) = |\hat{S}(k)|e^{j\theta_x(k)}, \tag{9.6.4}$$

where

$$e^{j\theta_x(k)} = \frac{X(k)}{|X(k)|} \tag{9.6.5}$$

and $\theta_x(k)$ is the phase of measured noisy signal. It is sufficient to use the noisy speech phase for practical purposes. Therefore we reconstructed the processed signal using the estimate of short-term speech magnitude spectrum $|\hat{S}(k)|$ and the phase of degraded speech, $\theta_x(k)$.

Substituting Equations (9.6.3) and (9.6.5) into Equation (9.6.4), the estimator can be expressed as

$$\hat{S}(k) = [|X(k)| - E|V(k)|]\frac{X(k)}{|X(k)|} = H(k)X(k), \qquad (9.6.6)$$

where

$$H(k) = 1 - \frac{E|V(k)|}{|X(k)|}. \qquad (9.6.7)$$

Note that the spectral subtraction algorithm given in Equations (9.6.6) and (9.6.7) avoids computation of the phase $\theta_x(k)$, which is too complicated to implement in a fixed-point hardware.

9.6.3 Implementation Considerations

A number of modifications are developed to reduce the auditory effect of spectral error. These methods are spectral magnitude averaging, half-wave rectification, and residual noise reduction. A detailed diagram for spectral subtraction algorithm is illustrated in Figure 9.21.

Spectral magnitude averaging

Since the spectral error is proportional to the difference between the noise spectrum and its mean, local averaging of the magnitude spectral

$$|X(k)| = \frac{1}{M}\sum_{i=1}^{M}|X_i(k)| \qquad (9.6.8)$$

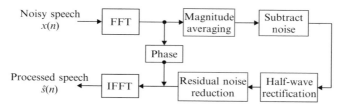

Figure 9.21 Detailed diagram of spectral subtraction algorithm

can be used to reduce the spectral error, where $X_i(k)$ is ith time-windowed transform of $x(n)$. One problem with this modification is that the speech signal has been considered as short-term stationary for a maximum of 30 ms. The averaging has the risk of some temporal smearing of short transitory sounds. From the simulation results, a reasonable compromise between variance reduction and time resolution appears to be averaging 2–3 frames.

Half-wave rectification

For each frequency bin where the signal magnitude spectrum $|X(k)|$ is less than the averaged noise magnitude spectrum $E|V(k)|$, the output is set to 0 because the magnitude spectrum cannot be negative. This modification can be implemented by half-wave rectifying the spectral subtraction filter $H(k)$. Thus Equation (9.6.6) becomes

$$\hat{S}(k) = \frac{H(k) + |H(k)|}{2} X(k). \tag{9.6.9}$$

The advantage of half-wave rectification is that any low variance coherent tonal noise is essentially eliminated. The disadvantage of half-wave rectification occurs in the situation where the sum of noise and speech at a frequency bin k is less than $E|V(k)|$. In this case the speech information at that frequency is incorrectly removed, implying a possible decrease in intelligibility.

As mentioned earlier, a small amount of noise improved the output speech quality. This idea can be implemented by using a software constraint

$$|S(k)| \geq 0.02E|V(k)|, \tag{9.6.10}$$

where the minimum spectrum floor is -34 dB respected to the estimated noise spectrum.

Residual noise reduction

For uncorrelated noise, the residual noise spectrum occurs randomly as narrowband magnitude spikes. This residual noise spectrum will have a magnitude between 0 and a maximum value measured during non-speech periods. When these narrowband components are transformed back to the time domain, the residual noise will sound like the sum of tones with random fundamental frequency which is turned on and off at a rate of about 20 ms. During speech frame the residual noise will also be perceived at those frequencies which are not masked by the speech.

Since the residual noise will randomly fluctuate in amplitude at each frame, it can be suppressed by replacing its current value with its minimum value chosen from the adjacent frames. The minimum value is used only when $|\hat{S}(k)|$ is less than the maximum residual noise calculated during non-speech periods. The motivation behind this replacement scheme is threefold: (1) If the $|\hat{S}(k)|$ lies below the maximum residual noise and it varies radically from frame to frame, there is a high probability that the spectrum at that frequency is due to noise. Therefore it can be suppressed by taking the minimum value. (2) If $|\hat{S}(k)|$ below the maximum but has a nearly constant value, there is a high probability that the spectrum at that frequency is due to low-energy speech. Therefore taking the minimum will retain the information. (3) If $|\hat{S}(k)|$ is greater than the maximum, the bias is sufficient. Thus the estimated spectrum $|\hat{S}(k)|$ is used to reconstruct the output speech. However, with this approach high-energy frequency bins are not averaged together. The disadvantages to the scheme are that more storage is required to save the maximum noise residuals and the magnitude values for three adjacent frames, and more computations are required to find the maximum value and minimum value of spectra for the three adjacent frames.

9.7 Projects Using the TMS320C55x

This section provides a list of experimental projects and applications that are related to communications at different levels. A large project can be partitioned into smaller modules and each portion may be simple in terms of algorithm development, simulation, and DSP implementation. The algorithms range from signal generation, error correction coding, filtering, to channel simulation.

9.7.1 Project Suggestions

Some DSP applications that can be used as the course projects for this book are listed in this section. Brief descriptions are provided, so that we can evaluate and define the scope of each project. The numbers in the parentheses indicate the level of difficulty of the projects, where the larger the number, the greater the difficulty.

Signal Generation and Simulation

1. White Gaussian noise generator (1)

2. Sinusoidal signal generator (1)

3. Telephone channel simulator (2)

4. Wireless fading channel simulator (2)

Adaptive Echo Canceler and Equalizer

5. Adaptive data echo canceler for dial-up modem applications (3)

6. Adaptive acoustic echo canceler for speakerphone applications (3)

7. Adaptive channel equalizer (2)

8. Adaptive equalizer for wireless communications (3)

Speech Codecs

9. A-law and μ-law companding (1)

10. International Telecommunications Union (ITU) G.726 ADPCM (3)

11. GSM full-rate vocoder (3)

Telecommunications

12. DTMF tone generation (1)

13. DTMF tone detection using Goertzel algorithm (2)

14. ITU V.21 FSK transmitter (2)

15. ITU V.22bis quadrature amplitude modulation – QAM (2)

16. ITU V.32bis trellis code modulation – TCM (2)

Error Coding

17. Cyclic redundancy code (1)

18. ITU V.32bis scrambler/de-scrambler (1)

19. ITU 32bis convolutional encoder (1)

20. ITU 32bis Viterbi decoder (3)

21. Interleaver/de-interleaver (1)

22. Reed–Solomon Encoder (3)

23. Reed–Solomon Decoder (3)

24. Turbo Decoder (3)

Image Processing

25. Discrete cosine transformation (DCT) (2)

26. Inverse discrete cosine transformation (IDCT) (2)

Signal generation and simulation are used widely for DSP algorithm development. They are an integrated part of many applications implemented using DSP processors. The most widely used signal generators are the sinusoid and random number generators. For designing modern communication applications, engineers often use channel simulations to study and implement DSP applications. The widely used channel models are telephone and wireless channels.

As discussed in previous sections, echoes exist in both the full-duplex dial-up telephone networks and hands-free telephones. For a full-duplex telephone communication, there exists the near-end and far-end data echoes. The adaptive data echo canceler is required for high-speed modems. The acoustic echo canceler is needed for speaker-phone applications used in conference rooms.

Speech codecs are the voice coder–decoder pairs that are used for transmitting speech (audio) signals more efficiently than passing the raw audio data samples. At the 8 kHz sampling rate, the 16-bit audio data requires the rate of 1 28 000 bits per second (= 128 kbps) to transmit through a given channel. By using speech codecs with speech compression techniques, many voice codecs can pass speech at a rate of 8 kbps or lower. Using lower bit rate vocoders, a channel with fixed capacity will be able to serve more users.

Telecommunication has changed our daily life dramatically. DSP applications can be found in various communication devices such as cellular phones and DSL modems. Modulation techniques are widely used in communications. Quadrature amplitude modulation (QAM) is one example used by modems to transmit information over the telephone lines.

Channel coding or error coding is becoming more and more important in telecommunication applications. Convolutional encoding and Viterbi decoding, also known as forward error correction (FEC) coding techniques, are used in modems to help improve the error-free transmission. Cyclic redundant check (CRC) is widely used to verify the data code correctness in the receiver.

Image processing is another important DSP application as a result of the increasing need for video compression for transmission and storage. Many standards exist, the JPEG (joint photographic experts group) is used for still images, while the MPEG (moving picture experts group) is designed to support more advanced video communications. The image compression is centered on the block-based DCT and IDCT technologies.

9.7.2 A Project Example – Wireless Application

Wireless communication is one of the most important technologies that has been developed and greatly improved in the past several years. Digital cellular phone systems include both the infrastructure such as the cellular base-stations and the handsets all use DSP processors. Some of the systems use general-purpose DSP processors, while others use DSP cores associating with ASIC technologies. A simplified wireless communication system is illustrated in Figure 9.22. The system can be divided into three sections: transmitter, receiver, and the communication channel. The system can also be distinguished as speech coding and decoding, channel coding and decoding, and finally, modulation and demodulation.

The speech (source) coding is important DSP applications in wireless communications. The vocoders are used to compress speech signals for bandwidth limited communication channels. The most popular vocoders for wireless communications compress speech samples from 64 kbps to the range of 6–13 kbps.

The FEC coding scheme is widely used in the communication systems as an important channel coding method to reduce the bit errors on noisy channels. The FEC used in the system shown in Figure 9.22 consists of the convolutional encoding and Viterbi decoding algorithms. Modern DSP processors such as the TMS320C55x have special instructions to aid efficient implementations for the computational intensive Viterbi decoders.

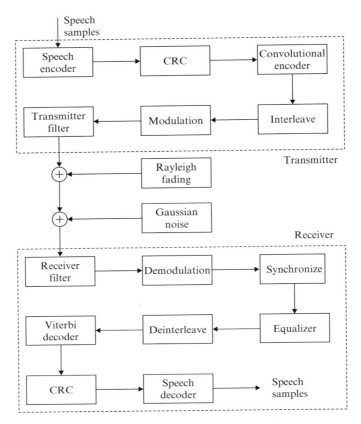

Figure 9.22   Simplified wireless communication system

Convolutional coding provides error correction capability by adding redundancy bits to the information bits. The convolutional encoding is usually implemented by either the table-lookup method or by the shift register method. Figure 9.23 shows a rate one-half (1/2) convolutional coder.

The convolutional encoder shown in Figure 9.23 can be expressed using the following two polynomial generators:

$$b_0 = 1 \oplus x \oplus x^3 \oplus x^5 \qquad\qquad (9.7.1)$$

and

$$b_1 = 1 \oplus x^2 \oplus x^3 \oplus x^4 \oplus x^5, \qquad\qquad (9.7.2)$$

where $\oplus$ denotes the modulo two adder, an XOR operation. For the rate 1/2 convolutional encoder, each input information bit has two encoded bits, where bit 0 is generated by (9.7.1) and bit 1 by generator (9.7.2). This redundancy enables the Viterbi decoder to choose the correct bits under noisy conditions. Convolutional encoding is often represented using the states. The convolutional encoder given in Figure 9.23 has a trellis of 32-state with each state connecting to two states. The basic block of this 32-state trellis diagram is illustrated in Figure 9.24.

For this encoding scheme, each encoding state at time n is linked to two states at time $n+1$ as shown in Figure 9.24. The Viterbi algorithm (see references for details) is used for decoding the trellis coded information bits by expanding the trellis over the received symbols. The Viterbi algorithm reduces the computational load by taking advantage of the special structure of the trellis codes. It calculates the 'distance' between the received signal path and all the accumulated trellis paths entering each state. After comparison, only the most likely path based on the current and past path history (called surviving path – the path with the shortest distance) is kept, and all other unlikely paths are discarded at each state. Such early rejection of the unlikely paths greatly reduces the computation needed for the decoding process. From Figure 9.24, each link from an old state at time n to a new state at time $n+1$ associates with a transition path. For example, the path m_x is the transition from state i to state j, and m_y is the transition path from state $i+16$ to state j. The accumulated path history is calculated as

$$\text{state}(j) = \min\{\text{state}(i) + m_x, \ \text{state}(i+16) + m_y\}, \tag{9.7.3}$$

where the new state history, state(j), is chosen as the smaller of the two accumulated pass history paths state(i) and state($i+16$) plus the transition paths m_x and m_y, respectively.

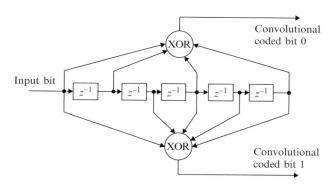

Figure 9.23   A rate 1/2, constraint length 5 convolutional encoder

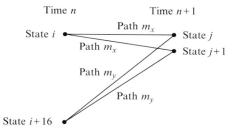

Figure 9.24   The trellis diagram of the rate 1/2 constraint length 5 convolutional encoder

In most communication systems, the cyclic redundancy check (CRC) is used to detect transmission errors. The implementation of CRC is usually done using the shift-register. For example, a 7-bit CRC can be represented using the following polynomial generator:

$$b_{CRC} = 1 \oplus x \oplus x^2 \oplus x^4 \oplus x^5 \oplus x^7. \tag{9.7.4}$$

Above CRC generator can produce a unique CRC code of a block sample up to 127 (2^7) bits. To generate the CRC code for longer data streams, use a longer CRC generator, such as the CRC16 and CRC32 polynomials, specified by the ITU.

At the front end of the system, transmit and receive filters are used to remove the out-of-band signals. The transmit filter and receive filter shown in Figure 9.22 are chosen to be square-root raised-cosine pulse-shape frequency response FIR filters. The detailed DSP implementation of FIR filters has been described in Chapter 5.

DSP processors are also often used to provide modulation and demodulation for digital communication systems. Modulation can be implemented in several ways. The most commonly used method is to arrange the transmitting symbols into the I (in-phase) and Q (quadrature) symbols. Figure 9.25 is a simplified modulation and demodulation scheme for communication systems, where w_c is the carrier frequency.

Other functions for the digital communication systems may include timing synchronization and recovery, automatic gain control (AGC), and match filtering. Although

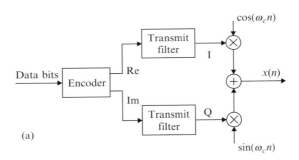

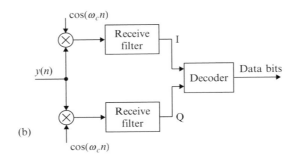

Figure 9.25 The simplified block diagram of modulator and demodulator for digital communication systems: (a) a passband transmitter modulator, and (b) a passband receiver demodulator

these functions can be implemented by either hardware or software, most modern digital communication systems use programmable DSP processors to implement these functions in software.

An important and challenging DSP task is channel equalization and estimation. The channel conditions of the wireless mobile communications are far more complicated than the dial-up channels, due to the deep channel fading characteristics and multipath interferences. Although some of the wireless communication devices use equalizers, most of them use channel estimation techniques. To combat the burst errors, interleave schemes are used. Although a severe fading may destroy an entire frame, it is unusual for the fading to last more than several frames. By spreading the data bits across a longer sequence succession, we can use the Viterbi algorithm to recover some of the lost bits at the receiver. As illustrated in Figure 9.26, a simple example of the interleaving technique is to read symbols in the order of row-by-row and write them out in the order of column-by-column.

In this example, the input data is coming in the order of 5 bits per frame as $\{b_0,\ b_1,\ b_2,\ b_3,\ b_4\}$, $\{b_5,\ b_6,\ b_7,\ b_8,\ b_9\}$, etc. as shown in Figure 9.26(a). These bits are written into a buffer column-by-column as shown in Figure 9.26(b), but are read-out row-by-row in the order of 5 bits $\{b_0,\ b_5,\ b_{10},\ b_{15},\ b_{20}\}$, etc. As a result, not all the bits from one frame will be lost by a bad receive slot due to the wireless channel fading.

Another important component in Figure 9.22 is the Rayleigh fading channel model. In a real-world mobile communication environment, the radio signals that a receiver antenna picks up comes from many paths caused by the surrounding buildings, trees, and many other objects. These signals can become constructive or destructive. As a mobile phone user travels, the relationship between the antenna and those signal paths changes, which causes the fading to be combined randomly. Such an effect can be modeled by the Rayleigh distribution, and hence, it is called Rayleigh fading. In order to provide a mobile communication environment for wireless design and research, effects of multipath fading must be considered. However, field driving tests or using hardware faders may not be the solutions for software development, debugging, and testing. A couple of Rayleigh fading models has been developed in the past; one was proposed by Jakes [34], and the second model using a second-order IIR filter. The Jakes fading channel model can be implemented in C as shown in Table 9.3. The channel noise is simulated using a white Gaussian noise generator. A simulation program of a simplified time-division multiple access (TDMA) wireless communication system is available in the software package.

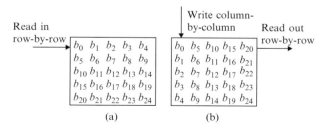

Figure 9.26 A simple example of interleave: (a) before interleave, and (b) after interleave

Table 9.3   Fading channel model proposed by Jakes

```
/*   PI = 3.14
     C = 30 00 00 000 m/s
     V = Mobile speed in mph
     Fc = Carrier frequency in Hz
     N = Number of simulated multi-path signals
     N0 = N/4 − 1/2, the number of oscillators
*/
wm = 2* PI* V* Fc/C;
xc(t) = sqrt(2)* cos(PI/4)* cos(wm* t);
xs(t) = sqrt(2)* sin(PI/4)* cos(wm* t);
for (n = 1; n <= N0; n++)
{
    wn = wm* cos(2* PI* n/N);
    xc(t) += 2* cos(PI* n/N0)* cos(wm* t);
    xs(t) += 2* sin(PI* n/N0)* cos(wm* t);
}
```

References

[1] S. M. Kuo and D. R. Morgan, *Active Noise Control Systems – Algorithms and DSP Implementations*, New York: Wiley, 1996.

[2] D. E. Knuth, *The Art of Computer Programming, vol. 2: Seminumerical Algorithms*, 2nd Ed., Reading, MA: Addison-Wesley, 1981.

[3] N. Ahmed and T. Natarajan, *Discrete-Time Signals and Systems*, Englewood Cliffs, NJ: Prentice-Hall, 1983.

[4] A. V. Oppenheim and R. W. Schafer, *Discrete-Time Signal Processing*, Englewood Cliffs, NJ: Prentice-Hall, 1989.

[5] S. J. Orfanidis, *Introduction to Signal Processing*, Englewood Cliffs, NJ: Prentice-Hall, 1996.

[6] J. G. Proakis and D. G. Manolakis, *Digital Signal Processing – Principles, Algorithms, and Applications*, 3rd Ed., Englewood Cliffs, NJ: Prentice-Hall, 1996.

[7] A. Bateman and W. Yates, *Digital Signal Processing Design*, New York: Computer Science Press, 1989.

[8] G. L. Smith, 'Dual-tone multifrequency receiver using the WE DSP16 digital signal processor,' Application note, AT&T.

[9] Analog Devices, *Digital Signal Processing Applications Using the ADSP-2100 Family*, Englewood Cliffs, NJ: Prentice-Hall, 1990.

[10] P. Mock, 'Add DTMF generation and decoding to DSP-uP designs,' in *Digital Signal Processing Applications with the TMS320 Family*, Texas Instruments, 1986, Chap. 19.

[11] D. O'Shaughnessy, 'Enhancing speech degraded by additive noise or interfering speakers,' *IEEE Communications Magazine*, Feb. 1989, pp. 46–52.

[12] B. Widrow et al., 'Adaptive noise canceling: principles and applications,' *Proc. of the IEEE*, vol. 63, Dec. 1975, pp. 1692–1716.

[13] M. R. Sambur, 'Adaptive noise canceling for speech signals,' *IEEE Trans. on ASSP*, vol. 26, Oct. 1978, pp. 419–423.

[14] S. F. Boll, 'Suppression of acoustic noise in speech using spectral subtraction,' *IEEE Trans. ASSP*, vol. 27, Apr. 1979, pp. 113–120.

[15] J. S. Lim and A. V. Oppenheim, 'Enhancement and bandwidth compression of noisy speech,' *Proc. of the IEEE*, vol. 67, 12, Dec. 1979, pp. 1586–1604.

[16] J. R. Deller, Jr., J. G. Proakis, and J. H. L. Hansen, *Discrete-Time Processing of Speech Signals*, New York: MacMillan, 1993.

[17] D. L. Duttweiler, 'A twelve-channel digital echo canceler,' *IEEE Trans. on Comm.*, vol. COM-26, May 1978, pp. 647–653.

[18] D. L. Duttweiler and Y. S. Chen, 'A single-chip VLSI echo canceler,' *Bell System Technical J.*, vol. 59, Feb. 1980, pp. 149–160.

[19] C. W. K. Gritton and D. W. Lin, 'Echo cancellation algorithms,' *IEEE ASSP Magazine*, Apr. 1984, pp. 30–38.

[20] 'Echo cancelers,' *CCITT Recommendation G.165*, 1984.

[21] M. M. Sondhi and D. A. Berkley, 'Silencing echoes on the telephone network,' *Proc. of IEEE*, vol. 68, Aug. 1980, pp. 948–963.

[22] M. M. Sondhi and W. Kellermann, 'Adaptive echo cancellation for speech signals,' in *Advances in Speech Signal Processing*, S. Furui and M. Sondhi, Eds., New York: Marcel Dekker, 1992, Chap. 11.

[23] Texas Instruments, Inc., *Acoustic Echo Cancellation Software for Hands-Free Wireless Systems*, Literature no. SPRA162, 1997.

[24] Texas Instruments, Inc., *Echo Cancellation S/W for TMS320C54x*, Literature no. BPRA054, 1997.

[25] Texas Instruments, Inc., *Implementing a Line-Echo Canceler Using Block Update & NLMS Algorithms-'C54x*, Literature no. SPRA188, 1997.

[26] Texas Instruments, Inc., *A-Law and mu-Law Companding Implementations Using the TMS320C54x*, Literature no. SPRA163A, 1997.

[27] Texas Instruments, Inc., *The Implementation of G.726 ADPCM on TMS320C54x DSP*, Literature no. BPRA053, 1997.

[28] Texas Instruments, Inc., *Cyclic Redundancy Check Computation: An Implementation Using the TMS320C54x*, Literature no. SPRA530, 1999.

[29] Texas Instruments, Inc., *DTMF Tone Generation and Detection on the TMS320C54x*, Literature no. SPRA096A, 1999.

[30] Texas Instruments, Inc., *IS-54 Simulation*, Literature no. SPRA135, 1994.

[31] Texas Instruments, Inc., *Implement High Speed Modem w/Multilevel Multidimensional Modulation-TMS320C542*, Literature no. SPRA321, 1997.

[32] Texas Instruments, Inc., *Viterbi Decoding Techniques in the TMS320C54x Family Application Report*, Literature no. SPRA071, 1996.

[33] A. J. Viterbi, 'Error bounds for convolutional codes and an asymptotically optimum decoding algorithm', *IEEE Trans. Information Theory*, vol. IT-13, Apr. 1967, pp. 260–269.

[34] W. C. Jakes, Jr., *Microwave Mobile Communications*, New York, NY: John Wiley & Sons, 1974.

Appendix A
Some Useful Formulas

This appendix briefly summarizes some basic formulas of algebra that will be used extensively in this book.

A.1 Trigonometric Identities

Trigonometric identities are often required in the manipulation of Fourier series, transforms, and harmonic analysis. Some of the most common identities are listed as follows:

$$\sin(-\alpha) = -\sin\alpha, \tag{A.1a}$$

$$\cos(-\alpha) = \cos\alpha, \tag{A.1b}$$

$$\sin(\alpha \pm \beta) = \sin\alpha\cos\beta \pm \cos\alpha\sin\beta, \tag{A.2a}$$

$$\cos(\alpha \pm \beta) = \cos\alpha\cos\beta \mp \sin\alpha\sin\beta, \tag{A.2b}$$

$$2\sin\alpha\sin\beta = \cos(\alpha - \beta) - \cos(\alpha + \beta), \tag{A.3a}$$

$$2\cos\alpha\cos\beta = \cos(\alpha + \beta) + \cos(\alpha - \beta), \tag{A.3b}$$

$$2\sin\alpha\cos\beta = \sin(\alpha + \beta) + \sin(\alpha - \beta), \tag{A.3c}$$

$$2\sin\beta\cos\alpha = \sin(\alpha + \beta) - \sin(\alpha - \beta), \tag{A.3d}$$

$$\sin\alpha \pm \sin\beta = 2\sin\left(\frac{\alpha \pm \beta}{2}\right)\cos\left(\frac{\alpha \mp \beta}{2}\right), \tag{A.4a}$$

$$\cos\alpha + \cos\beta = 2\cos\left(\frac{\alpha + \beta}{2}\right)\cos\left(\frac{\alpha - \beta}{2}\right), \tag{A.4b}$$

$$\cos\alpha - \cos\beta = -2\sin\left(\frac{\alpha + \beta}{2}\right)\sin\left(\frac{\alpha - \beta}{2}\right), \tag{A.4c}$$

$$\sin(2\alpha) = 2\sin\alpha\cos\alpha, \tag{A.5a}$$

$$\cos(2\alpha) = 2\cos^2\alpha - 1 = 1 - 2\sin^2\alpha, \tag{A.5b}$$

$$\sin\left(\frac{\alpha}{2}\right) = \sqrt{\frac{1}{2}(1 - \cos\alpha)}, \tag{A.6a}$$

$$\cos\left(\frac{\alpha}{2}\right) = \sqrt{\frac{1}{2}(1 + \cos\alpha)}, \tag{A.6b}$$

$$\sin^2\alpha + \cos^2\alpha = 1, \tag{A.7a}$$

$$\sin^2\alpha = \frac{1}{2}[1 - \cos(2\alpha)], \tag{A.7b}$$

$$\cos^2\alpha = \frac{1}{2}[1 + \cos(2\alpha)], \tag{A.7c}$$

$$e^{\pm j\alpha} = \cos\alpha \pm j\sin\alpha, \tag{A.8a}$$

$$\sin\alpha = \frac{1}{2j}(e^{j\alpha} - e^{-j\alpha}), \tag{A.8b}$$

$$\cos\alpha = \frac{1}{2}(e^{j\alpha} + e^{-j\alpha}). \tag{A.8c}$$

In Euler's theorem (A.8), $j = \sqrt{-1}$. The basic concepts and manipulations of complex number will be reviewed in Section A.3.

A.2 Geometric Series

The geometric series is used in discrete-time signal analysis to evaluate functions in closed form. Its basic form is

$$\sum_{n=0}^{N-1} x^n = \frac{1 - x^N}{1 - x}, \quad x \neq 1. \tag{A.9}$$

This is a widely used identity. For example,

$$\sum_{n=0}^{N-1} e^{-j\omega n} = \sum_{n=0}^{N-1}(e^{-j\omega})^n = \frac{1 - e^{-j\omega N}}{1 - e^{-j\omega}}. \tag{A.10}$$

If the magnitude of x is less than 1, the infinite geometric series converges to

$$\sum_{n=0}^{\infty} x^n = \frac{1}{1 - x}, \quad |x| < 1. \tag{A.11}$$

A.3   Complex Variables

A complex number z can be expressed in rectangular (Cartesian) form as

$$z = x + jy = \text{Re}[z] + j\text{Im}[z]. \tag{A.12}$$

Since the complex number z represents the point (x, y) in the two-dimensional plane, it can be drawn as a vector illustrated in Figure A.1. The horizontal coordinate x is called the real part, and the vertical coordinate y is the imaginary part.

As shown in Figure A.1, the vector z also can be defined by its length (radius) r and its direction (angle) θ. The x and y coordinates of the vector are given by

$$x = r\cos\theta \quad \text{and} \quad y = r\sin\theta. \tag{A.13}$$

Therefore the vector z can be expressed in polar form as

$$z = r\cos\theta + jr\sin\theta = re^{j\theta}, \tag{A.14}$$

where

$$r = |z| = \sqrt{x^2 + y^2} \tag{A.15}$$

is the magnitude of the vector z and

$$\theta = \tan^{-1}\left(\frac{y}{x}\right) \tag{A.16}$$

is its phase in radians.

The basic arithmetic operations for two complex numbers, $z_1 = x_1 + jy_1$ and $z_2 = x_2 + jy_2$, are listed as follows:

$$z_1 \pm z_2 = (x_1 \pm x_2) + j(y_1 \pm y_2), \tag{A.17}$$

$$z_1 z_2 = (x_1 x_2 - y_1 y_2) + j(x_1 y_2 + x_2 y_1) \tag{A.18a}$$

$$= (r_1 r_2)e^{j(\theta_1 + \theta_2)}, \tag{A.18b}$$

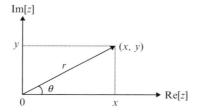

Figure A.1   Complex numbers represented as a vector

$$\frac{z_1}{z_2} = \frac{(x_1 x_2 + y_1 y_2) + j(x_2 y_1 - x_1 y_2)}{x_2^2 + y_2^2} \tag{A.19a}$$

$$= \frac{r_1}{r_2} e^{j(\theta_1 - \theta_2)}. \tag{A.19b}$$

Note that addition and subtraction are straightforward in rectangular form, but is difficult in polar form. Division is simple in polar form, but is complicated in rectangular form.

The complex arithmetic of the complex number x can be listed as

$$z^* = x - jy = re^{-j\theta}, \tag{A.20}$$

where * denotes complex-conjugate operation. In addition,

$$zz^* = |z|^2, \tag{A.21}$$

$$z^{-1} = \frac{1}{z} = \frac{1}{r} e^{-j\theta}, \tag{A.22}$$

$$z^N = r^N e^{jN\theta}. \tag{A.23}$$

The solution of

$$z^N = 1 \tag{A.24}$$

are

$$z_k = e^{j\theta_k} = e^{j(2\pi k/N)}, \quad k = 0, 1, \ldots, N-1. \tag{A.25}$$

As illustrated in Figure A.2, these N solutions are equally spaced around the unit circle $|z| = 1$. The angular spacing between them is $\theta = 2\pi/N$.

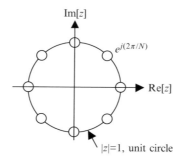

Figure A.2 Graphical display of the Nth roots of unity, $N = 8$

A.4 Impulse Functions

The unit impulse function $\delta(t)$ can be defined as

$$\delta(t) = \begin{cases} 1, & \text{if } t = 0 \\ 0, & \text{if } t \neq 0. \end{cases} \tag{A.26}$$

Thus we have

$$\int_{-\infty}^{\infty} \delta(t)dt = 1 \tag{A.27}$$

and

$$\int_{-\infty}^{\infty} \delta(t - t_0)x(t)dt = x(t_0), \tag{A.28}$$

where t_0 is a real number.

A.5 Vector Concepts

Vectors and matrices are often used in signal analysis to represent the state of a system at a particular time, a set of signal values, and a set of linear equations. The vector concepts can be applied to effectively describe a DSP algorithm. For example, define an $L \times 1$ coefficient vector as a column vector

$$\mathbf{b} = [b_0 \ b_1 \ldots b_{L-1}]^T, \tag{A.29}$$

where T denotes the transpose operator and the bold lower case character is used to denote a vector. We further define an input signal vector at time n as

$$\mathbf{x}(n) = [x(n) \ x(n-1) \ldots x(n-L+1)]^T. \tag{A.30}$$

The output signal of FIR filter defined in (3.1.16) can be expressed in vector form as

$$y(n) = \sum_{l=0}^{L-1} b_l x(n-l) = \mathbf{b}^T \mathbf{x}(n) = \mathbf{x}^T(n)\mathbf{b}. \tag{A.31}$$

Therefore, the linear convolution of an FIR filter can be described as the inner (or dot) product of the coefficient and signal vectors, and the result is a scalar $y(n)$.

If we further define the coefficient vector

$$\mathbf{a} = [a_1 \ a_2 \cdots a_M]^T \tag{A.32}$$

and the previous output signal vector

$$\mathbf{y}(n-1) = [y(n-1) \; y(n-2) \ldots y(n-M)]^T, \tag{A.33}$$

then the input/output equation of IIR filter given in (3.2.18) can be expressed as

$$y(n) = \mathbf{b}^T \mathbf{x}(n) + \mathbf{a}^T \mathbf{y}(n-1). \tag{A.34}$$

A.6 Units of Power

Power and energy calculations are important in circuit analysis. Power is defined as the time rate of expending or absorbing energy, and can be expressed in the form of a derivative as

$$P = \frac{dE}{dt}, \tag{A.35}$$

where P is the power in watts, E is the energy in joules, and t is the time in seconds. The power associated with the voltage and current can be expressed as

$$P = vi = \frac{v^2}{R} = i^2 R, \tag{A.36}$$

where v is the voltage in volts, i is the current in amperes, and R is the resistance in ohms.

The unit bel, named in honor of Alexander Graham Bell, is defined as the common logarithm of the ratio of two power, P_x and P_y. In engineering applications, the most popular description of signal strength is decibel (dB) defined as

$$N = 10 \log_{10} \left(\frac{P_x}{P_y} \right) \text{dB}. \tag{A.37}$$

Therefore the decibel unit is used to describe the ratio of two powers and requires a reference value, P_y for comparison.

It is important to note that both the current $i(t)$ and voltage $v(t)$ can be considered as an analog signal $x(t)$, thus the power of signal is proportional to the square of signal amplitude. For example, if the signal $x(t)$ is amplified by a factor g, that is, $y(t) = gx(t)$. The signal gain can be expressed in dB as

$$\text{gain} = 10 \log_{10} \left(\frac{P_x}{P_y} \right) = 20 \log_{10}(g), \tag{A.38}$$

since the power is a function of the square of the voltage (or current) as shown in (A.36). As the second example, consider that the sound-pressure level, L_p, in decibels corresponds to a sound pressure P_x referenced to $p_y = 20\mu P_a$ (pascals). When the reference signal $y(t)$ has power P_y equal to 1 milliwatt, the power unit of $x(t)$ is called dBm (dB with respect to 1 milliwatt).

Reference

[1] Jan J. Tuma, *Engineering Mathematics Handbook*, New York, NY: McGraw-Hall, 1979.

Appendix B

Introduction of MATLAB for DSP Applications

MATLAB (MATrix LABoratory) is an interactive technical computing environment for scientific and engineering applications. It integrates numerical analysis, matrix computation, signal processing, and graphics in an easy-to-use environment. By using its relatively simple programming capability, MATLAB can be easily extended to create new functions. MATLAB is further enhanced by numerous toolboxes such as the Signal Processing Toolbox. The version we use in this book is based on MATLAB for Windows, version 5.1.

Several reference books provide a concise tutorial on MATLAB and introduce DSP using MATLAB. However, the best way to learn a new tool is by using it. A useful command for getting started is intro, which covers the basic concepts in the MATLAB language. MATLAB has an extensive on-line help system, which can be used to answer any questions not answered in this appendix. Also, there are many demonstration programs that illustrate various capabilities of MATLAB, which can be viewed by using the command demo. In this appendix, MATLAB is briefly introduced with emphasis on DSP concepts introduced in Chapters 3 and 4.

B.1 Elementary Operations

This section briefly introduces the MATLAB environment for numerical computation, data analysis, and graphics.

B.1.1 Initializing Variables and Vectors

The fundamental data-type of MATLAB is array. Vectors, scalars, matrices are handled as special cases of the basic array. A finite-duration sequence can be represented by MATLAB as a row vector. To declare a variable, simply assign it a value at the MATLAB prompt. For example, a sequence $x(n) = \{2, 4, 6, 3, 1\}$ for $n = 0, 1, 2, 3, 4$ can be represented in MATLAB by two row vectors n and xn as follows:

```
n = [0, 1, 2, 3, 4];
xn = [2, 4, 6, 3, 1];
```

Note that the MATLAB command prompt '≫' in the command window is ignored throughout this book.

The above commands are examples of the MATLAB assignment statement, which consists of a variable name followed by an equal sign and the data values to assign to the variable. The data values are enclosed in brackets, which can be separated by commas and/or blanks. A scalar does not need brackets. For example,

```
Alpha = 0.9999;
```

MATLAB statements are case sensitive, for example, `Alpha` is different from `alpha`. There is no need to declare variables as integer, real (`float` or `double` in C), or complex because MATLAB automatically sets the variables to be real with double precision. The output of every command is displayed on the screen, however, a semicolon '`;`' at the end of a command suppresses the screen output, except for graphics and on-line help commands.

The `xn` vector itself is sufficient to represent the sequence $x(n)$, since the time index n is trivial when the sequence begins at $n = 0$. It is important to note that MATLAB assumes all vectors are indexed starting with 1, and thus $xn(1) = 2$, $xn(2) = 4$, ... and $xn(5) = 1$. We can check individual values of the vector `xn`. For example, typing

```
xn(3)
```

will display the value of `xn (3)`.

MATLAB saves previously typed commands in a buffer. These commands can be recalled with the up-arrow key '↑' and down-arrow key '↓'. This helps in editing previous commands with different arguments. Terminating a MATLAB session will delete all the variables in the workspace. These variables can be saved for later use by using the MATLAB command

```
save
```

This command saves all variables in the file `matlab.mat`. These variables can be restored to the workspace using the `load` command. The command

```
save file_name xn yn
```

will save only selective variables `xn` and `yn` in the file named `file_name`.

MATLAB provides an on-line help system accessible by using the `help` command. For example, to get information about the function `save`, we can enter the following statement at the command prompt:

```
help save
```

The `help` command will return the text information on how to use `save` in the command window. The `help` command with no arguments will display a list of directories that contains the MATLAB related files. A more general search for information is provided by `lookfor`.

B.1.2 Graphics

MATLAB provides a variety of sophisticated techniques for presenting and visualizing data as 2D and 3D graphs and annotating these graphs. The most useful command for generating a simple 2D plot is

```
plot(x, y, 'options');
```

where x and y are vectors containing the *x*- and *y*-coordinates of points on the graph. The `options` are optional arguments that specify the color, the line style, which will be discussed later. The data that we plot are usually read from data files or computed in our programs and stored in vectors. For example, to plot the sequence $x(n)$, we can use a simple plot from data stored in two vectors, with vectors n (*x*-axis) and xn (*y*-axis) as

```
plot(n, xn);
```

This command produces a graph of xn versus n, a connected plot with straight lines between the data points $[n, x(n)]$. The outputs of all graphics commands given in the command window are flushed to the separated graphics window.

If xn is a vector, `plot(xn)` produces a linear graph of the elements of xn versus the index of the elements of xn. For a causal sequence, we can use *x*-vector representation alone as

```
plot(xn);
```

In this case, the plot is generated with the values of the indices of the vector xn used as the n values.

The command `plot(x, y)` generates a line plot that connects the points represented by the vectors x and y with line segments. We can pass a character string as an argument to the plot function in order to specify various line styles, plot symbols, and colors. Table B.1 summarizes the options for lines and marks, and the color options are listed in Table B.2.

For example, the following command:

```
plot(x, y, 'r--' );
```

will create a line plot with the red dashed line.

Table B.1 Line and mark options

Line type	Indicator	Point type	Indicator
solid	–	plus	+
dashed	– –	star	*
dotted	:	circle	o
dash-dot	– .	x-mark	x

Table B.2 Color options

Symbol	Color
y	yellow
m	magenta
c	cyan
r	red
g	green
b	blue
w	white
k	black

Plots may be annotated with `title`, `xlabel`, and `ylabel` commands. For example,

```
plot(n, xn);
title('Time-domain signal x(n)');
xlabel('Time index');
ylabel('Amplitude');
```

where `title` gives the plot with the title 'Time-domain signal x(n)', `xlabel` labels the x-axis with 'Time index' and `ylabel` labels the y-axis with 'Amplitude'. Note that these commands can be written in the same line.

By default, MATLAB automatically scales the axes to fit the data values. However, we can override this scaling with the `axis` command. For example, the `plot` statement followed by

```
axis([xmin xmax ymin ymax]);
```

sets the scaling limits for the x- and y-axes on the current plot. The `axis` command must follow the plot command to have the desired effect. This command is especially useful when we want to compare curves from different plots using the identical scale. The `axis` command may be used to zoom-in (or zoom-out) on a particular section of the plot. There are some predefined string-arguments for the `axis` command. For example,

```
axis('equal');
```

sets equal scale on both axes, and

```
axis('square');
```

sets the default rectangular graphic frame to a square.

The command `plot(x, y)` assumes that the x and y axes are divided into equally spaced intervals; these plots are called linear plots. The MATLAB commands can also generate a logarithmic scale (base 10) using the following commands:

```
semilogx(x,y) — using a logarithmic scale for x and a linear scale for y
semilogy(x,y) — using a linear scale for x and a logarithmic scale for y
loglog(x,y) — using a logarithmic scales for both x and y
```

Generally, we use the linear plot to display a time-domain signal, but we prefer to use the logarithmic scale for y to show the magnitude response in the unit of decibels, which will be discussed in Chapter 4.

There are many other specialized graphics functions for 2D plotting. For example, the command

```
stem(n,xn);
```

produces the 'lollipop' presentation of the same data. In addition, `bar` creates a bar graph, `contour` makes contour plots, `hist` makes histograms, etc.

To compare different vectors by plotting the latter over the former, we can use the command

```
hold on
```

to generate overlay plots. This command freezes the current plot in the graphics window. All subsequent plots generated by the `plot` command are simply added to the existing plot. To return to normal plotting, use

```
hold off
```

to clear the `hold` command. When the entire set of data is available, the `plot` command with multiple arguments can be used to generate an overlay plot. For example, if we have two sets of data (x1, y1) and (x2, y2), the command

```
plot(x1, y1, x2, y2, ':' );
```

plots (x1, y1) with a solid line and (x2, y2) with a dotted line on the same graph.

Multiple plots per window can be done with the MATLAB `subplot` function. The `subplot` command allows us to split the graph window into sub-windows. The possible splits can be either two sub-windows or four sub-windows. Two windows can be arranged as either top-and-bottom or left-and-right. The arguments to the `subplot(m,n,p)` command are three integers m, n, and p. The integer m and n specify that the graph window is to be split into an m-by-n grid of smaller windows, and the digit p specifies the pth window for the current plot. The windows are numbered from left to right, top to bottom. For example,

```
subplot(2,1,1), plot(n), subplot(2,1,2), plot(xn);
```

will split the graph window into a top plot for vector n and a bottom plot for vector xn.

B.1.3 Basic Operators

MATLAB is an expression language. It interprets and evaluates typed expression. MATLAB statements are frequently of the form

```
variable = expression
```

or simply

```
expression
```

Since MATLAB supports long variable names (up to 19 characters, start with a letter, followed by letters, or digits, or underscores), we should take advantage of this feature to give variables descriptive names.

The default operations in MATLAB are matrix (including vector and scalar) operations. The arithmetic operations between two scalars (1×1 matrix) a and b are: $a + b$ (addition), $a - b$ (subtraction), $a * b$ (multiplication), a/b (division), and $a^\wedge b$ (a^b). An array operation is performed element-by-element. Suppose A and B vectors are row vectors with the same number of elements. To generate a new row vector C with values that are the operations of corresponding values in A and B element-by-element, we use $A + B$, $A-B$, $A.*B$, $A./B$, and $A.^\wedge B$. For example,

```
x = [1, 2, 3]; y = [4, 5, 6];
```

then

```
z = x.*y
```

results in

```
z = 4 10 18
```

A period preceding an operator indicates an array or element-by-element operation. For addition and subtraction, array operation and scalar operation are the same. Array (element-by-element) operations apply not only to operations between two vectors of the same size, but also to operations between a scalar and vector. For example, every element in a vector A can be multiplied by a scalar b in MATLAB as $B = b*A$ or $B = b.*A$. In general, when 'point' is used with another arithmetic operator, it modifies that operator's usual matrix definition to a pointwise one.

Six relational operators: < (less than), <= (less than or equal), > (greater than), >= (greater than or equal), == (equal), and ~= (not equal), are available for comparing two matrices of equal dimensions. MATLAB compares the pairs of corresponding elements. The result is a matrix of ones and zeros, with one representing 'true' and zero representing 'false.' In addition, the operators & (AND), | (OR), ~ (NOT), and xor (exclusive OR) are the logical operators. These operators are particularly useful in if statements. For example,

```
if a > b
  do something
end
```

The colon operator ';' is useful for creating index arrays and creating vectors of evenly spaced values. The index range can be generated using a start (initial value), a skip (increment), and an end (final value). Therefore, a regularly spaced vector of numbers is obtained by means of

```
n = [start:skip:end]
```

Note that no brackets are required if a vector is generated this way. However, brackets are required to force the concatenation of the two vectors. Without the skip parameter, the default increment is 1. For example,

```
n = 0:2:100;
```

generates the vector n = [0 2 4 ... 100], and

```
m = [1:10 20:2:40];
```

produces the vector m = [1 2 ... 10 20 22 ... 40].

In DSP application, the vector form of the impulse response $h(n) = 0.8^n$ for $n = 0,1, \ldots,127$ can be generated by the commands

```
n = [0:127]; hn = (0.8).^n;
```

where $h(n)$ is stored in the vector hn.

We also can use build-in function linspace(start, end, length) instead of colon operator ':'. For example,

```
n = linspace(0,10,6);
```

generates n = [0 2 4 6 8 10], which is the same as using

```
n = 0:2:10;
```

Program flow can be controlled in MATLAB using if statement, while loop, for loop, and switch statement. These are similar to any high-level language such as C, which will be reviewed in Appendix C. Since MATLAB is an interpreted language, certain common programming habits are intrinsically inefficient. The primary one is the use of for loops to perform simple operation over an entire vector. Rather than writing a loop, try to find a vector function or the nested composition of a few vector functions that will accomplish the desired result. For example, the following for-loop:

```
for n = 0:127
  x(n + 1) = sin(2*pi*100/8000*n);
end
```

can be replaced by the much efficient vector operation as follows:

```
n = 0:127;
x = sin(2*pi*100/8000*n);
```

B.1.4 Files

MATLAB provides three types of files for storing information: M-files, Mat-files, and Mex-files. M-files are text files, with a '.m' extension. There are two types of M-files: script files and function files. A script file is a user-created file with a sequence of MATLAB commands. The file must be saved with a .m extension. A script file can be executed by typing its name (without extension) at the command prompt in the command window. It is equivalent to typing all the commands stored in the script file at the MATLAB prompt. A script file may contain any number of commands, including those built-in and user-written functions. Script files are useful when we have to repeat a set of commands and functions several times. It is important to note that we should not name a script file the same as the name of a variable in the workspace and the name of a variable it created. In addition, avoid names that clash with built-in functions. We can use any text editor to write, edit, and save M-files. However, MATLAB provides its own text editor. On PC, select New→M-file from the File menu. A new edit window will appear for creating a script file.

A function file is also an M-file, just like a script file, except it has a function definition line on the top that defines the input and output explicitly. We will discuss function files later. Mat-files are binary data files with a '.mat' extension. Mat-files are created by MATLAB when we save data with the `save` command. The data is written in a special format that MATLAB can read. Mat-files can be loaded into MATLAB with the `load` command. Mex-files are MATLAB callable C programs with `.mex` extension. We do not use and discuss this type of files in this book.

B.2   Generation and Processing of Digital Signals

Arithmetic expressions often require computations other than addition, subtraction, multiplication, division, and exponentiation. For examples, many expressions require the use of logarithms, trigonometric functions, etc. MATLAB provides hundreds of built-in functions. With so many available functions, it is important to know how to look for functions and how to use them. Typing `help` in the command window brings out a list of categories. We can get help on one of these categories by typing the selected category name after `help`. For example, typing `help graph2d` gives a list of 2D graphs with a very brief description of each function. Further help can be obtained by typing `help` followed by the exact name of the function.

A set of elementary mathematical functions is listed in the last section. All function names must be in lowercase. As discussed in Chapter 3, a MATLAB function `sin` (or `cos`) can be used to generate a digital sinusoidal signal

$$x(n) = A\sin(2\pi fnT + \phi), \quad n = 0, 1, \ldots, N - 1. \tag{B.1}$$

For example, to generate $x(n)$ for $A = 1.5$, $f = 100\,\mathrm{Hz}$, $T = 0.001$ second (1 ms), $\phi = 0.25\pi$, and $N = 100$, we can easily use the following MATLAB script (`figb1.m` in the software package):

```
n = [0:99];
xn = 1.5* sin(2* pi* 100* n* 0.001+0.25* pi);
```

where the function `pi` returns the value of π. To view the generated sinewave, we can use

```
plot(n, xn); title('Sinewave');
xlabel('Time index'); ylabel('Amplitude');
```

The waveform of the generated sinewave is shown in Figure B.1.

In Figure B.1, a trivial integer index n is used for the x-axis instead of an actual time index in seconds. To better represent the time-domain signal, we can use the colon operator to generate values between the first and third numbers, using the second number as the increment. For example, if we wish to view $x(n)$ generated in the previous example with the actual time index $t = 0, 0.001, \ldots, 0.099$, we can use the following script (`figb2.m` in the software package):

```
n = [0:99];
xn = 1.5*sin(2*pi*100*n*0.001+0.25*pi);
t = [0:0.001:0.099];
plot(t,xn); title('Sinewave');
xlabel('Time in second'); ylabel('Amplitude');
```

The result is shown in Figure B.2.

In addition to these `sin`, `cos`, `rand`, and `randn` functions discussed in Chapter 3, MATLAB provides many other functions, such as `abs(x)`, `log(x)`, etc. The arguments or parameters of the function are contained in parentheses following the name of the

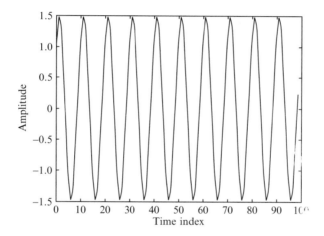

Figure B.1   Sinewave using integer index

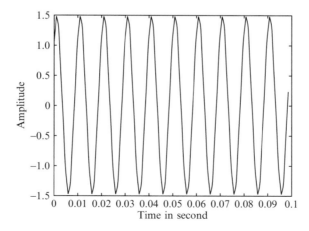

Figure B.2   Sinewave using time index

function. If a function contains more than one argument, it is very important to list the arguments in the correct order. Some functions also require that the arguments be in specific units. For example, the trigonometric functions assume that the arguments are in radians. It is possible to nest a sequence of function calls. For example, the following equation:

$$y = \sum_{n=1}^{L} \log(|x(n)|) \tag{B.2}$$

can be implemented as

```
y = sum(log(abs(xn)));
```

where xn is the vector containing the elements $x(n)$.

The built-in functions are optimized for vector operations. Writing efficient MATLAB code (scripts or user-written functions) requires a programming style that generates small functions that are vectorized. The primary way to avoid loops is to use MATLAB functions as often as possible. The details of user-written functions will be presented in Section B.4.

Two sequences $x_1(n)$ and $x_2(n)$ can be added sample-by-sample to form a new sequence

$$y(n) = x_1(n) + x_2(n). \tag{B.3}$$

Adding their corresponding sample sums these two sequences. The summation of two sequences can be implemented in MATLAB by the arithmetic operator '+' if sequences are of equal length. For example, we can add a random noise with a sinewave as follows:

```
n = [0:127];
x1n = 1.5*sin(2*pi*100*n*0.001+0.25*pi);
x2n = 1.2*randn(1,128);
yn = x1n+x2;
```

A given sequence $x(n)$ multiplied by a constant α can be implemented in MATLAB by the scaling operation. For example, $y(n) = \alpha x(n)$, where each sample in $x(n)$ is multiplied by a scalar $\alpha = 1.5$, can be implemented as

```
yn = 1.5*xn;
```

Consider the discrete-time linear time-invariant system. Let $x(n)$ be the input sequence and $y(n)$ be the output sequence of the system. If $h(n)$ is the impulse response of the system, the output signal of the system can be expressed as

$$y(n) = \sum_{k=-\infty}^{\infty} x(k)h(n-k) = \sum_{k=-\infty}^{\infty} h(k)x(n-k). \tag{B.4}$$

As discussed in Chapter 3, a digital system is called a causal system if $h(n) = 0$, for $n < 0$, a digital signal is called a causal signal if $x(n) = 0$ for $n < 0$. If the sequences $x(n)$

and $h(n)$ are finite causal sequences, MATLAB has a function called `conv(a,b)` that computes the convolution between vectors a and b. For example,

```
xn = [1, 3, 5, 7, 9] ; hn = [2, 4, 6, 8, 10];
yn = conv(xn, hn)
yn = 2   10   28   60   110   148   160   142   90
```

B.3   DSP Applications

As discussed in Chapter 3, the differential equation of an IIR system can be expressed as

$$
\begin{aligned}
y(n) &= b_0 x(n) + b_1 x(n-1) + \cdots + b_{L-1} x(n-L+1) - a_1 y(n-1) \\
&\quad - \cdots - a_M y(n-M) \\
&= \sum_{l=0}^{L-1} b_l x(n-l) - \sum_{m=1}^{M} a_m y(n-m).
\end{aligned}
\tag{B.5}
$$

As discussed in Chapter 4, the transfer function of this IIR filter can be expressed as

$$
H(z) = \frac{\displaystyle\sum_{l=0}^{L-1} b_l z^{-l}}{\displaystyle\sum_{m=0}^{M} a_m z^{-m}},
\tag{B.6}
$$

where $a_0 = 1$. The MATLAB function

```
y = filter(b, a, xn);
```

performs filter operation to the data in vector xn that contains signal $x(n)$ with a filter described by the vector b that contains the coefficients b_l and the vector a that contains the coefficients a_m, producing the filter result y. For example, the difference equation of an IIR filter is given as

$$
y(n) = 0.0305 x(n) - 0.0305 x(n-2) + 1.5695 y(n-1) - 0.9391 y(n-2),
$$

the MATLAB script (`figb3.m` in the software package) to compute the filter output yn with input sequence xn is given as follows:

```
n = [0:139];
x1n = 1.5* sin(2* pi* 100* n* 0.001+0.25* pi);
x2n = 1.2* randn(1,140);
xn = x1n+x2n;
b = [0.0305, 0 -0.0305];
a = [1, - 1.5695, 0.9391];
yn = filter(b, a, xn);
plot(n, xn, ':', n, yn, '-' );
```

Figure B.3 shows the MATLAB plots of input and output signals. Note that the vector a is defined based on coefficients a_m used in Equation (B.6), which have different signs than the coefficients used in Equation (B.5). An FIR filter can be implemented by setting a = [1] or using

```
filter(b,1,x);
```

where the vector b consists of FIR filter coefficients b_l.

MATLAB can interface with three different types of data files: Mat-files, ASCII files, and binary files. The Mat-file and binary file contain data stored in a memory-efficient binary format, whereas an ASCII file contains information stored in ASCII characters. Mat-files are preferable for data that is going to be generated and used by MATLAB programs.

Data saved in both the Mat-file and ASCII file can be loaded (retrieved) from the disk file into an array in workspace using the load command. For example,

```
Load xn;
```

will load the Mat-file xn.mat, and

```
load xn.dat;
```

will read the data from the ASCII file with the name xn.dat in the disk into an array with the name xn.

To load the file xn.bin stored in the binary format into the array xn, we have to use the following C-like commands:

```
fid = fopen('xn.bin' , 'r');
xn = fread(fid, 'float32');
```

where the fopen command open the file xn.bin, 'r' indicates to open the file for reading, and fid is a file identifier associated with an open file. The second command

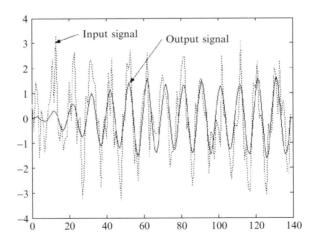

Figure B.3   Filter input (dotted line) and output (solid line) waveforms

fread reads binary data from the specified file and writes it into the array xn, and 'float32' indicates each data in the file is a 32-bit floating-point value. Actually, MATLAB supports all the C and FORTRAN data types such as int, long, etc. Other related MATLAB commands are fclose, fscanf, sscanf, fseek, and ftell.

Mat-files are generated by MATALB program using the save command, which contains a file name and the array to be stored in the file. For example,

```
save out_file yn;
```

will save samples in the array yn into a disk file out_file.mat, where the .mat extension is automatically added to the filename. To save variables in the ASCII format, we can use a save command with an additional keyword as follows:

```
save out_file yn –ascii;
```

To save data in the binary format, we can use the fwrite command, which writes the elements of an array to the specified file.

B.4   User-Written Functions

As discussed in the previous section, MATLAB has an intensive set of functions. Some of the functions are intrinsic (built-in) to the MATLAB itself. A function file is a script file. Details about individual functions are available in the on-line help facility and the MATLAB Reference Guide. Other functions are available in the library of external M-files called the MATLAB toolbox such as Signal Processing Toolbox. Finally, functions can be developed by individual users for more specialized applications. This is an important feature of MATLAB.

Each user-written function should have a single purpose. This will lead to short, simple modules that can be linked together by functional composition to produce more complex operations. Avoid the temptation to build super functions with many options and outputs. The function M-file has very specific rules that are summarized as follows.

An example of the user-written function dft.m given in Chapter 4 is listed as follows:

```
function [Xk] = dft(xn, N)
%   Discrete Fourier transform function
%     [Xk] = dft(xn, N)
%   where xn is the time-domain signal x(n)
%      N is the length of sequence
%      Xk is the frequency-domain X(k)
n = [0:1:N−1];
k = [0:1:N−1];
WN = exp(−j*2pi/N);   % Twiddle factor
nk = n' *k;            % N by N matrix
WNnk = WN.^nk;         % Twiddle factor matrix
Xk = xn* WNnk;         % DFT
```

A function file begins with a function definition line, which has a well-defined list of inputs and outputs as follows:

```
function [output variables] = facction_name(input variables);
```

The function must begin with a line containing the word `function`, which is followed by the output arguments, an equal sign, and the name of the function. The input arguments to the function follow the function name and are enclosed in parentheses. This first line distinguishes the function file from a script file. The name of the function should match the name of the M-file. If there is a conflict, it is the name of the M-file located on the disk that is known to the MATLAB command environment.

A line beginning with the `%` (the percent symbol) sign is a comment line and a '`%`' sign may be put anywhere. Anything after a `%` in a line is ignored by MATLAB during execution. The first few lines immediately following the function definition line should be comments because they will be displayed if `help` is requested for the function name. For example, typing

        help dft

will display five comment lines from `dft.m` M-file on the screen as help information if this M-file is in the current directory. These comment lines are usually used to provide a brief explanation, define the function calling sequence, and then a definition of the input and output arguments.

The only information returned from the function is contained in the output arguments, which are matrices. For example, `Xk` in the `dft.m`. Multiple output arguments are also possible if square brackets surround the list of output arguments. MATLAB returns whatever value is contained in the output matrix when the function completes.

The same matrix names can be used in both a function and the program that references it since the function makes local copies of its arguments. However, these local variables disappear after the function completes, thus any values computed in the function, other than the output arguments, are not accessible from the program.

It is possible to declare a set of variables to be accessible to all or some functions without passing the variables in the input list. For example,

        global input_x;

The `global` command declares the variable `input_x` to be global. This command goes before any executable command in the functions and scripts that need to access the value of the global variables. Be careful with the names of the global variables to avoid conflict with other local variables.

B.5   Summary of Useful MATLAB Functions

`sin` – sine
`cos` – cosine
`tan` – tangent
`abs` – absolute value
`angle` – phase angle
`sqrt` – square root
`real` – real part of complex variable
`imag` – imaginary part of complex variable
`conj` – complex conjugate
`round` – round towards the nearest integer

fix – round towards zero
floor – round towards $-\infty$
ceil – round towards ∞
sign – signum function
rem – remainder or modulus
exp – exponential, base e
log – natural logarithm
log10 – logarithm, base 10
max – maximum value
min – minimum value
mean – mean value
median – median value
std – standard deviation
sort – sorting
sum – sum of elements
corrcoef – correlation coefficients
cov – covariance matrix
conv – convolution
cov – covariance
deconv – deconvolution
fft – fast Fourier transform
ifft – inverse fast Fourier transform

In addition, the *Signal Processing Toolbox* provides many additional signal-processing functions.

References

[1] D. M. Etter, *Introduction to MATLAB for Engineers and Scientists*, Englewood Cliffs, NJ: Prentice-Hall, 1996.
[2] V. K. Ingle and J. G. Proakis, *Digital Signal Processing Using MATLAB V.4*, Boston: PWS Publishing, 1997.
[3] E. W. Kamen and B. S. Heck, *Fundamentals of Signals and Systems Using MATLAB*, Englewood Cliffs, NJ: Prentice-Hall, 1997.
[4] J. H. McClellan, et al., *Computer-Based Exercises for Signal Processing Using MATLAB 5*, Englewood Cliffs, NJ: Prentice-Hall, 1998.
[5] R. Pratap, *Getting Started with MATLAB 5*, New York: Oxford University Press, 1999.
[6] *MATLAB User's Guide*, Math Works, 1992.
[7] *MATLAB Reference Guide*, Math Works, 1992.
[8] *Signal Processing Toolbox for Use with MATLAB*, Math Works, 1994.

Appendix C

Introduction of C Programming for DSP Applications

C has become the language of choice for many DSP software developments not only because of its powerful commands and data structures but also because of its portability to migrate between DSP platforms and devices. In this appendix, we will cover some of the important features of C for DSP applications.

The processes of compilation, linking/loading, and execution of C programs differ slightly among operating environments. To illustrate the process we use a general UNIX system C compiler shown in Figure C.1 as an example. C compiler translates high-level C programs into machine language that can be executed by computers or DSP processors such as the TMS320C55x. The fact that C compilers are available for a wide range of computer platforms and DSP processors makes C programming the most portable software for DSP applications. Many C programming environments include debugger programs, which are useful for identifying errors in source programs. Debugger programs allow us to see values stored in variables at different points in a program and to step through the program line by line.

The purpose of DSP programming is to manipulate digital signals for a specific signal processing application. To achieve this goal, DSP programs must be able to organize the variables (different data types), describe the actions (operators), control the operations (program flow), and move data back and forth between the outside world and the program (input/output). This appendix provides a brief overview of the elements required for efficient programming of DSP algorithms in C language and introduces fundamental C programming concepts using C examples, but does not attempt to cover all the elements in detail. C programming language used throughout this book is conformed to the ANSI C standard (American National Standard Institute C Standard).

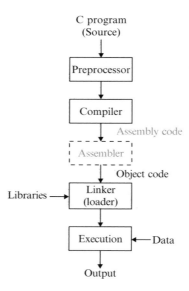

Figure C.1   Program compilation, linking, and execution

C.1   A Simple C Program

In this section, we will present a simple C program and use it as an example to introduce C language program components. As discussed in Section 3.1, an *N*-point unit-impulse sequence can be written as

$$\delta(n) = \begin{cases} 1, & n = 0 \\ 0, & n = 1, 2, \ldots N - 1. \end{cases} \tag{C.1}$$

The following C program (`impulse.c` in the software package) can be used to generate this unit-sample sequence

```
/*****************************************************************
*   IMPULSE.C — Unit impulse sequence generator                 *
*****************************************************************/
#include <stdio.h>
#include <stdlib.h>
#include <math.h>
#define  K 1024

void main()
{
   float y[K];
   int k;
   int N = 256;

   /* Generate unit impulse sequence */
```

```
    for (k = 1; k < N; k++)
    {
        y[k] = 0.0;    //Clear array
    }
    y[0] = 1.0;        //y(0)=1
}
```

A program written in C must have a few basic components. We now briefly discuss these components used in this example C program.

C program comments may contain any message beginning with the characters sequence /* and ending with the characters sequence */. The comments will be ignored by the compiler. Program comments may be interspersed within the source code as well as at the beginning of a function. In above example, the extra asterisks around the program comments in lines one through three are there only to enhance the appearance of the comments; they are not necessary. Most of the C compiler nowadays also accepts the C++ programming language comments sequence, //. In our example, we mixed both comment sequences for demonstration purpose. Although program comments are optional, good programming style requires that they be used throughout a program to document the operations and to improve its readability. Detailed comments are very important in maintaining complicated DSP software for new readers, even for the original programmers after time has passed.

The preprocessor is the first pass of the C compiler. It reads in C source files as input and produces output files for the C compiler. The tasks of preprocessor are to remove comments, expand macro definition, interpret include files, and check conditional compilation. Preprocessor directives give instructions to the compiler that are performed before the program is compiled. Each preprocessor directive begins with a pound sign '#' followed by the preprocessor keyword. For example, the program impulse.c contains the following directives:

```
#include <stdio.h>
#include <stdlib.h>
#include <math.h>
#define  K 1024
```

The #include <file> directive copies the file from a directory inside the standard C compiler, while #include 'file' directly copies the file from current working directory for compilation unless a path has been specified. Thus the first three directives specify that the statements in the files of stdio.h, stdlib.h, and math.h should be inserted in place of the directives before the program is compiled. Note that these files are provided with the C compiler. The #define directs the preprocessor to replace subsequent occurrences of K with the constant value 1024.

A C program may consist of one or more functions and one and only one function is called main() with which the operating system will start executing. After starting the application, the main function is executed first, and it is usually responsible for the main control of the application program. This function can be written to return a value, or it can be written as a void function that does not return a value. The body of the function is enclosed in braces as follows:

```
void main()
{
    variable declarations; /* Statements define variables */
    executable statements; /* Statements execute program  */
}
```

As shown in the example, all C statements end with a semicolon. The function contains two types of statements – statements that define memory locations that will be used in the program and statements that specify actions to be taken.

C.1.1 Variables and Assignment Operators

Before a variable can be used in a C program, it must be declared to inform the compiler the name and type of the variable. A variable name is defined by declaring a sequence of characters (the variable identifier or name) as a particular predefined type of data. C constants are specific values that are included in the C statements, while variables are memory locations that are assigned a name or identifier. The variable declaration uses the following syntax:

```
data_type name;
```

For example, in the simple example we have

```
int k;
```

The term `int` indicates that the variable named k will store as integer data value. C also allows multiple variables to be defined within one statement by separating them with the commas. For example,

```
int i,j,k;
```

An identifier may be any sequence of characters (usually with some length restrictions) that starts with a letter or an underscore, and cannot be any of the C compiler reserved keywords. Note that C is case sensitive; making the variable k different from the variable K. C language supports several data types that represent: integers numbers, floating-point numbers, and text data. Arrays of each variable type and pointers of each type may also be declared and used. Once variables are defined to be a given size and type, some sort of manipulation can be performed using the variables.

Memory locations must be defined before other statements use them. Initial values can also be specified at the same time when memory locations are defined. For example,

```
int N = 256;
```

defines the variable N as an integer, and assigns it with the value 256.

An assignment statement is used to assign a value to an identifier. The most basic assignment operator in C is the single equal sign, =, where the value to the right of the equal sign is assigned to the variable on the left. The general form of the assignment statement is

```
identifier = expression;
```

where the expression can be a constant, another variable, or the result of an operation. C also allows multiple expressions to be placed within one statement by separating them with the commas. Each expression is evaluated left to right, and the entire expression assumes the value of the last expression which is evaluated. Multiple assignments are also allowed in C, for example,

```
int i = j = k = 0;
```

In this case, the statement is evaluated from right to left, so that 0 is assigned to k, j, and i.

C.1.2 Numeric Data Types and Conversion

Numeric data types are used to specify the types of numbers that will be contained in variables. There are several types of data used depending on the format in which the numbers are stored and the accuracy of the data. In C, numeric numbers are either integers (short, int, long) or floating-point (float, double, long double) values. The specific ranges of values are system dependent, which means that the ranges may vary from one computer to another. Table C.1 contains information on the precision and range of integers represented by a 32-bit machine and a 16-bit machine. Thus the size of a variable declared as just int depends on the compiler implementation and could make the program behave differently on different machine. To make a program truly portable, the program should contain only short and long declarations. In practice, explicit defined data types are often used, such as:

```
#define Word16 short
#define Word32 long
main()
{
   Word16 k;   /* Declare as 16-bit variable */
   Word32 x;   /* Declare as 32-bit variable */

   statements;
}
```

Table C.1 Example of integer type limits

Data type	Value on 32-bit machine	Value on 16-bit machine
Short	[−32 768, 32 767]	[−32 768, 32 767]
unsigned short	[0, 65 535]	[0, 65 535]
Int	[−2 14 74 83 648, 2 14 74 83 647]	[−32 768, 32 767]
unsigned int	[0, 4 29 49 67 295]	[0, 65 535]
Long	[−2 14 74 83 648, 2 14 74 83 647]	[−2 14 74 83 648, 2 14 74 83 647]
unsigned long	[0, 4 29 49 67 295]	[0, 4 29 49 67 295]

Instead of using `short` and `long` data type, the example code uses `Word16` for the 16-bit integer data type and `Word32` for the 32-bit data type. In addition, the three integer types (`int`, `short`, and `long`) can be declared as unsigned by preceding the declaration with `unsigned`. For example,

```
unsigned int counter;
```

where `counter` has a value range from 0 to 65 535.

Statements and expressions using the operators should normally use variables and constants of the same type. If data types are mixed, C uses two basic rules to automatically make type conversions:

1. If an operation involves two types, the value with a lower rank is converted to the type of higher rank. This process is called promotion, and the ranking from highest to lowest type is `double`, `float`, `long`, `int`, `short`, and `char`.

2. In an assignment statement, the result is converted to the type of the variable that is being assigned. This may result in promotion or demotion when the value is truncated to a lower ranking type.

Sometimes the conversion must be stated explicitly in order to demand that a conversion be done in a certain way. A cast operator places the name of the desired type in parentheses before the variable or expression, thus allowing us to specify a type change in the value. For example, the data casting (`int`) used in the following expressions treats the floating-point number z as an integer:

```
int x, y;
float z = 2.8;
x = (int)z;          /* Truncate z to an integer x */
y = (int)(z+0.5);    /* Rounding z to an integer y */
```

The casting result will truncate 2.8 to 2 and store it in x, and allows rounding of the floating variable z to an integer 3 and stores it in y.

C.1.3 Arrays

An array groups distinct variables of the same type under a single name. A one-dimensional array can be visualized as a list of values arranged in either a row or a column. We assign an identifier to an array, and then distinguish between elements or values in the array by using subscripts. The subscripts always start with 0 and are incremented by 1. In C, all data types can be declared as an array by placing the number of elements to be assigned to an array in brackets after its name. One-dimensional array is declared as

```
data_type array_name[N];
```

where the `array_name` is the name of an array of N elements of data type specified. For example,

```
float y[5];
```

where an integer expression 5 in brackets specifies there are five `float` (floating-point) elements in the array `y[]`. The first value in the `y` array is referenced by `y[0]`, the second value in the `y` array is referenced by `y[1]`, and the last element is indexed by `y[K−1]`.

Multidimensional arrays can be defined simply by appending more brackets containing the array size in each dimension. For example,

```
int matrix_a[4][2];
```

defines a 4×2 matrix called `matrix_a`. The matrix array would be referenced as `matrix_a[i][j]`, where `i` and `j` are row and column indices respectively.

An array can be initialized when it is defined, or the values can be assigned to it using program statements. To initialize the array at the same time when it is defined, the values are specified in a sequence that is separated by commas and enclosed with braces. For example,

```
float y[5] = {1.0, 0.0, 0.0, 0.0, 0.0};
```

initializes a 5-point unit impulse response sequence in the floating-point array `y[]`. Arrays can also be assigned values by means of program statements. For example, the following example generates an *N*-point unit impulse sequence.

```
for (k = 1; k < N; k++)
{
    y[k] = 0.0;    /* Clear array */
}
y[0] = 1.0;        /* y(0) = 1    */
```

A more detailed discussion on arrays and loops will be given later.

C.2   Arithmetic and Bitwise Operators

Once variables are defined to be a given size and type, a certain manipulation (operator) can be performed using the variables. We have discussed assignment operators in C.1.1. This section will introduce arithmetic and bitwise operators. Logical operators will be introduced later.

C.2.1   Arithmetic Operators

C supplies arithmetic operators on two operands: + (add), − (subtract), * (multiply), / (divide), and % (modulus, integer remainder after division). The first four operators are defined for all types of variables, while the modulus operator is only defined for integer operands. C does not have an exponential operator. However, the library function `pow(x, y)` may be used to compute x^y. Note that in C, `a = pow(x, y)` is an expression, while `a = pow(x, y);` is a statement. Thus `c = b+(a=pow(x, y));` is a statement. The result of this statement would be that the result returned by the function `pow(x, y)` is assigned to `a` and `b+a` is assigned to `c`.

The modulus operator is useful in implementing a circular pointer for signal processing. For example,

```
k = (k+1)%128;
```

makes k a circular pointer of range from 0 to 127.

C also includes increment (++) and decrement (– –) operators for incrementing and decrementing variables. For example, i++ is equal to the statement i = i+1. These operators can be applied either in a prefix position or in a postfix position. If the increment or decrement operator is in a prefix position, the identifier is modified first, and then the new value is used to evaluate the rest of the expression. If the increment or decrement operator is in a postfix position, the old value of the identifier is used to evaluate the rest of the expression, and then the identifier is modified. These unary operators (require only one operand) are often used for updating counters and address pointers.

C also allows operators to be combined with the assignment operator '=' so that almost any statement of the form

```
variable = variable operator expression;
```

can be replaced with

```
variable operator = expression;
```

For example,

```
x = x+y;
```

is equal to the statement

```
x += y;
```

Some compiler implementations may generate code that is more efficient if the combined operator is used. The combined operators include +, –, *, /, %, and other logical operators.

C.2.2 Bitwise Operators

C supplies the binary bitwise operators: & (bitwise AND), | (bitwise OR), ^ (bitwise exclusive OR), ≪ (arithmetic shift left), and ≫ (arithmetic shift right), which are performed on integer operands. The unary bitwise operator, which inverts all the bits in the operand, is implemented with the ~ symbol. These bitwise operators make C programming an efficient programming language for DSP applications.

C.3 An FIR Filter Program

To introduce more features of C programming, an example C program firfltr.c that implements an FIR filter is included in the software package.

C.3.1   Command-Line Arguments

The function `main` can have two parameters, `argc` and `argv[]`, to catch arguments passed to `main` from the command line when the program begins executing. These arguments could be file names on which the program is to act or options that influence the logic of the program. The parameter `argv[]` is an array of pointers to strings, and `argc` is an `int` whose value is equal to the number of strings to which `argv[]` points to. The command-line arguments are passed to the main( ) function as follows:

```
void main (int argc, char * argv[])
```

Suppose that we compile the `firfltr.c` such that the executable program is generated and saved as `firfltr.exe`. We can run the program on a PC under MS-DOS Prompt by typing

```
firfltr infile coefile outfile <enter>
```

The operating system passes the strings on the command line to `main`. More precisely, the operating system stores the strings on the command line in memory and sets `argv[0]` to the address of the first string (`firfltr`), the name of the file that holds the program to be executed on the command line. `argv[1]` points to the address of the second string (`infile`) on the command line, `argv[2]` to `coefile`, and `argv[3]` to `outfile`. The argument `argc` is set to the number of strings on the command line. In this example, `argc = 4`.

The use of command-line arguments makes the executable program flexible, because we can run the program with different arguments (data files, parameter values, etc.) specified at the execution time without modifying the program and re-compiling it again. For example, the file `firfltr.exe` can be used to perform FIR filtering function for different FIR filter with their coefficients defined by `coefile`. This program can also be used to filter different input signals contained in the `infile`. The flexibility is especially convenient when the parameter values used in the program need to be tuned based on given data.

C.3.2   Pointers

A pointer is a variable that holds the address of data, rather than the data itself. The use of pointers is usually closely related to manipulating the elements in an array. Two special pointer operators are required to effectively manipulate pointers. The indirection operator `*` is used whenever the data stored at the address pointed to by a pointer (indirect addressing) is required. The address operator `&` is used to set the pointer to the desired address. For example,

```
int i = 5;
int * ptr;
ptr = &i;
* ptr = 8;
```

The first statement declares i as an integer of value 5, the second statement declares that ptr is a pointer to an integer variable, and the third statement sets the pointer ptr to the address of the integer variable i. Finally, the last statement changes the data at the address pointed by ptr to 8. This results in changing the value of variable i from 5 to 8.

An array introduced in Section C.1.3 is essentially a section of memory that is allocated by the compiler and assigned the name given in the declaration statement. In fact, the name given is just a fixed pointer to the beginning of the array. In C, the array name can be used as a pointer or it can be used to reference elements of the array. For example, in the function shift in firfltr.c, the statement

```
float * x;
```

defines x as a pointer to floating-point variables. Thus *x and x[0] are exactly equivalent, although the meaning of x[0] is often more clear.

C.3.3 C Functions

As discussed earlier, all C programs consist of one or more functions, including the main(). In C, functions (subroutines) are available from libraries such as the standard C library and programmer-defined routines that are written specifically to accompany the main function, such as:

```
void shift(float *, int, float);
float fir(float *, float *, int);
```

These functions are sets of statements that typically perform an operation, such as shift to update data buffers for FIR filters, or as fir to compute an output of the FIR filter.

To maintain simplicity and readability for more complicated applications, we develop programs that use a main() function plus additional functions, instead of using one long main function. In C, any function can call any other function, or be called by any other function. Breaking a long program into a set of simple functions has following advantages:

1. A function can be written and tested separately from other parts of the program. Thus module development can be done in parallel for large projects. Several engineers can work on the same project if it is separated into modules because the individual modules can be developed and tested independently of each other.

2. A function is smaller than the complete program, so testing it separately is easier.

3. Once a function has been carefully tested, it can be used in other programs without being retested. This reusability is a very important issue in the development of large software systems because it greatly reduces development time.

4. The use of functions frequently reduces the overall length of a program, because many solutions include steps that are repeated several places in the program.

A function consists of a definition statement followed by the function body. The first part of the definition statement defines a function name and the type of value that is returned by the function. A pair of parentheses containing an optional argument list and a pair of braces containing the optional executable statements. For example,

```
float fir(float *x, float *h, int ntap)
{
    float yn = 0.0;   /* Output of FIR filter      */
    int i;            /* Loop index                */

    more statements

    return(yn);       /* Return y(n) value to main */
}
```

The first line declares a function called `fir` that will return the floating-point value `yn` to the main program. The variable declaration represents the arguments (`float *x`, `float *h`, and `int ntap`) passed to the function. Note that the variables `x` and `h` are actually the pointers to the beginning of floating-point arrays that are allocated by the calling function. By passing the pointer, only one value (address of the first element in the array) is passed to the function instead of the large floating-point array. Additional local variables (`yn` and `i`) used by the function are declared in the function body, which is enclosed in braces. The return statement, such as `return(yn)`, passes the result back to the calling function. Note that the expression type should match the return-type indicated in the function definition. If the function does not return a value, the type is `void`.

The definition statement of a function defines the parameters that are required by the function and these parameters are called formal parameters. Any statement that calls the function must include values that correspond to the parameters and these are called actual parameters. The formal parameters and the corresponding actual parameters must match in number, type, and order. For example, the definition statement of `fir` function is

```
float fir(float *x,   float *h,   int ntap);
```

The statement from the main program that calls the function is

```
yn = fir(xn_buf, bn_coef, len_imp);   /* FIR filtering */
```

Thus the variables `x`, `h`, `ntap` are the formal parameters and the variables `xn_buf`, `bn_coef`, `len_imp` are the corresponding actual parameters. When the references to the `fir` function is executed in `main`, the values of the actual parameters are copied to the formal parameters, and the statements in the `fir` function are executed using the new values in `x`, `h`, `ntap`.

It is important to note that since C always pass function arguments by value rather than by reference, the values of the formal parameters are not moved back to the actual parameters when the function completes its execution. When the function needs to modify the variables in the calling program, we must specify the function arguments as pointers to the beginning of the variables in the calling program's memory. For example, the `main` function calls the function `shift` as

```
shift(xn_buf, len_imp, xn);
```

where the address (value) of the first element in xn_buf is passed to the function shift. Thus in shift, we use

```
float *x;
```

which defines the pointer $*x$ points to the first element of xn_buf in the main function. Therefore the signal buffer array x[i] used in the function shift is actually the same array xn_buf used by the main function. Consequently, it can be modified by the function shift.

C.3.4 Files and I/O Operations

To use standard input/output functions provided by the C compiler, we must have the statement #include <stdio.h> that includes the standard I/O header file for function declaration. Some functions that perform standard input/output identify the file to read or write by using a file pointer to store the address of information required to access the file. To define file pointers, we use

```
FILE   *fpin;    /* File pointer to x(n) */
FILE   *fpimp;   /* File pointer to b(n) */
FILE   *fpout;   /* File pointer to y(n) */
```

We can open a file with the function fopen and close it with the function fclose. For example,

```
fpin = fopen(argv[1],"rb");
fpout = fopen(argv[3],"wb");
```

The function fopen requires two arguments – the name of the file and the mode. The name of the file is given in the character string argv[], or simply use 'file_name' where file_name is a file name that contains the desired data. For example, to open the ASCII data file infile.dat for reading, we can use

```
fpin = fopen("infile.dat","r");
```

For data files in ASCII format, the mode 'r' opens the file for reading and the mode 'w' opens a new file for writing. Appending a character b to the mode string, such as 'rb' and 'wb', is used to open a binary formatted data file. If the file is successfully opened, fopen returns a pointer to FILE that references the opened file. Otherwise, it returns NULL. The function fclose expects a pointer to FILE, such as the following statement:

```
fclose(fpin);
```

will close the file pointed by the pointer fpin while the statement of fcloseall() used in our previous example firfltr will close all the open files. When a program terminates, all open files are automatically closed.

Functions such as scanf and printf perform formatted input/output. Formatting input or output is to control where data is read or written, to convert input to the desired type (int, float, etc.), and to write output in the desired manner. The function scanf, fscanf, and sscanf provides formatted input (scanning or reading). For example, the statement

```
fscanf(fpimp, "%f", &bn);
```

reads from an arbitrary file pointed by the file pointer fpimp to a variable of address &bn, and %f indicates the number is floating-point data. In addition, the formatted I/O functions also recognize %d for decimal integers, %x for hexadecimals, %c for characters, and %s for character strings.

The function fwrite writes binary data. That is, fwrite writes blocks of data without formatting to a file that has been opened in binary mode. The data written may be integers or floating-point numbers in binary form (they may represent digital sounds or pictures). Conversely, the function fread is used for reading unformatted binary data from a file. The function fread requires four arguments. For example, the statement

```
fread(&xn, sizeof(float), 1, fpin);
```

reads 1 item, each item of size float data type, into the array xn from the file pointed to by the file pointer fpin. The fread function returns the number of items successfully read. The operator sizeof(object) has a value that is the amount of storage required by an object. The values of sizeof for different data types may vary from system to system. For example, in a workstation, the value of sizeof(int) is 4 (bytes), whereas on fixed-point DSP systems, the value of sizeof(int) is typically 2 (bytes).

The function fwrite expects four arguments. For example, the statement

```
fwrite(&yn, sizeof(float), 1, fpout);
```

writes the binary form float array yn of size sizeof(float) to the file pointed to by the pointer fpout. The difference between using fwrite to write a floating-point number to a file and using fprintf with the %f format descriptor is that fwrite writes the value in binary format using 4 bytes, whereas fprintf writes the value as ASCII text which usually need more than 4 bytes.

C.4 Control Structures and Loops

C language has a complete set of program control features that allow conditional execution or repetition of statements in loops based on the result of an expression.

C.4.1 Control Structures

The C language provides two basic methods for executing a statement or series statements conditionally: the if statement and the switch-case statement. The if statement allows us to test a condition and then execute statements based on whether the given condition is true or false. The if statement has the following general format:

```
if (condition)
    statement1;
```

If the condition is true, the `statement1` is executed; otherwise, this statement will be skipped. When more than one statement needs to be executed if the condition is true, a compound statement that consists of a left brace {, some number of statements, and a right brace } is used as follows:

```
if (condition)
{
    statements;
}
```

If the condition is true, those statements enclosed in braces are executed; if the condition is false, we skip these statements. Figure C.2 shows flowchart of the program control with the simple `if` statement.

An `if/else` statement allows us to execute one set of statements if the condition is true and a different set if the condition is false. The simplest form of an `if/else` statement is

```
if (condition)
{
    statements A;
}
else
{
    statements B;
}
```

A flowchart for this `if/else` statement is illustrated in Figure C.3. By using compound statements, the if/else control structure can be nested.

When a program must choose between several alternatives, the `if/else` statement becomes inconvenient and somewhat inefficient. When more than four alternatives from a single expression are chosen, the `switch-case` statement is very useful. The basic syntax of the `switch-case` statement is

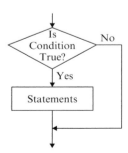

Figure C.2   Flowchart for if statement

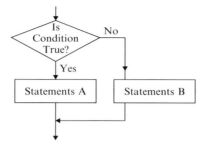

Figure C.3 Flowchart for if/else statement

```
switch(integer expression)
{
   case constant_1:
     statements;
     break;
   case constant_2;
     statements;
     break;
   ...
   default:
     statements;
}
```

Program control jumps to the statement if the case label with the constant (an integer or single character in quotes) matches the result of the integer expression in the `switch` statement. If no constant matches the expression value, the control goes to the statement following the `default` label. When a match occurs, the statements following the corresponding case label will be executed. The program execution will continue until the end of the switch statement is reached, or a `break` statement that redirects control to the end of the `switch-case` statement is reached. A `break` statement should be included in each case not required after the last `case` or `default` statement.

C.4.2 Logical Operators

The condition used in an `if` or `if/else` statement is an expression that can be evaluated to be true or false. It is composed of expressions combined with relational and sometimes logical operators. A logical operator is any operator that gives a result of true or false. Table C.2 gives the relational operators that can be used to compare two expressions in C. C also supports three logical operators listed in Table C.3 that compare conditions.

Table C.2   Relational operators

Relational Operator	Interpretation
<	less than
<=	less than or equal to
>	greater than
>=	greater than or equal to
==	equal to
!=	not equal to

Table C.3   Logical operators

Logical Operator	Interpretation
!	logical NOT
&&	logical AND
||	logical OR

C.4.3 Loops

C contains three different loop structures that allow a statement or group of statements to be repeated for a fixed (or variable) number of times, and they are `for` loop, `while` loop, and `do/while` loop.

Many DSP operations require loops that are based on the value of the variable (loop counter) that is incremented (or decremented) by the same amount each time it passes through the loop. When the counter reaches a specified value, we want the program execution to exit the loop. This type of loop can be implemented with `for` loop, which combines an initialization statement, an end condition statement, and an action statement into one control structure. The most frequent use of `for` loop is indexing an array through its elements. For example, the simple C program listed in Section C.1 uses the following `for` loop:

```
for (k = 1; k < N; k++)
{
    y[k] = 0.0;      /* Clear array */
}
```

This `for` statement sets k to 1 (one) first and then checks if k is less than the number N. If the test condition is true, it executes the statement $y[k] = 0.0$; increments k, and then repeats the loop until k is equal to N. Note that the integer k is incremented at the end of the loop if the test condition statement $k < N$ is true. When the loop is completed, the elements of the array y are set to zero from $y[1]$ up to $y[N-1]$.

The `while` loop repeats the statements until a test expression becomes false or zero. The general format of a `while` loop is

```
while(condition)
{
    statements;
}
```

The `condition` is evaluated before the statements within the loop are executed. If the condition is false, the loop statements are skipped, and execution continues with the statement following the `while` loop. If the condition is true, then the loop statements are executed, and the condition is evaluated again. This repetition continues until the condition is false. Note that the decision to go through the loop is made before the loop is ever started. Thus it is possible that the loop is never executed. For example, in the FIR filter program `firfltr.c`, the `while` loop

```
while((fscanf(fpimp, "%f", &hn)) != EOF)
{
    bn_coef[len_imp] = (float)bn;   /* Read coefficients */
    xn_buf[len_imp] = 0.;            /* Clear x(n) vector */
    len_imp++;                       /* Order of filter   */
}
```

will be executed until the file pointer reaches the end_of_file (`EOF`).

The `do/while` loop is used when a group of statements needs to be repeated and the exit condition should be tested at the end of the loop. The general format of `do/while` loop is

```
do {
    statements;
} while(condition);
```

The decision to go through the loop again is made after the loop is executed so that the loop is executed at least once. The format of `do-while` is similar to the `while` loop, except that the `do` key word starts the statement and `while` ends the statement.

C.5 Data Type Used by the TMS320C55x

The TMS320C55x supports the ANSI C standard. It identifies certain implementation-defined features that may differ from other C compilers, depending on the type of processor, the DSP run-time environment, and the host environment. In this section, we will present the data types that are used by the TMS320C55x C compiler.

The C55x is designed for real-time DSP applications. The most common data type is the 16-bit fixed-point integer representation of data variables. An important difference in data type definition used by the C55x and some other C compiler is the size between data types of `char` and `int`, `float` and `double`, and `long double`. Table C.4 lists the data type supported by the C55x C compiler.

From this table, it is clear that all the integer data types (`char`, `short` and `int`), either signed or unsigned, have the equivalent type and are all in 16-bit size. The `long`

Table C.4 Data type supported by the TMS320C55x C compiler

Data Type	Size	Representation	Range
Char	16-bit	ASCII	$[-32\,768, 32\,767]$
unsigned char	16-bit	ASCII	$[0, 65\,535]$
short	16-bit	2s complement	$[-32\,768, 32\,767]$
unsigned short	16-bit	Binary	$[0, 65\,535]$
int	16-bit	2s complement	$[-32\,768, 32\,767]$
unsigned int	16-bit	Binary	$[0, 65\,535]$
long	32-bit	2s complement	$[-2\,147\,483\,648, 2\,147\,483\,647]$
unsigned long	32-bit	Binary	$[0, 42\,949\,67\,295]$
float	32-bit	IEEE 32-bit	$[1.175\,494\,4\ 10^{-38}, 3.402\,82\,346\ 10^{38}]$
double	32-bit	IEEE 32-bit	$[1.175\,494\ 10^{-38}, 3.402\,82\,346\ 10^{38}]$
long double	32-bit	IEEE 32-bit	$[1.175\,494\ 10^{-38}, 3.402\,82\,346\ 10^{38}]$
enum	16-bit	2s complement	$[-32\,768, 32\,767]$
data pointer	16-bit	Small memory	$[0x0, 0xFFFF]$
data pointer	23-bit	Large memory	$[0x0, 0x7FFFFF]$
program pointer	24-bit	Function	$[0x0, 0xFFFFFF]$

data type uses 32-bit binary values. The signed data types use 2's-complement notation. Finally, all floating-point data types are the same and represented by the IEEE single-precision format. It is the programmer/engineer's responsibility to correctly define and use the data types while writing the program. When porting applications from one platform to the other, it is equally important that the correct data type conversion is applied, such as convert all the data defined as char from 8-bit to 16-bit integer.

References

[1] P. M. Embree, *C Algorithms for Real-Time DSP*, Englewood Cliffs, NJ: Prentice-Hall, 1995.
[2] P. M. Embree and B. Kimble, *C Language Algorithms for Digital Signal Processing*, Englewood Cliffs, NJ: Prentice-Hall, 1991.
[3] D. M. Etter, *Introduction to C++ for Engineers and Scientists*, Englewood Cliffs, NJ: Prentice-Hall, 1997.
[4] S. P. Harbison and G. L. Harbison, *C: A Reference Manual*, Englewood Cliffs, NJ: Prentice-Hall, 1987.
[5] R. Johnsonbaugh and M. Kalin, *C for Scientists and Engineers*, Englewood Cliffs, NJ: Prentice-Hall, 1997.
[6] B. W. Kernighan and D. M. Ritchie, *The C Programming Language*, Englewood Cliffs, NJ: Prentice-Hall, 1978.

Appendix D
About the Software

The web sites, http:/www.ceet.niu.edu/faculty/kuo/books/rtdsp.html and http://pages. prodigy.net/sunheel/web/dspweb.htm, that accompany this book include program and data files that can be used to demonstrate the DSP concepts introduced in this book and to conduct the experiments at the last section of each chapter. MATLAB scripts with extension .m are developed and tested with MathWorks MATLAB version 5.1 (http://www.mathworks.com). Assembly programs with extension .asm and C programs with extension .c or .h have been compiled and tested with Texas Instruments' Code Composer Studio IDE C5000 version 1.2. Data files with extension .dat are used for C code, and the files with extension .inc are used for the C55x assembly programs. The files in the software package are divided into the 10 groups, one for each chapter.

Chapter 1
exp1.c
exp1.cmd

Chapter 2
exp2a.c
exp2b.c
exp2.cmd
exp2_sum.asm
exp2b_1.asm
exp2b_2.asm
exp2b_3.asm
exp2b_4.asm

Chapter 3
exp3a.c
exp3b.c
exp3c.c
exp3d.c
exp3e.c
exp3.cmd

sine_cos.asm
ovf_sat.asm
exp3d_iir.asm
timit1.asc
timit1.dat

Chapter 4
exp4a.c
exp4.cmd
exp4b.asm
sine_cos.asm
mag_128.asm
dft_128.asm
vectors.asm
input.dat
input.inc
dft.m
exam4_15.m
exam4_16.m
exam4_17.m
Imp.dat

Imprtf.m
magrtf.m
all_pole.m
notch.m
notch1.m
notch2.m

Chapter 5
exp5a.c
exp5b.c
exp5c.c
exp5.cmd
signal_gen.c
fir.asm
fir2macs.asm
firsymm.asm
input5.dat

Chapter 6
exp6a.c
exp6b.c

exp6c.c
exp6d.c
iir.c
iir_i1.c
iir_i2.c
signal_gen2.c
exp6.cmd
iirform2.asm
sine.asm
exam6_12.m
exam6_13.m
s671.m

Chapter 7
exp7a.c
exp7b.c
exp7c.c
exp7d.c
fbit_rev.c
fft_a.c
fft_float.c
ibit_rev.c
test_fft.c
w_table.c
exp7.cmd
icomplex.h
fcomplex.h
bit_rev.asm
fft.asm
freqflt.asm
olap_add.asm
fir.m
input7.dat
input7_f.dat

input7_i.dat
firlp256.dat
firlp512.dat
firlp8.dat
firlp16.dat
firlp32.dat
firlp64.dat
firlp128.dat
firlp48.dat

Chapter 8
exp8a.c
exp8b.c
init.c
random.c
signal.c
exp8.cmd
adaptive.asm
alp.asm
fir_filt.asm
randdata.dat
LP_coef.dat
noise.dat
exam8_7a.m
exam8_7b.m

Chapter 9
ex9_1.c
ex9_2.c
ex9_3.c
pn_squ.c
uran.c
ex9.cmd
cos_sin.asm

gen.asm
pn_gen.asm
jkch.c
crcgen.c
diff.c
diffenc.c
errorhnd.c
interlev.c
convenc.c
noise.c
rx_src.c
sim.c
slotbits.c
sync_cor.c
tx_src.c
vitham.c
vselp_io.c
vselp.h
speech.dat
cosine.c
rand_num.c
project.cmd

Appendices
figb1.m
figb2.m
figb3.m
impulse.c
firfltr.c

Index

R2's complement, 96, 103, 118
3-dB bandwidth, 188, 281
3-dB cut-off frequency, 249–50
60 Hz hum, 6, 184
6-dB signal-to-quantization-noise gain, 8, 100, 105

accumulator (AC0–AC3), 27, 38, 57, 62, 97, 169, 382
acoustic
  echo, 427
  echo cancellation, 163, 409, 417, 426, 427
  echo path, 427
active noise control system, 373
adaptive
  algorithm, 351, 360, 368, 377, 385, 409, 420, 427
  channel equalization, 379
  echo cancellation, 417, 418, 419, 427
  filter, 351, 359, 369, 419, 427
  linear prediction (predictor), 373, 374, 390
  noise cancellation (ANC), 375, 430
  notch filter, 377
  step size, 370, 371
  system identification, 372, 373, 385
ADC, see analog-to-digital converter
adders, 80
address-data flow unit (AU), 36, 38, 174
addressing modes, 50
aliasing, 6, 156, 260, 401
allpass filter, 185
all-pole model (system), 163
all-zero model (system), 145, 163
amplitude spectrum, see magnitude spectrum
analog
  filter, 181, 241, 247
  interface chip (AIC), 10
  signal, 1, 241
analog-to-digital converter (ADC), 3, 6, 8, 10, 100
anti-aliasing filter, 6, 9, 156

anti-imaging filter, 9
anti-resonances, 163
anti-symmetry
  filter, 230
  properties, 195
application specific integrated circuits (ASIC), 11, 437
arithmetic and logic unit (ALU), 36, 38, 39
arithmetic
  error, 98, 382
  instruction, 63
ASIC, see Application specific integrated circuits
ASCII (text) file, 18, 19
assembler directives, 44
assembly statement syntax, 49
autocorrelation, 352, 356, 361, 362
  properties, 353
  matrix, 361
autocovariance, 352
automatic gain controller (AGC), 4, 440
autopower spectrum, 356
auxiliary register (AR0–AR7), 36, 38, 62
averaging technique, 410

bandlimited, 5, 155, 259
bandpass filter, 184, 185, 254
bandstop (band-reject) filter, 185, 254
bandwidth, 2, 155, 188, 254
barrel shifter, 38
Bessel
  filter, 253
  function, 212
bilinear transform, 259
binary
  file, 18
  point, 95
bit manipulation instruction, 64
bit-reversal, 318, 343
Blackman window, 210

block
    FIR filter, 225
    processing, 198, 225, 331
block-repeat
    counter, 168
    instruction (RPTB), 168
breakpoint, 26
built-in parallel instruction, 60
bus contention, 232, 234
buses, 39
butterfly network (computation), 316, 317, 335
Butterworth filter, 249, 271

Cascade (series) form, 142, 199, 266, 269
Cauchy's integral (residue) theorem, 139
causal
    filter, 182, 203
    signal, 78, 134
    system, 88
C compiler, 19, 42
CD player, 3, 9
center
    clipper, 408, 425
    frequency, 187, 254
central limit theorem, 95
channel
    coding, 437
    equalization, 359, 399, 441
characteristic equation, 150
Chebyshev
    approximation, 118, 219
    filter, 252, 271
circular
    buffer, 200, 217
    convolution, 311, 330
    pointer, 400
    shift, 310
circular addressing
    mode, 36, 58, 227
    buffer size registers (BK03, BK47, BKC), 58, 62
    starting address registers (BSA01, BSA23,
        BSA45, BSA67, BSAC), 58, 62
clipping threshold, 408, 425
coefficient
    data pointer (CDP), 38
    quantization, 101, 115, 223, 276
code composer studio (CCS), 19, 48
code conversion process, 41
code division multiple access (CDMA), 3, 353
code optimizer, 42
coder/decoder (CODEC), 10
comb filter, 192, 200, 216
combined parallel instruction, 61
comfort noise, 408, 431

common object file format (COFF), 41, 44
companding, 8
compiler optimization, 43, 294
complex
    arithmetic, 380, 414, 448
    LMS algorithm, 380
    variable, 165, 447
complex-conjugate property, 153, 309
complexity, 2
continuous-time signal, 1
convergence, 366, 368
    factor, see step size
    speed, 368
convolution, 135, 330
    sum, 87–8, 189
convolutional encoding, 438
correlation function, 94, 352
cosine and sine power series expansion, 117–18
CRC, see cyclic redundant check
critical frequencies, 257, 261
crosscorrelation, 352, 354, 361
    vector, 361
crosscovariance, 352–3
cross-power spectrum, 357
crosstalk effect, 376
C-to-ASM interlister, 42
cumulative probability distribution function, 90
cutoff frequency, 6, 184
cyclic redundant check (CRC), 437, 440

DAC, see digital-to-analog converter
data
    buses, 39
    computation unit (DU), 36, 38, 174
    generation unit (DAGEN), 60
    page pointer (DP), 52
dBm, 451
DC
    component, 128
    offset, 6, 184
DCT, see discrete cosine transformation
debugger, 19, 25
decibel (dB), 450
decimation, 8
decimation-in-frequency, 315, 319
decimation-in-time, 315
decision-directed algorithm, 380
delay
    equalizer, 185
    property, 135
    units, 81
delta function, 131, 207, 449
design of FIR filter, 201
deterministic signal, 77, 351

DFT, see discrete Fourier transform
DFT matrix, 306
differentiator, 220
digital building block, 11, 12
digital
   filter, 143, 181
   filter design, 181
   resonator, 150
   signal, 1
   signal processing (DSP), 1
   signal processors (DSP chips), 11, 35
digital-to-analog converter (DAC), 3, 9
digit decoder, 416
digitization, 4
direct addressing mode, 52
direct form, 198, 263–4
discrete
   cosine transformation (DCT), 436
   Fourier transform, 152, 157, 171, 303, 308, 411
discrete-time
   Fourier transform (DTFT), 153, 304, 356
   frequency, 79
   signal, 1, 5, 152, 245
   system, 77
dispersive delay, 419
dot product, 73
double-talk, 423
   detector, 423, 427, 429
DSP, see digital signal processing
   chip selection, 16
   hardware, 11
   software development, 16
   software development tool, 18
   start kit (DSK), 48
   system design, 14
DTMF (dual-tone multi-frequency)
   frequencies, 411
   tone detection, 410
   tone generator, 403
dual slope, see analog-to-digital converter

echo
   cancellation, 399
   path, 417, 419, 421
   return loss (ERL), 424
   return loss enhancement (ERLE), 421
eigenvalue spread, 369
elementary digital signal, 77
elliptic filter, 253–54, 271
encoding process, 7
energy, 85, 416, 450
ensemble average, 352, also see expectation
   operation
equalizer, 379

ergodic, 352, 355
error
   coding, 437
   contours, 363–64
   surface, 363
Euler's
   formula, 129
   theorem, 446
evaluation module (EVM), 48, 49
even function, 128, 131, 149, 309, 353, 356
excess mean-square error (MSE), 369, 370, 383,
   420, 429
excitation signal, 372, 419
expectation
   operation, 92, 352
   value, see mean

far-end, 418, 419
fast
   convolution, 198, 330, 344
   Fourier transform, 159, 314, 321, 336, 411
feedback coefficients, 144
feedforward coefficients, 144
FFT, see fast Fourier transform
file header format, 111–12
filter, 181
   coefficients (weights or taps), 83
   design procedure, 201
   specifications, 185
finite impulse response, see FIR filter
finite precision (wordlength) effects, 107, 267, 334,
   382
FIR filter, 83, 84, 89, 143, 144, 181, 189, 221, 227,
   360
fixed-point, 13, 95, 223, 284
flash ADCs, see analog-to-digital converter
flat delay, 419
flexibility, 1
floating-point, 13, 223, 284
folding
   frequency, see Nyquist frequency
   phenomenon, 156
forced response, 147, 245
forward error correction (FEC), 437
Fourier
   coefficient, 128, 152
   series, 128, 152
      series (window) method, 202
   transform, 130, 132, 133, 153, 157, 158, 242
four-wire facility, 418
fractional number, 95, 97
frequency, 78, 127
   offset test, 416
   resolution, 159, 307, 322, 326

frequency (*cont.*)
  response, 148, 153
  sampling method, 202, 214, 255
    transform, 253
  warping, 261
fundamental frequency, 129, 130, 156

G.165 (G.168), 422
gain, 4
Gaussian distribution function, 95
general extension language, 32
geometric series, 215, 446
Gibbs phenomenon, 205
Goertzel algorithm, 411
gradient, 365
  estimate, 366
granulation noise, 100
group delay, 188, 196
guard bits, 106

half-wave rectification, 434
Hamming window, 209, 221
hands-free telephone, see speakerphone
Hann (Hanning) window, 209
Hanning filter, 198
harmonics, 128, 192
Harvard architecture, 12
highpass filter, 184, 249, 253, 254
Hilbert transformer, 185, 197, 220, 378
hybrid, 418, 427
  loss, see echo return loss

ideal
  filter, 183
  oscillator, 129
  sampler, 4, 5
  system, 107
IDCT, see inverse discrete cosine transformation
IDFT, see inverse fast Fourier transform
IFFT, see inverse fast Fourier transform
IIR Filters, 88, 89, 143, 241, 255, 263, 271, 279, 360
image
  coding, 399
  recognition, 2
impedance, 248, 417
implementation procedure, 107
implied parallel instructions, 60
impulse
  functions, 449
  response, 83, 87, 202, 203, 255
  train, 177, 245
impulse-invariant method, 255, 256
indirect addressing mode, 53

infinite impulse response, see IIR filters
in-circuit emulator (XDS), 48, 49
in-place FFT algorithms, 315
input
  autocorrelation matrix, 361
  channel, 3
  quantization, 98, 277
  signal conditioning, 3
  vector, 360, 381
instantaneous
  normalized frequency, 402
  power, 424
  squared error, 366
instruction
  buffer queue (IBQ), 60, 341
  buffer unit (IU), 36, 37
  set, 63
interfacing C with assembly code, 71
interleave, 441
interpolation, 8, 401
  filter, 198
interrupt, 174
  service routine (ISR), 174
  vector table, 174
intersymbol interference (ISI), 379
intrinsics, 289
inverse
  discrete cosine transformation (IDCT), 436
  discrete Fourier transform (IDFT), 153, 158, 306
  fast Fourier transform (IFFT), 320
  filter, 198
  Fourier transform, 130
  Laplace transform, 242
  z-transform, 136

JPEG (Joint Photographic Experts Group), 1, 437

Kaiser window, 212
Kronecker delta function, 78

Laplace transform, 133, 241
leakage (or smearing), 208
  factor, 371, 372
leaky LMS algorithm, 371, 382, 390
learning curve, 368
least-mean-square, see LMS algorithm
least significant bit (LSB), 8, 40, 96
limit cycle oscillation, 98
line
  echo, 417
  spectrum, 129
linear
  chirp signal, 402, 409

congruential method, 404
convolution, 87–8, 135, 189, 198, 243–4, 314, 330
  interpolation, 401
  phase, 188
  phase filter, 183, 194
  prediction, 373, 374
  system, 87
  time-invariant (LTI) system, 133
linearity (superposition), 134, 182, 308
linker, 46
  command file, 23
LMS algorithm, 351, 366
logarithmic quantizer, 10
logic instructions, 64
long division, 136, 137
lookup-table method, 400, 403
loop unrolling, 228
lowpass filter, 184, 204, 249, 250
LPC, 165

MAC unit, see multiply-accumulate
magnitude, 447
  bit, 95, 96, 382
  distortion, 185
  response, 148–9, 151, 154, 160, 164, 182
  spectrum, 129, 130–31, 153–54, 308
  test, 414
mainlobe, 207, 325
maintainability, 17
mapping properties, 246, 260, 261
marginally stable, 148, 245, 274
MATLAB (matrix laboratory), 2, 202, 453
  basic Operators, 457
  built-in function, 465
  files, 459
  functions, 466
  graphics, 455
  M-files, 459
  Mat-files, 460
  Mex-files, 460
  script files, 459
  user-written functions, 465
maximum-length sequence, 163
mean, 92
mean-square error (MSE), 361, 383
mean-squared value, 93, 353, 356
MEMORY directive, see linker
memory map, 40
memory-mapped register addressing mode, 56
memory mapped registers (MMRs), 40
microprocessor and microcontrollers (μP), 11
minimum
  MSE, 362, 383

  step size, 384
mixture of C and assembly programming, 68
MMR, see memory mapped registers
mnemonic assembly code, 42
modulo operation, 293, 404–5
most significant bit (MSB), 95
move instruction, 65
moving average filter, 84, 89, 145, 149, 193, 194
MPEG (Moving Picture Experts Group), 437
MSE, see mean-square error
  surface, 363
  time constant, 368–9
multipliers, 80
multiply–accumulate, 36, 227
multi-rate, 8

natural response, 147
near-end, 418
negative symmetry, see anti-symmetric
network echo, 417
noise
  generators, 404
  reduction, 399, 429
  subtraction, 430
nonlinear phase, 194
non-causal filter, 182, 198, 203
non-parametric methods, 322
non-real-time DSP, 2
non-recursive filter, 89
non-uniform quantization, 8
normal distribution function, 9
normalized
  digital frequency, 155
  frequency, 79, 271
  LMS algorithm, 368, 371, 428
  step size, 371
notch filter, 160, 185, 377
Nyquist
  frequency, 6, 156, 197
  interval, 6
  rate, 6

odd function, 128, 129, 131, 149, 309
off-line system modeling, 373, 409, 428
one-sided z-transform, 134
operand types, 51
optimum
  filter, 362
  step size, 383
  weight vector, 362
oscillatory behavior, 205
output channel, 3
overflow, 103, 112
overlap-add, 332

overlap-save, 331
oversampling, 6
overshoot, 182

parallel
    connection, 141
    converter, 9
    execution, 60
    form, 268
parallelism, 59
parametric equalizer, 283
parametric methods, 322
Park–McClellan algorithm, 202, 219
Parseval's theorem, 129, 131, 328
partial-fraction expansion, 137
passband, 183, 271
    edge (cutoff) frequency, 185
    ripple, 186
PDS, see power density spectrum
peak value, 85
performance
    (or cost) function, 361
    surface, 363
periodic signal, 127, 152, 153, 177, 304
periodicity, 305, 310
periodogram, 328
peripheral data-page pointer (PDP), 52
phase, 432–3, 447
    distortion, 185
    response (shift), 149, 154, 160, 182, 188
    spectrum, 129, 131, 154, 188, 308
pipeline, 59
    break down, 60
    protection unit, 38
PN sequence, see pseudo-random number
polar form, 447
pole, 134, 144, 150, 163, 244, 251, 269, 276, 277
pole–zero
    cancellation, 146, 147, 193, 216
    plot (diagram), 134, 145, 147, 150
    system (model), 145, 163, 165
polynomial approximation, 117, 249, 399
positive symmetric, see symmetric
power, 85, 93, 128, 129, 354, 356, 368, 450
    density spectrum (PDS), 129, 131, 328, 356
    estimator, 84, 147, 424
    spectral density, see PDS
    spectrum, see PDS
practical system, 107
prediction delay, 374
pre-warping, 261
probability, 90
    density function, 91
probe point, 28

processing time, 381
profiler (profiling), 29
program
    address generator, 57
    counter (PC), 38, 174
    execution pipeline, 59
    fetch pipeline, 59
    flow control instructions, 66
    flow unit (PU), 36, 38
program-read address bus (PAB), 39
program-read data bus (PB), 39
Prony's method, 165
pseudo-random
    binary sequence generator, 406
    numbers (noise), 94, 404

Q format, 97, 118, 291, 296
quadrants, 118, 401
quadrature amplitude modulation (QAM), 380, 437
quadrature phase, 402
quality factor, 188
quantization, 4
    effect, 223
    errors (noise), 7, 98, 99, 100, 107, 223, 276, 278, 335, 401
    process, 7, 99, 109, 111, 223
    step (interval, width, resolution), 98

radix-2 FFT, 159, 336
raised cosine function, 209
random
    number generation, 399, 404
    process (signals), 78, 89, 351
    variable, 90, 352
Rayleigh fading, 441
realization of IIR filter, 263
real-time
    constraint, 14
    emulator, 18
    signal processing, 2, 381
reconstruction filter, 9, 10
rectangular
    pulse train, 128
    window, 207, 324, 327
recursive
    oscillator, 281
    power estimator, 147
region of convergence (ROC), 134
register, 95, 406
reliability, 2
Remez algorithm, 219
repeat operation, 36, 293
reproducibility, 2

residual
  echo, 420
  echo suppressor, 408, 425
  method, 139
  noise reduction, 434
resonator (peaking) filter, 185, 217, 280
reverberation, 163, 426, 427
rise time, 182
room transfer function (RTF), 162, 165
roots, 244
  of polynomials, 145, 146
rounding and saturation control logic, 38
rounding and truncation, 98
roundoff error (noise), 102, 267, 382, 383, 429
running averaging filter
  see moving averaging filter
run-time support library, 24, 293

sample
  autocorrelation function, 355
  function, 245, 352
  mean, 355
  space, 90
  variance, 355
sample and hold, 4
sampling
  frequency (rate), 5, 6, 283, 330, 413
  function, 4, 245
  process, 4
  theorem, 5, 6, 156
saturation arithmetic, 103, 112
scaling
  factor, 106, 193, 224, 278, 382
  signal, 105
Schur–Cohn stability test, 275
second harmonic test, 416
SECTION directive, see linker
serial converter, 9
settling time, 182
Shannon's sampling theorem, 5
sidelobes, 207, 325
sigma-delta ADC, see analog-to-digital converter
signal buffer, 199, 222, 227
signal-to-quantization-noise ratio, 7–8, 100
sign bit, 96, 382
simulator, 17, 48
sinc function, 204
sinewave
  generation, 167, 226, 281, 399
  table address increment, 400
single-repeat instruction (RPT), 168
sinusoidal
  signal, 78
  steady-state response, 245

sirens, 403
slicer, 380
smearing, 208, 325
smoothing filter, 9, 198
software development tools, 40
SP, see stack pointer
speakerphone, 426
spectral
  dynamic range, 369
  estimator, 433
    leakage, 324
  magnitude averaging, 433
  resolution, 324
  smearing, 326
  subtraction, 375, 431
spectrogram, 332
spectrum, 131, 161, 309, 322, 325
speech
  enhancement, 429
  (source) coding, 375, 399, 437
  recognition, 2, 375
square wave, 176
squared-magnitude response, 149
stability, 182, 274
  condition, 148, 245, 274
  constraint, 367, 369
  triangle, 275
stack, 174
stack pointer (SP), 52
stalling, 384
standard deviation, 93
stationary, 369
statistical independence, 94
status registers, 57
steady-state response, 147, 154, 182, 188, 245
steepest descent, 365
step size, 365, 370, 383, 421, 429
stochastic gradient algorithm, see LMS algorithm
stochastic process, see random process
stopband, 183
  edge frequency, 185
  ripple (or attenuation), 186
subband acoustic echo canceler, 427
successive approximation, see analog-to-digital
    converter
sum of products, 73
superposition, 182
switched capacitor filter, 9
symmetry, 195, 230, 305
synthesis equation, 158
synthesizes sine function, 117
system
  gain, 149, 154
  identification, 372

system (*cont.*)
   stack pointer (SSP), 175

tail delay, 420
talk-off, 411
tapped-delay-line, 84, 200
Taylor series expansion, 117
temporal, 352
temporary register (T0–T3), 38
time
   average, 352
   delay, 188, 353
   shifting, 135
time-invariant filter (system), 181
time-quantization error, 401
TMS320C3x, 13
TMS320C55x, 13, 35
tolerance (or ripple) scheme, 185
tone generation and detection, 399, 403
tone-to-total energy test, 416
total harmonic distortion, 399
trace, 368
training
   signal, 409
   stage, 380, 409
transfer function, 141, 143, 189, 198, 244
transformer, 192
transient response, 147, 182, 191, 245
transition
   band, 185, 187
   discontinuity, 205
transversal filter, see FIR filter
trellis, 439
trigonometric
   function, 399
   identities, 445
truncation, 203, 205
twiddle
   factor, 158, 165, 167, 305, 314, 334

matrix, 306
twist, 411
   test, 415
two-sided z-transform, 134
two-wire facility, 418

uncorrelated, 94
uniform density function, 91, 92, 93
unit
   circle, 133
   delay, 81, 135
unit-impulse sequence, 78, 449
units of power, 450
unit-step sequence, 78
unstable condition, 148, 245
user-built parallel instruction, 61

variable watch, 20
variance, 93, 100
vector concept, 449
Viterbi decoding, 437, 439
voice activity detector (VAD), 430, 431
von Neumann architecture, 12

weight vector, 220, 361, 365, 366
white noise, 357, 409
wide-sense stationary (WSS), 353
window functions, 208, 326
wireless application, 437
wordlength, 2, 7, 96, 224

zero, 134, 144, 151, 163, 244
   padding, 313, 326, 330
zero-mean, 92, 357
zero-order-hold, 9
zero-order modified Bessel function, 212
zero-overhead looping, 36, 227
z-plane, 133, 246
z-transform, 133, 136, 153, 246, 311, 356